DYNAMICS AND MODELLING OF REACTIVE SYSTEMS

Publication No. 44
of the Mathematics Research Center
The University of Wisconsin—Madison

Academic Press Rapid Manuscript Reproduction

DYNAMICS AND MODELLING OF REACTIVE SYSTEMS

Edited by

WARREN E. STEWART
W. HARMON RAY
CHARLES C. CONLEY

Mathematics Research Center
University of Wisconsin
Madison, Wisconsin

Proceedings of an Advanced Seminar
Conducted by the Mathematics Research Center
The University of Wisconsin—Madison
October 22–24, 1979

ACADEMIC PRESS 1980
A Subsidiary of Harcourt Brace Jovanovich, Publishers
NEW YORK LONDON TORONTO SYDNEY SAN FRANCISCO

ACADEMIC PRESS, INC.
111 Fifth Avenue, New York, New York 10003

United Kingdom Edition published by
ACADEMIC PRESS, INC. (LONDON) LTD.
24/28 Oval Road, London NW1 7DX

LIBRARY OF CONGRESS CATALOG CARD NUMBER: 80-19714

PRINTED IN THE UNITED STATES OF AMERICA

80 81 82 83   9 8 7 6 5 4 3 2 1

This volume is dedicated to

PROFESSOR JOSEPH O. HIRSCHFELDER

in recognition of his pioneering contributions to the theory of reactive systems, and his valued counsel and support of MRC from its inception.

Contents

Contributors

Numbers in parentheses indicate the pages on which authors' contributions begin.

Neal R. Amundson (353), Department of Chemical Engineering, University of Houston, Houston, Texas 77004

Rutherford Aris (1), Department of Chemical Engineering and Materials Science, University of Minnesota, Minneapolis, Minnesota 55455

Donald G. Aronson (161), School of Mathematics, University of Minnesota, Minneapolis, Minnesota 55455

George F. Carrier (333), Division of Applied Sciences, Harvard University, Cambridge, Massachusetts 02138

Martin Feinberg (59), Department of Chemical Engineering, University of Rochester, Rochester, New York 14627

Philip S. Feldman (333), Engineering Sciences Laboratory, TRW Defense and Space Systems Group, Redondo Beach, California 90278

Francis E. Fendell (333), Engineering Sciences Laboratory, TRW Defense and Space Systems Group, Redondo Beach, California 90278

E. Dieter Gilles (37), Institut für Systemdynamik und Regelungstechnik, Universität Stuttgart, Pfaffenwaldring 9, D-7000 Stuttgart 80, Germany

David M. Golden (315), Department of Chemical Kinetics, SRI International, Menlo Park, California 94025

Paul S. Gough (375), Paul Gough Associates, Inc., Portsmouth, New Hampshire 03801

Louis N. Howard (195), Massachusetts Institute of Technology, Department of Mathematics, Cambridge, Massachusetts 02139

Joseph B. Keller (211), Department of Mathematics and Mechanical Engineering, Stanford University, Stanford, California 94305

Dan Luss (131), Department of Chemical Engineering, University of Houston, Houston, Texas 77004

Eduardo Mon (353), Department of Chemical Engineering, University of Houston, Houston, Texas 77004

Gary T. Renola (177), Exxon Research and Engineering, Baytown, Texas 77520

John Rinzel (259), Mathematical Research Branch, NIAMMD, National Institutes of Health, Bethesda, Maryland 20205

Willi Ruppel (37), Institut für Systemdynamik und Regelungstechnik, Universität Stuttgart, Pfafenwaldring 9, D-7000 Stuttgart 80, Germany

Roger A. Schmitz (177), Department of Chemical Engineering, University of Notre Dame, Notre Dame, Indiana 46556

John H. Seinfeld (225), California Institute of Technology, Pasadena, California 91125; Mathematics Research Center, University of Wisconsin—Madison, Madison, Wisconsin 53706

Forman A. Williams (293), Department of Applied Mechanics and Engineering Sciences, University of California, San Diego, La Jolla, California 90293; Department of Mechanical and Aerospace Engineering, Princeton University, Princeton, New Jersey 08544

Anthony P. Zioudas (177), Department of Chemical Engineering, University of Illinois, Urbana, Illinois 61801

Preface

This volume represents the proceedings of the Advanced Seminar on Dynamics and Modelling of Reactive Systems, held at the University of Wisconsin on October 22–24, 1979. This seminar, conducted by the Mathematics Research Center of the University, brought together distinguished engineers and applied mathematicians to discuss their research in this broad and rapidly developing field.

The goals of the advanced seminar were (i) to assess the current level of understanding of dynamics of chemically reacting systems and (ii) to provide a forum for exchange of ideas between engineers and mathematicians working in this important area. Thanks to the breadth of the formal lectures, and the intensity of the many informal discussions, these objectives were achieved.

Because of the diverse disciplines of the contributors to this volume, no attempt has been made to enforce a uniform style. Rather, each author has been encouraged to give an impression of the current state of his particular field, while providing new results for the initiated reader. Thus, in some chapters the material has the flavor of mathematics while in others the flavor is definitely that of chemistry or engineering. We hope that the reader shares our belief that this variety of viewpoints enhances the value of the volume.

We extend our thanks to the authors, and to John Nohel and the staff of MRC, for their cooperation in all aspects of the conference. We are especially grateful to Gladys Moran, who handled many organizational details of the conference, and to Dorothy Bowar who worked with us in editing this volume. Finally, we thank the following agencies for their support that made the advanced seminar possible:

1. The United States Army Research Office (Contract DAAG29-75-C-0024);
2. The National Science Foundation (Grant ENG-7918040);
3. The Office of Naval Research (Grant N00014-79-G-0061).

Warren E. Stewart
W. Harmon Ray
Charles C. Conley

Hierarchies of Models in Reactive Systems

Rutherford Aris

1. INTRODUCTION

J. Maynard Smith in his valuable introduction to models in ecology [30] has drawn a distinction between two aspects of mathematical systems that represent non-mathematical prototypes. On the one hand they are useful when their outcome reproduces the behavior of the prototype and may be developed and extended, often by incorporating more and more detail, toward greater fidelity and comprehensiveness. This he suggests might usefully be called a 'simulation' leaving the word 'model' for the mathematical system that gains its validity, not from its particularity and insight into a special system, but from its limited generality and ability to handle certain broad classes of systems and to explain their unifying and differentiating features. The one advances actual understanding, the other contributes to conceptual progress. Of course the distinction is not a hard and fast one, nor are the two to be put in opposition to each other. Like theory and experiment - and indeed almost any pair of contrasts - they gain their vitality from their interaction. (Cf. however other usages of 'model' and 'simulation', e.g. [37]).

A second general point that may be made is that models do not come in isolation but in hierarchies or, perhaps it would be better to say, in families and their kinship relations often shed as much light on the situation as does the

ISBN 0-12-669550-4

exploration of the individual model. This has been discussed in a variety of ways elsewhere [1]; here I wish to concentrate on the physical system consisting of a fluid flowing through a tube with reaction taking place on the wall.

Immediately a range of possibilities leaps to mind and though we shall have to specify the system more straitly later on there are a large number of questions that may be asked. These may be geometrical (G_1, is the tube straight? G_2, what is its cross-section? G_3, is it finite length? - to the last of which the answer is, "Of course!"), hydrodynamical (H_1, is the fluid Newtonian? H_2, is the flow developed? H_3, is the flow laminar or turbulent?) or physicochemical (P_1, are the diffusion coefficients constant? P_2, are there several reacting species? P_3, what are the kinetics of the reaction?). When these are answered specifically we shall have our system, but we should always keep an eye on the neighboring cases and ask if our methods can be extended to them or whether they are relying on specific features. In formulating the model the questions that arise can be answered in various ways; definitely (G_1 or H_1, for example, might be answered with an unequivocal, "yes"; G_2, by "circular to within 0.001%"), assumptively (P_1 - "In the absence of any information we assume constancy"; H_2 - "We'll ignore the entrance region") or provisionally (G_3 - "It is of course finite but, being long, we model it by a infinite tube"; P_3 - "For this model the kinetics are linear"). Sometimes the answer specifies the system, S, sometimes the model, Σ.

2. The systems

We consider a long, straight tube of circular cross-section through which the reacting stream passes in Newtonian laminar flow with negligible entrance effect. There is no homogeneous reaction but as the R chemical species diffuse to the wall they react in a system of first order reactions. The diffusion coefficients are constant. We call this general system S_R and have the equations:

$$\frac{\partial c_r}{\partial t} = D_{rs} \{\frac{1}{r} \frac{\partial}{\partial r} (r \frac{\partial c_s}{\partial r}) + \frac{\partial^2 c_s}{\partial x^2}\} - 2U(1 - \frac{r^2}{a^2}) \frac{\partial c_r}{\partial x}, \qquad (1)$$

$$\frac{\partial c_r}{\partial r} = 0,\ r = 0;\ D_{rs}\frac{\partial c_s}{\partial r} + k_{rs}c_s = 0,\ r = a. \tag{2}$$

The summation convention on repeated dummy indices has been invoked and the notation, though transparent, is given below. It should be remarked that $k_{rs}c_s$ is the rate of <u>dis</u>appearance of the rth species as a function of all concentrations.

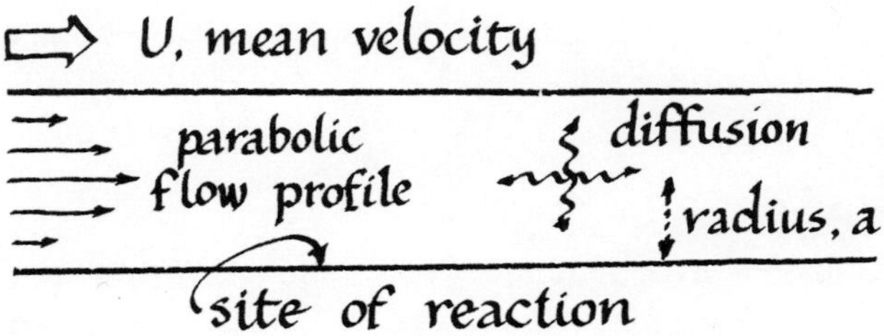

Fig. 1. The System.

When a single reactant is involved we may drop suffixes and refer to S_1.

The further conditions are:

$$c_r(r,x,0) = c_{ro}(r,x); \tag{3}$$

S_{R1}, infinite tube $c_r(r,x,t) \to 0$ as $x \to \pm\infty$; (4)

S_{R2}, semi-infinite tube $c_r(r,x,t) \to 0$ as $x \to \infty$ (5)

$$-D_{rs}\frac{\partial c_s}{\partial x} + 2U(1 - \frac{r^2}{a^2})c_r = f_r(r,t),\ x = 0 \tag{6}$$

S_{R3}, finite tube $D_{rs}\dfrac{\partial c_s}{\partial r} = 0,\ x = L$ (7)

where equation (6) is common to both S_{R2} and S_{R3}. The last two are patient of steady forms if the inlet flux is constant; these will be denoted by S_{R4} and S_{R5} respectively. It is sometimes assumed the inlet concentration can be prescribed so that for the semi-infinite case

S_{R6} $c_r(r,0,t) = c_{ri}(r,t),\ c_r(r,x,t) \to 0,\ x \to \infty$. (8)

The corresponding models are:

Σ_{R1}: Equations (1), (2), (3), (4); (10), (11), (12), (13);

Σ_{R2}: Equations (1), (2), (3), (5), (6); (10), (11), (12), (13), (14);

Σ_{R3}: Equations (1), (2), (3), (6), (7); (10), (11), (12), (14), (15);

Σ_{R4}: Equations (1), with 0 on the left, (2), (5), (6) with $f_r = f_r(r)$; (10), (11), (13), (14);

Σ_{R5}: Equations (1) with 0 on the left, (2), (6) with $f_r = f_r(r)$, (7); (10), (11), (14), (15).

Σ_{R6}: Equations (1), (2), (3), (8); (10), (11), (12), (16).

We will write the dimensionless equations in vector form taking

$$(\underset{\sim}{u})_r = c_r/c_o, \ (\underset{\sim}{D})_{rs} = D_{rs}/D, \ \tau = Dt/a^2, \ \xi = x/a, \ \rho = r/a$$

$$\tilde{\omega} = Va/D, \ (\underset{\sim}{\kappa})_{rs} = k_{rs}\, a/D \tag{9}$$

where c_o and D are a characteristic concentration and diffusion coefficient respectively. Thus

$$\underset{\sim}{u}_\tau = \underset{\sim}{D}(L\underset{\sim}{u} + \underset{\sim}{u}_{\xi\xi}) - 2\tilde{\omega}(1-\rho^2)u_\xi \tag{10}$$

$$\underset{\sim}{u}_\rho(0,\xi,\tau) = 0, \ \underset{\sim}{D}\underset{\sim}{u}_\rho(1,\xi,\tau) + \underset{\sim}{\kappa}\underset{\sim}{u}(1,\xi,\tau) = 0 \tag{11}$$

$$\underset{\sim}{u}(\rho,\xi,0) = \underset{\sim}{u}_o(\rho,\xi) \tag{12}$$

$$\underset{\sim}{u}(\rho,\pm\infty,\tau) = 0 \tag{13}$$

$$-\underset{\sim}{D}\underset{\sim}{u}_\xi(\rho,0,\tau) + 2\tilde{\omega}(1-\rho^2)\ \underset{\sim}{u}(\rho,0,\tau) = \underset{\sim}{\rho}(\rho,\tau) \tag{14}$$

$$\underset{\sim}{D}\underset{\sim}{u}_\xi(\rho,L/a,\tau) = 0 \tag{15}$$

$$\underset{\sim}{u}(\rho,0,\tau) = \underset{\sim}{u}_i(\rho,\tau), \ \underset{\sim}{u}(\rho,\infty,\tau) = 0 \tag{16}$$

L denotes the radial Laplacian $Lu = \rho^{-1}(\rho u_\rho)_\rho$. For Σ_1 - models $\underset{\sim}{u}$ is simply treated as a scalar and $\underset{\sim}{D} = 1$. In many cases $\underset{\sim}{D} = \mathrm{diag}(D_1, D_2, \ldots D_r)$. $\underset{\sim}{K}$ will be singular for reversible systems, but is lower triangular for a sequence of irreversible reactions.

3. The non-reactive cases

When there is no reaction we have the well-known case of Taylor diffusion which has considerable literature. Though this cannot be reviewed in detail, it is worth pausing to get it in perspective. The models will be denoted by T. In [32] Taylor started with $T_1 \equiv \Sigma_{11}$ with $\kappa = 0$, but he immediately remarks that the effect of longitudinal diffusion will be small compared with that of convection and arrives at T_2 which is T_1 with $u_{\xi\xi}$ removed. What follows is not a complete solution of T_2, though this can be done with confluent hypergeometric functions [22], but the typically Taylorian insight that the transfer across planes moving with the mean speed of the stream will depend on the radial variation of c but that c_ξ would be virtually independent of ρ. This permitted a very simple calculation to show that the mean concentration $\bar{c}$ should satisfy

$$\bar{c}_\tau + \tilde{\omega}\bar{c}_\xi = D_e \bar{c}_{\xi\xi} \tag{17}$$

where the effective diffusion coefficient (or Taylor diffusion coefficient) is

$$D_e = \tilde{\omega}^2/48 \ . \tag{18}$$

We observe that a new model has arisen in which the complex interaction of velocity profile and diffusion is wrapped up in an equivalent dispersion coefficient superimposed on a uniform mean flow. This simple equivalent model will appear again in other forms: here it will be denoted by T_e.

Historically, Taylor's next work was applied to turbulent flow [33] and to the conditions under which diffusivities might be measured [34]. However Aris [2] returned to T_1 and found that the assumptions of Taylor's approach need not be made if the evolution of the moments of the solute distribution was considered. Thus if

$$v_p(\rho,\tau) = \int_{\infty}^{\infty} \xi^p \, u(\rho,\xi,\tau)d\xi \tag{19}$$

and

$$w_p(\tau) = \int_0^1 2\rho \ v_p(\rho,\tau)d\tau \tag{20}$$

it was found that

$$w_0(\tau) = \text{const.} = 1(\text{say}), \ w_1(\tau) \sim \tilde{\omega}\tau + w_{100} \ , \tag{21}$$

$$w_2(\tau) \sim 2(1 + \tilde{\omega}^2/48)\tau + \tilde{\omega}^2\tau^2 + w_{200} \ .$$

Complete expressions were of course found for $v_p(\rho,\tau)$ and $w_p(\tau)$ but these depend on the initial distribution of solute as do w_{100} and w_{200} in eqn. (21). The beauty of the asymptotic model, T_a, in the form

$$w_0 = 1, \ w_1 \sim \tilde{\omega}\tau, \ w_2 - w_1^2 \sim 2D_e\tau \tag{22}$$

$$D_e = (1 + \tilde{\omega}^2/48) \tag{23}$$

is that it is free from the particularities of the initial distribution. Coupled with the fact, conjectured in [2] but proved by Chatwin [5], that any distribution approaches normality the model T_e with D_e given by eqn. 23 is given validity in an asymptotic sense. Notice that this is a model in Smith's sense for we have gained insight into the dispersive process but have not been able to compute the details of a particular distribution.

In passing it may be mentioned that further work on the calculation of moments was done by Horn [17] and that Brenner [3] has recently generalized Taylor dispersion very considerably. His model, T_b, takes local and global spaces (in T_1, ρ is the coordinate in local space and ξ in the global) and, approaching from the viewpoint of a single particle using a probability density rather than a concentration field, he obtains an equivalent velocity vector and dispersion dyadic by the method of moments. T_b may thus be regarded as an overarching model under which the following scheme may be adumbrated.

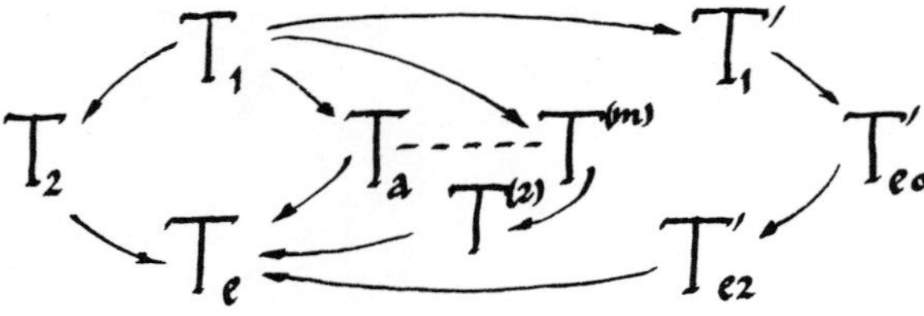

Fig. 2. Relationships of the models.

Mention of asymptotic behavior raises the question of the initial behavior during the period before equilibrium in the cross-section has been attained. Implicit in Taylor's first paper [32] is an estimate of the duration of this initial period, as also in [2], but the subject has also been studied in detail by Lighthill [23] and Chatwin [6,7]. An asymptotic analysis by Fife and Nicholes [10] has provided the best justification for T_a. If ℓ is a measure of the length over which the initial distribution is spread-and, for this analysis, it must be smoothly spread-T_a is valid when $\varepsilon = \tilde{\omega}a/\ell$ is small. They also show agreement with the approach of Gill, to which we now turn, for $\tau > 0.1$. The advantage of Fife's method (a non-standard two-timing technique) is that it can be proved that the difference between the exact solution and his expansion to ε^m terms is bounded by $K\varepsilon^{m+1}$; K does however depend on the smoothness of the initial data.

Gill's approach, adumbrated in [11] and developed in a series of papers with Sankarasubramanian [13-16,29], was first to transform to a coordinate system moving with the mean speed of flow

$$\zeta = \xi - \tilde{\omega}\tau \tag{24}$$

giving

$$u_\tau = Lu + u_{\xi\xi} - \tilde{\omega}(1-2\rho^2)u_\xi \tag{25}$$

and then to seek a solution of the form

$$u(\rho;\zeta,\tau) = \bar{u}(\zeta,\tau) + \sum_1^\infty f_k(\rho,\tau)\frac{\partial^k\bar{u}}{\partial\zeta^k} \tag{26}$$

(Such an expression can be obtained formally by expanding the Fourier transform of u with respect to ζ). It is then assumed that

$$\bar{u}_\tau = \sum_1^\infty K_k(\tau)\frac{\partial^k\bar{u}}{\partial\zeta^k} \tag{27}$$

and combining eqns. (25) - (27) and equating to zero the coefficients of the $\partial^k\bar{u}/\partial\zeta^k$ gives a sequence of equations:

$$f_{1,\tau} = Lf_1 - \{\tilde{\omega}(1-2\rho^2) + K_1\}, \tag{28}$$

$$f_{2,\tau} = Lf_2 - \{\tilde{\omega}(1-2\rho^2) + K_1\}f_1 + \{1-K_2\}, \tag{29}$$

$$f_{k+2,\tau} = Lf_{k+2} - \{\tilde{\omega}(1-2\rho^2)+K_1\}f_{k+1} + \{1-K_2\}f_k - \sum_1^k K_{i+2}f_{k-i}, \quad (f_o=1) \tag{30}$$

In addition, by eqn. (26),

$$\bar{f}_k = 0 \tag{29}$$

Thus averaging eqn. (28) gives

$$K_1 = 0$$

while the same operation on eqn. 29 gives

$$K_2 = 1 - 2\tilde{\omega}\int_o^1 \rho(1-2\rho^2)\, f_1(\rho,\tau)d\rho \tag{30}$$

$$= 1 + 4\tilde{\omega}\int_o^1 \rho^3 f_1(\rho,\tau)$$

This shows that the equation for f_1 must be solved and if the initial distribution across the tube is uniform $f_1(\rho,0) = 0$. Then (28) and (30) give

$$K_2 = 1 + \frac{\tilde{\omega}^2}{48}\{1 - 768 \sum_1^\infty \frac{J_3(\lambda_n)J_2(\lambda_n)}{\lambda_n^5[J_0(\lambda_n)]^2} e^{-\lambda_n^2\tau}\} \tag{31}$$

where $J_1(\lambda_n) = 0$. Thus the asymptotic value of D_e, namely $1+\tilde{\omega}^2/48$ is reached, for practical purposes, by the time τ exceeds 0.25 if the initial distribution is uniform. If the series in eqn. (27) is truncated at k=2 we have a new model, T_e', which is T_e with D_e given by eqn. (31).

Gill and his colleagues explored the accuracy of this model for certain standard initial distributions [12,15]; cf. also [1] where the relationship of various models is discussed. The validity of the truncation of the series in eqn. (27) is not easy to assess and the principal advance in clarifying this has recently been made by de Gance and Johns [8,9]. Though their work bears more immediately on the reactive case to be considered next one or two points should be mentioned here. Their key idea is to separate the construction of the solution of the equations for T_1 from the construction of the effective dispersion coefficient i.e. of T_e or, as the model in eqn. (27) might be called, $T_{e\infty}'$. They define an mth Taylor-Gill model, T_{em}' by

$$\bar{u}_\tau^{(m)} = \sum_{k=0}^{m} (-)^k X_k^{(m)}(\tau) \frac{\partial^k \bar{u}^{(m)}}{\partial \xi^k} \tag{32}$$

where $X_k^{(m)}$ are obtained by equating the first (m+1) Hermite moments of $\bar{u}$ and $\bar{u}^{(m)}$. They show the advantage of using these Hermite moments

$$V_p(\rho,\tau) = \int_{-\infty}^{\infty} h_p(\xi)\, u(\rho,\xi,\tau)d\xi \tag{33}$$

and

$$W_p(\tau) = \overline{V_p(\rho,\tau)} \tag{34}$$

where

$$h_p(\xi) = \frac{1}{\pi^{\frac{1}{2}} 2^p p!} \sum_{q=0}^{[p/2]} (-)^q \frac{p!}{q!(p-2q)!} (2\xi)^{p-2q} \tag{35}$$

and $[p/2] = p/2$ (p,even) or $(p-1)/2$ (p,odd). The Hermite moment satisfies

$$(V_p)_\tau = L V_p + 2(2p)^{\frac{1}{2}} \tilde{\omega}(1-\rho^2) V_{p-1} + 2\{p(p-1)\}^{\frac{1}{2}} V_{p-2} \tag{36}$$

which can be solved recursively for $p = 0,1,2...$ By using the Hermite moments, they ensure that the kth dispersion coefficient does not change in any model T'_{em} once $m \geq k$ i.e. $X_k^{(m)} = X_k^{(m')}$ for $m,m' \geq k$. This is a very important result. DeGance and Johns also show that there is a mathematically preferred weight function for averaging over the cross-section, but, since this is the area average when $\kappa = 0$, we will reserve mention of this to the next section. The whole work is a very elegant and penetrating analysis bringing together the approach of Taylor and Gill with the idea of matching moments.

4. A single reactant

Returning now to the reactive case with a single species we must ask how far the lessons of Taylor diffusion can be expected to carry over. The great conceptual lesson is that initial and asymptotic developments can be nicely discriminated. The initial development of the concentration field depends on the details of the reactant distribution at time $t=0$. The asymptotic development is a random dispersion qualitatively described by constant dispersion coefficients - the limiting values of Gill's K_k or Johns' $X_k^{(m)}$. This is implicit in Taylor's initial treatment, becomes increasingly clear in the succeeding work of Aris, Chatwin, Gill and Subramanian and De Gance and Johns and is explicit in Fife and Nicholes' two time scales. Calculations for various initial conditions show that the asymptotic value is achieved by the time $\tau=\frac{1}{2}$. It is of course this asymptotic value that gives the dispersion coefficient its conceptual validity, a validity enhanced by its independence of the averaging process in the cross-section, for to solve (27) or (32) with the time-dependent coefficients dictated by a particular initial distribution is merely to solve that particular initial value problem. And this is what raises problems in the reactive case since an important part of the process may be over before the asymptotic values are attained. In fact to anticipate some of the results, if we want to ensure that no more than 10% of the reactant be consumed before $\tau=\frac{1}{2}$, we cannot allow κ to be greater than about 0.1. Thus as simulations our Σ-series will not succeed, but as models they retain their validity.

Bearing this in mind let us go forward with a brief exposition of De Gance and Johns' approach. In the scalar case the equations are

$$u_\tau = Lu + u_{\xi\xi} - 2\tilde{\omega}(1-\rho^2)u_\xi, \tag{37}$$

$$u_\rho + \kappa u = 0, \rho = 1; \ u_\rho = 0, \rho = 0; \tag{38}$$

$$u(\rho,\xi,0) = u_o(\rho,\xi) \ . \tag{39}$$

They require the condition that $\int_{-\infty}^{\infty} u_o(\rho,\xi)\, e^{\xi^2} d\xi$ be finite, but this is no problem since an initial distribution confined to a finite region is conceptually appropriate. The Hermite moments defined by (33) and (35) satisfy

$$(V_p)_\tau = LV_p + 2(2p)^{\frac{1}{2}} \tilde{\omega}(1-\rho^2)V_{p-1} + 2\{p(p-1)\}^{\frac{1}{2}}V_{p-2} \tag{36 bis}$$

with

$$(V_p)_\rho = 0, \ \rho = 0; \ (V_p)_\rho + \kappa V_p = 0, \rho = 1 \tag{40}$$

$$V_p(\rho,0) = \int h_p(\xi)\, u_o(\rho,\xi)d\xi \tag{41}$$

Thus V_o satisfies the radial diffusion equation and is

$$V_o(\rho,\tau) = \sum_1^\infty a_n\, e^{-\lambda_n^2 \tau} J_o(\lambda_n\rho) \tag{42}$$

with

$$a_n = \lambda_n^2 \int_o^1 2\rho\, V_o(\rho,0)\, J_o(\lambda_n\rho)d\rho / J_o^2(\lambda_n)\ \{\kappa^2+\lambda_n^2\} \tag{43}$$

and λ_n is the nth root of

$$\lambda J_1(\lambda) + \kappa J_o(\lambda) = 0 \ . \tag{44}$$

The disposition of the roots is shown in the next figure and λ_1^2 is given in Fig. 4 and Table 1. When κ is small

$$\lambda_1^2 = 2\kappa - \frac{1}{2}\kappa^2 + \ldots \tag{45}$$

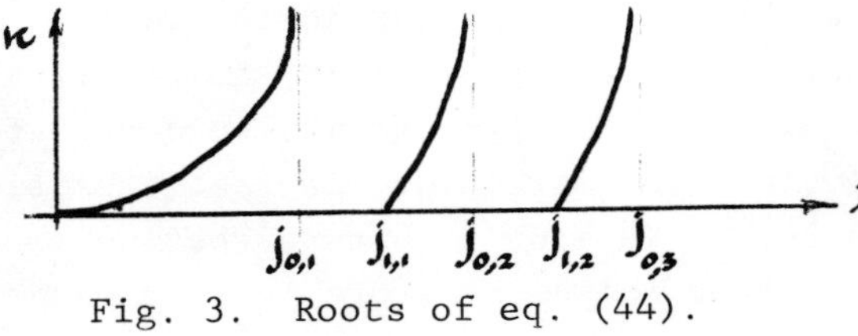

Fig. 3. Roots of eq. (44).

when large

$$\lambda_1 \sim j_{01}\ (1 - \frac{1}{\kappa} + \ldots) \tag{46}$$

where j_{01} = 2.4048 is the first zero of $J_o(\lambda)$. In the limit of infinite κ the denominator in eqn. (43) is, of course, $\lambda_n^2 J_1^2(\lambda_n)$; $\lambda_1^2 J_1^2(\lambda_1) = 1.56$. Asymptotically V_o is given by the first term of the series

$$V_o(\rho,\tau) = a_1 J_o(\lambda_1 \rho)\ e^{-\lambda_1^2 \tau}$$

Accordingly the total amount of the reactant that remains is

$$\int_o^1 2\rho\ V_o(\rho,\tau) d\rho \sim A_1 e^{-\lambda_1^2 \tau} \tag{47}$$

and the reactant appears to be reacting with a rate constant $\lambda_1^2 D/a^2$. This makes good sense for in the limit of small κ, $\lambda_1^2 D/a^2 \sim 2\pi ak/\pi a^2$; while as $\kappa \to \infty$ the whole process is controlled by diffusion at a rate $5.76D/a^2$.

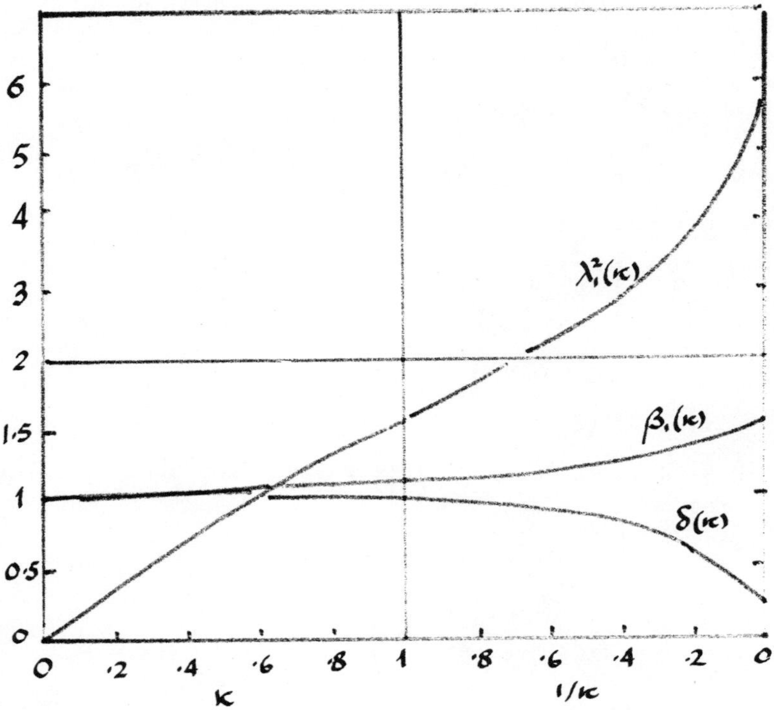

Fig. 4. Values of λ_1^2, β_1 and δ as functions of κ.

Johns and DeGance introduce a rather elegant refinement at this point. They observe that, though asymptotically

$$\dot{W}_o \sim -\lambda_1^2 W_o$$

for any average

$$W_p(\tau) = \int_o^1 w(\rho)\, V_p(\rho,\tau)d\rho / \int_o^1 w(\rho)d\rho, \quad w \geq 0 \ , \tag{48}$$

there is a preferred average when $w(\rho) = J_o(\lambda_1 \rho)$. It is mathematically preferable in that it simplifies all averages giving, from eqn. 42

$$W_o(\tau) = A_1 e^{-\lambda_1^2 \tau} \tag{49}$$

not asymptotically but always. From now on we will use this preferred average.

Proceeding to V_1 we have

$$(V_1) = LV_1 + 2^{3/2}\, \tilde{\omega}(1-\rho^2)V_o \tag{50}$$

with

$$(V_1)_\rho = 0, \ \rho = 0; \ (V_1)_\rho + \kappa(V_1) = 0, \ \rho = 1 \ . \tag{51}$$

This solution takes the form

$$V_1(\rho,\tau) = \tilde{\omega} \sum_1^\infty \{b_n + 2^{\frac{1}{2}} \beta_n \tau + f_n(\rho)\}\, a_n\, e^{-\lambda_n^2 \tau} J_o(\lambda_n \rho) \tag{52}$$

where on substitution and comparison of terms we find

$$\rho J_o(\lambda_n \rho) f_n'' + \{J_o(\lambda_n \rho) - 2\lambda_n \rho\, J_1(\lambda_n \rho)\} f_n' =$$

$$\rho\, \{2^{\frac{1}{2}}\beta_n - 2^{3/2}(1-\rho^2)\}\, J_o(\lambda_n \rho) \tag{53}$$

with

$$f_n'(0) = f_n'(1) = 0 \quad . \tag{54}$$

Multiplying by $J_o(\lambda_n \rho)$ and integrating from 0 to 1, using (54), shows that

$$\beta_n = 2 \int_o^1 \rho(1-\rho^2)\, J_o^2(\lambda_n \rho)d\rho / \int_o^1 \rho\, J_o^2(\lambda_n \rho)d\rho$$

$$= \frac{4}{3} \quad \frac{\kappa^2 - \kappa\lambda_n^2 + \lambda_n^4 + \kappa^2\lambda_n^2}{\lambda_n^2(\lambda_n^2 + \kappa^2)} \tag{55}$$

Because $J_o(\lambda_n\rho)$ has zeros in (0,1) for $n > 1$, the calculation of f_2, f_3 etc. is not trivial though Johns gives formulae that are equivalent to solving (53). Fortunately we are only interested in the dominant term and so need only consider $f_1(\rho)$ which is given by

$$2^{-\frac{1}{2}}f_1(\rho) = \int_o^\rho \frac{d\rho'}{\rho' J_o^2(\lambda_1\rho')} \int_o^{\rho'} \{\beta_n - 2(1-\rho''^2)\}\rho'' J_o^2(\lambda_1\rho'')d\rho'' \tag{56}$$

Now the dominant term of V_1 is

$$\tilde{\omega}\beta_1 \tau a_1 e^{-\lambda_1^2\tau} J_o(\lambda_1\rho)$$

and

$$W_1 = 2^{\frac{1}{2}} \tilde{\omega}\beta_1 \tau W_o . \tag{57}$$

The square root of 2 is an artefact of the Hermite polynomial and equation (57) merely says that the center of gravity of the reactant moves at β_1 times the mean speed of the stream. That β_1 should be greater than 1 does not surprise us, since the reactant is being destroyed at the wall where the flow is less than the average. However it does suggest that there may be some interesting couplings in the multicomponent case. β_1 is shown in Fig. 4 and Table 1: $\beta_1\tilde{\omega}$ is Gill's - K_1(eq. 43 of [29]) or Johns X_1 (Table 3 of [8]).

We could now proceed V_2 and from the asymptotic form of W_2 deduce a dispersion coefficient, X_2 or K_2. Instead let us take advantage of a most elegant formula of DeGance & Johns [8] that in the preferred mean the dispersion coefficient is (in our notation)

$$1 + \frac{\tilde{\omega}^2}{2} \{\frac{2(1-\rho^2)V_1}{W_o} - \beta_1 \frac{W_1}{W_o}\} . \tag{58}$$

The beauty of this formula is that it does not require a knowledge of V_2 but rests solely on suitable averages of V_1. What is more, any term proportional to $J_o(\lambda_1\rho)$ drops out since β_1 is such that $2(1-\rho^2)\ J_o(\lambda_1\rho) = \beta_1\ J_o(\lambda_1\rho)$. Thus the only term that survives is that with $f_1(\rho)$. After some manipulation we have a dimensionless dispersion coefficient of

$$\Delta = \Delta(\kappa,\tilde{\omega}) = 1 + \frac{\tilde{\omega}^2}{48}\,\delta(\kappa) \tag{59}$$

where

$$\delta(K) = 96\,\{2\int_0^1 \rho(1-\rho^2)\, f_1(\rho)\, J_0^2(\lambda_1\rho)d\rho - \beta_1\int_0^1\rho\, f_1(\rho)$$

$$J_0^2(\lambda_1\rho)d\rho\,\}/\{J_0^2(\lambda_1) + J_1^2(\lambda_1)\} \quad . \tag{60}$$

This has been written in such a way that $\delta(0) = 1$, for as $\kappa \to 0$, $\lambda_1^2 \sim 2\kappa$, $\beta_1 \to 1$, $f_1 \to \frac{1}{8}\rho^2(\rho^2-2)$ and we recover the non-reactive Taylor dispersion coefficient. Thus eqns. (56), (59) and (60) define Gill's $K_2(Pe)^2$ or Johns X_2. $\delta(\kappa)$ is also given in Fig. 4 and Table 1.

The second order Taylor-Gill model is thus

$$\bar{u}_\tau = \{1 + \frac{\tilde{\omega}^2}{48}\,\delta(\kappa)\}\,\bar{u}_{\xi\xi} - \beta_1(\kappa)\tilde{\omega}\,\bar{u}_\xi - \lambda_1^2(\kappa)\bar{u} \tag{60}$$

where the dependence of the equivalent coefficients on the two parameters is quite explicit. It should be recalled at this point that the asymptotic values are attained for $\tau > 0.5$.

5. Comparison of steady states

If we consider the steady state in a long reactor we seek solutions of

$$\{1 + \frac{\tilde{\omega}^2}{48}\,\delta(\kappa)\}\,\bar{u}_{\xi\xi} - \beta_1(\kappa)\tilde{\omega}\bar{u}_\xi - \lambda_1^2(\kappa)\bar{u} = 0 \tag{61}$$

with

$$\bar{u}(0) = 1,\ \bar{u}(\xi) \to 0 \text{ as } \xi \to \infty\ . \tag{62}$$

Thus

$$\bar{u}(\xi) = e^{-\mu\xi} \tag{63}$$

where μ is the positive root of

$$\{1 + \frac{\tilde{\omega}^2}{48}\,\delta(\kappa)\}\mu^2 + \beta_1(\kappa)\tilde{\omega}\mu - \lambda_1^2(\kappa) = 0 \quad . \tag{64}$$

This constant μ gives us a better idea of the performance of the reactor. This model is $\Sigma_{e6}^{(2)}$.

It is instructive to compare this with the first term in the solution of Σ_{16}, of which the equations are:

$$Lu + u_{\xi\xi} - 2\tilde{\omega}(1-\rho^2)u_\xi = 0$$

$$u_\rho(0,\xi) = 0,\ u_\rho(1,\xi) + \kappa u(1,\xi) = 0 \tag{65}$$

$$u(\rho,0) = 1$$

Table 1

κ	κ^{-1}	λ_1^2	β_1	δ
0		0	1	1
.1		.1951	1.0165	1
.2		.3807	1.0325	1
.3		.5572	1.0480	1
.4		.7252	1.0632	.9987
.5		.8851	1.0778	.9942
.6		1.0372	1.0920	.9884
.7		1.1821	1.1057	.9815
.8		1.3201	1.1189	.9735
.9		1.4516	1.1317	.9647
1	1	1.5770	1.1440	.9551
1.11	.9	1.7095	1.1572	.9437
1.25	.8	1.8656	1.1729	.9287
1.43	.7	2.0522	1.1919	.9083
1.67	.6	2.2786	1.2153	.8801
2	.5	2.5582	1.2446	.8402
2.50	.4	2.9105	1.2820	.7824
3.33	.3	3.3630	1.3305	.6976
5	.2	3.9593	1.3938	.5743
10	.1	4.7502	1.4740	.4095
∞	0	5.7831	1.5639	.2397

If we seek a solution of the form $e^{-\mu\xi}\ g(\rho)$ then

$$Lg + \{2\mu\tilde{\omega}(1-\rho^2) + \mu^2\}g = 0 \tag{66}$$

and g can be expressed in terms of Kummer's confluent hypergeometric function

$$M(a,c,x) = {}_1F_1(a;c;x) = \sum_0^\infty \frac{(a)_n}{(c)_n} \frac{x^n}{n!} \tag{67}$$

In fact

$$g(\rho) = e^{-\omega\rho^2}\ M(\tfrac{1-\omega}{2},\ 1,\ 2\omega\rho^2) \tag{68}$$

where

$$\omega = (2\mu\tilde{\omega} + \mu^2)/2\ \sqrt{2\mu\tilde{\omega}}\ . \tag{69}$$

We are interested in the smallest value of μ that makes $g' + \kappa g = 0$ and this after some manipulation leads to the equation

$$\kappa = (1 \pm \omega) - (1 \mp \omega)\,\frac{M(\frac{1-\omega}{2} \pm 1,\ 1,\ 2\omega)}{M(\frac{1-\omega}{2},\ 1,\ 2\omega)} \tag{70}$$

where the upper or lower signs must be taken consistently.

One can also compare even simpler models. The plug-flow, non-diffuse model would be Σ_7:

$$\tilde{\omega} u_\xi + 2\kappa u = 0$$

giving

$$\mu = 2\kappa / \tilde{\omega}$$

The plug-flow, non-diffusive model allowing for the speed up of flow and diffusion to the wall would be Σ_8:

$$\beta_1(\kappa)\tilde{\omega}\, u_\xi + \lambda_1^2(\kappa) u = 0$$

or

$$\mu = \lambda_1^2(\kappa)/\beta_1(\kappa)\tilde{\omega}$$

The plug-flow model with Taylor diffusion uncorrected for longitudinal diffusion or reaction would be Σ_9:

$$(\tilde{\omega}^2/48)u_{\xi\xi} - \beta_1(\kappa)\tilde{\omega}u_\xi - \lambda_1^2(\kappa)u = 0$$

giving

$$\mu = 24\ \{-\beta_1 + \sqrt{\beta_1^2 + \lambda_1^2/12}\}/\tilde{\omega}$$

When longitudinal diffusion is allowed for but the Taylor term is not corrected for reaction Σ_{10}:

$$(1 + \frac{\tilde{\omega}^2}{48})u_{\xi\xi} - \beta_1\tilde{\omega}u_\xi - \lambda_1^2 u = 0$$

and

$$\mu = \{-\beta_1\tilde{\omega} + \beta_1^2\tilde{\omega}^2 + (4 + \frac{\tilde{\omega}^2}{12})\lambda_1^2\}/(2 + \frac{\tilde{\omega}^2}{24}) .$$

It would also be mentioned that there have been a number of studies of the steady state equations usually neglecting longitudinal diffusion. Thus Solomon and Hudson [31] used a Galerkin method for simultaneous homogeneous and wall reaction; Ulrichson and Schmitz [35] used a finite scheme to discuss the effect of entrance length; Lupa and Dranoff [24] used the eigen function expansion for annular reactors; while Pancharatnam and Homsey [25] were concerned with the asymptotic case of large $\tilde{\omega}$.

These are compared for μ = 0.1 and 0.5 in Fig. 5 and it is clear that the δ κ) refinement has little influence for this value of μ. Since $\mu < 2.4048$ it is concluded that it never has great significance.

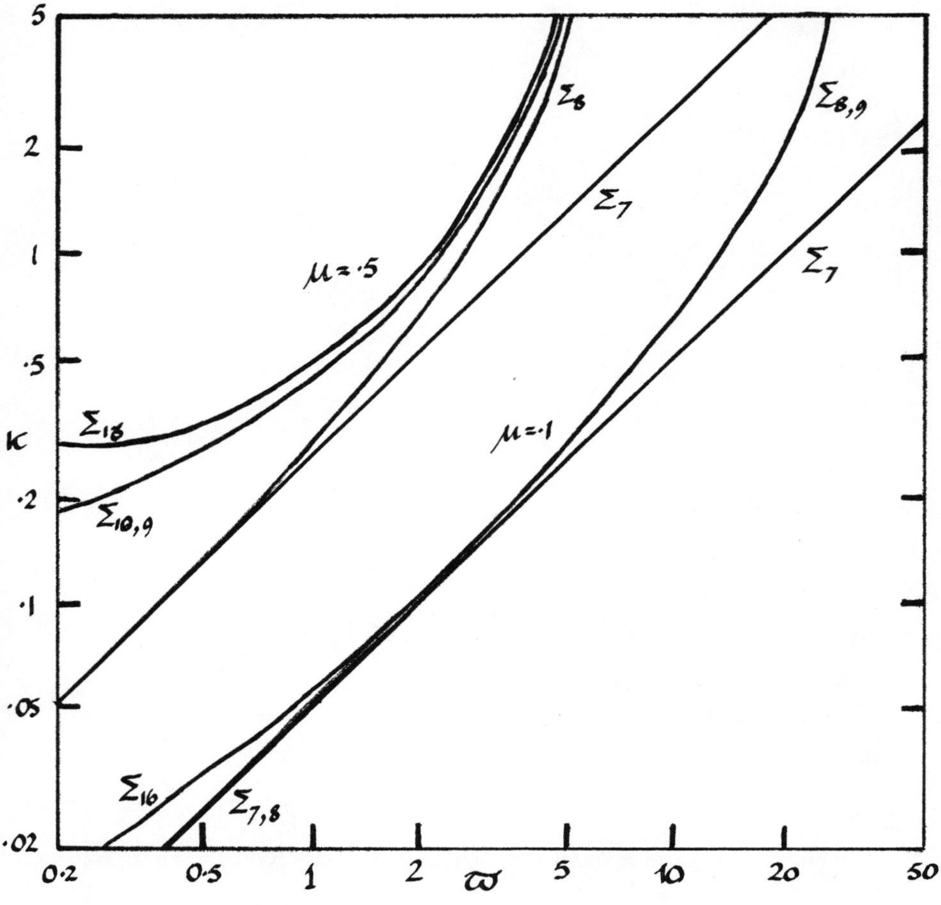

Fig. 5. Contours of μ=.5 and .1 in the R,$\tilde{\omega}$-plane.

6. Systems of first order reactions

Before seeing how the G-S-DeG-J analysis can be used with systems let us follow Johns and DeGance [20] in a very illuminating extension of Taylor's intuitive analysis. We start out from eqns. (10) and (11) and let a bar denote the area average, i.e.

$$\bar{\underset{\sim}{u}}(\xi,\tau) = \int_0^1 2\rho \underset{\sim}{u}(\rho,\xi,\tau)d\rho \tag{71}$$

We are aiming to get an equivalent model for $\bar{u}$, namely

$$\bar{\underset{\sim}{u}}_\xi = \underset{\sim}{D}_e \bar{\underset{\sim}{u}}_{\xi\xi} - \tilde{\omega}\underset{\sim}{\Omega}_e \bar{\underset{\sim}{u}}_\xi - \underset{\sim}{K}_e \bar{u} \tag{72}$$

Averaging eqn. (10) gives

$$\bar{\underset{\sim}{u}}_\tau = \underset{\sim}{D}\bar{\underset{\sim}{u}}_{\xi\xi} - 2\underset{\sim}{\kappa}\underset{\sim}{u}_1 - 2\tilde{\omega}\left[\overline{(1-\rho^2)u}\right]_\xi \tag{73}$$

where $\underset{\sim}{u}_1$ denotes $\underset{\sim}{u}(1,\xi,\tau)$.

If the radial variation of $\underset{\sim}{u}$ and its derivatives is slight we may hope that it could be approximated by $\underset{\sim}{u}'$ satisfying

$$\bar{\underset{\sim}{u}}'_\tau = \underset{\sim}{D}L\underset{\sim}{u}' + \underset{\sim}{D}\bar{\underset{\sim}{u}}'_{\xi\xi} - 2\tilde{\omega}(1-\rho^2)\bar{u}'_\xi \tag{74}$$

$$u'_\rho = 0,\ \rho = 0; \quad \underset{\sim}{D}\underset{\sim}{u}'_\rho + \underset{\sim}{\kappa}\underset{\sim}{u}' = 0,\ \rho = 1 \tag{75}$$

Now (74) can be written

$$L\underset{\sim}{u}' = 2\tilde{\omega}(1-\rho^2)\underset{\sim}{D}^{-1}\bar{\underset{\sim}{u}}'_\xi + (\underset{\sim}{D}^{-1}\bar{\underset{\sim}{u}}^1_\tau - \bar{\underset{\sim}{u}}'_{\xi\xi}) \tag{76}$$

Since the right hand side is merely a quadratic in ρ the solution can be obtained by quadratures. The condition at $\rho = 0$ rules out the logarithmic term and the mean fixes the constant to give

$$\underset{\sim}{u}' = \bar{\underset{\sim}{u}}' - \frac{1}{8}(1-2\rho^2)(\underset{\sim}{D}^{-1}\bar{\underset{\sim}{u}}'_\tau - \bar{\underset{\sim}{u}}'_{\xi\xi})$$
$$- \frac{1}{24}(5 - 12\rho^2 + 3\rho^4)\tilde{\omega}\ \underset{\sim}{D}^{-1}\underset{\sim}{u}'\xi \tag{77}$$

Thus on $\rho = 1$

$$\underset{\sim}{u}'_1 = \bar{\underset{\sim}{u}}' + \frac{1}{8}(\underset{\sim}{D}^{-1}\bar{\underset{\sim}{u}}'_\tau - \bar{\underset{\sim}{u}}'_{\xi\xi}) + \frac{1}{6}\tilde{\omega}\,\underset{\sim}{D}^{-1}\bar{\underset{\sim}{u}}'_\xi \tag{78}$$

$$\underset{\sim}{u}'_\rho = \frac{1}{2}(\underset{\sim}{D}^{-1}\bar{\underset{\sim}{u}}'_\tau - \bar{\underset{\sim}{u}}'_{\xi\xi}) + \frac{1}{2}\tilde{\omega}\,\underset{\sim}{D}^{-1}\bar{\underset{\sim}{u}}'_\xi$$

so that by the boundary condition (75)

$$\bar{\underset{\sim}{u}}'_\tau - \underset{\sim}{D}\bar{\underset{\sim}{u}}'_{\xi\xi} = -(\underset{\sim}{I} + \frac{1}{4}\underset{\sim}{\kappa}\underset{\sim}{D}^{-1})^{-1}[2\underset{\sim}{\kappa}\,\bar{\underset{\sim}{u}}' + \tilde{\omega}(\underset{\sim}{I} + \frac{1}{3}\underset{\sim}{\kappa}\underset{\sim}{D}^{-1})\bar{u}'_\xi] \tag{79}$$

and substituting this back into eqn. (77) gives

$$\begin{aligned}\underset{\sim}{u}' = {} & \{\underset{\sim}{I} + \frac{1}{4}(1-2\rho^2)\underset{\sim}{D}^{-1}(\underset{\sim}{I} + \frac{1}{4}\underset{\sim}{\kappa}\underset{\sim}{D}^{-1})^{-1}\underset{\sim}{\kappa}\}\,\bar{\underset{\sim}{u}}' \\ & + \frac{\tilde{\omega}}{24}\{3(1-2\rho^2)\underset{\sim}{D}^{-1}(\underset{\sim}{I} + \frac{1}{4}\underset{\sim}{\kappa}\underset{\sim}{D}^{-1})^{-1}(\underset{\sim}{I} + \frac{1}{3}\underset{\sim}{\kappa}\underset{\sim}{D}^{-1}) \\ & -(5 - 12\rho^2 + 3\rho^4)\underset{\sim}{D}^{-1}\}\bar{\underset{\sim}{u}}'_\xi\end{aligned} \tag{80}$$

Thus

$$\overline{[(1-\rho^2)\underset{\sim}{u}']} = \{\frac{1}{2}\underset{\sim}{I} + \underset{\sim}{C}\underset{\sim}{\kappa}\}\bar{\underset{\sim}{u}}' + \frac{\tilde{\omega}}{96}\{2\underset{\sim}{C}\underset{\sim}{B} - 3\underset{\sim}{D}^{-1}\}\bar{\underset{\sim}{u}}'_\xi$$

$$u'_\xi = \{I - \frac{1}{4}\underset{\sim}{C}\kappa\}\bar{\underset{\sim}{u}}_1 + \frac{\tilde{\omega}}{24}\{4\underset{\sim}{D}^{-1} - 3\underset{\sim}{C}\underset{\sim}{B}\}\bar{\underset{\sim}{u}}'_\xi$$

where

$$\underset{\sim}{C} = \underset{\sim}{D}^{-1}(I + \frac{1}{4}\underset{\sim}{\kappa}\underset{\sim}{D}^{-1})^{-1}, \quad B = (\underset{\sim}{I} + \frac{1}{3}\underset{\sim}{\kappa}\underset{\sim}{D}^{-1})$$

Using these estimates in eq. (73) gives

$$\begin{aligned}\bar{u}_\tau = {} & [\underset{\sim}{D} + \frac{\tilde{\omega}^2}{48}\{3\underset{\sim}{D}^{-1} - 2\underset{\sim}{C}\underset{\sim}{B}\}]\,\bar{u}'_{\xi\xi} \\ & - [\{\underset{\sim}{I} + \frac{1}{12}\underset{\sim}{C}\underset{\sim}{\kappa}\} + \frac{1}{12}\underset{\sim}{\kappa}\{4D^{-1} - 3\underset{\sim}{C}\underset{\sim}{B}\}]\,\tilde{\omega}\bar{u}'_\xi \\ & - 2\underset{\sim}{\kappa}\{\underset{\sim}{I} - \frac{1}{4}\underset{\sim}{C}\underset{\sim}{\kappa}\}\bar{u}'\end{aligned}$$

Thus

$$\begin{aligned}\underset{\sim}{D}_e & = \underset{\sim}{D} + \frac{\tilde{\omega}^2}{48}\underset{\sim}{D}^{-1}(I + \frac{1}{4}\underset{\sim}{\kappa}\underset{\sim}{D}^{-1})^{-1}(\underset{\sim}{I} + \frac{1}{12}\underset{\sim}{\kappa}\underset{\sim}{D}^{-1}) = \underset{\sim}{D} + \frac{\tilde{\omega}^2}{48} \\ & \quad \underset{\sim}{C}(I + \frac{1}{12}\underset{\sim}{\kappa}\underset{\sim}{D}^{-1}) = \underset{\sim}{D} + \frac{\tilde{\omega}^2}{144}(2\underset{\sim}{C} + \underset{\sim}{D}^{-1})\end{aligned} \tag{81}$$

$$\begin{aligned}\underset{\sim}{\Omega}_e & = \underset{\sim}{I} + \frac{1}{12}\underset{\sim}{D}^{-1}(I + \frac{1}{4}\underset{\sim}{\kappa}\underset{\sim}{D}^{-1})^{-1}\underset{\sim}{\kappa} + \frac{1}{12}\underset{\sim}{\kappa}\underset{\sim}{D}^{-1}(\underset{\sim}{I} + \frac{1}{4}\underset{\sim}{\kappa}\underset{\sim}{D}^{-1})^{-1} \\ & = \underset{\sim}{I} + \frac{1}{12}(\underset{\sim}{C}\underset{\sim}{\kappa} + \underset{\sim}{\kappa}\underset{\sim}{C})\end{aligned} \tag{82}$$

and

$$\begin{aligned}\underset{\sim}{K}_e & = 2\underset{\sim}{\kappa}\,\underset{\sim}{D}^{-1}(\underset{\sim}{I} + \frac{1}{4}\underset{\sim}{\kappa}\underset{\sim}{D}^{-1})^{-1}\underset{\sim}{D} = 2(\underset{\sim}{I} + \frac{1}{4}\underset{\sim}{\kappa}\underset{\sim}{D}^{-1})^{-1}\underset{\sim}{\kappa} \\ & = 2\underset{\sim}{D}\underset{\sim}{C}\underset{\sim}{\kappa} = 2\underset{\sim}{\kappa}\underset{\sim}{C}\underset{\sim}{D}\end{aligned} \tag{83}$$

where

$$\underset{\sim}{C} = \underset{\sim}{D}^{-1}(\underset{\sim}{I} + \frac{1}{4}\underset{\sim}{\kappa}\underset{\sim}{D}^{-1})^{-1} \tag{84}$$

In the scalar case when κ is small

$$K_e = 2\kappa(1 + \frac{1}{4}\kappa)^{-1} \sim \lambda_1^2(\kappa), \quad \Omega_e = \{1 + \frac{1}{6}\kappa(1 + \frac{1}{4}\kappa)^{-1}\} \sim \beta_1(\kappa)$$

$$D_e = 1 + \frac{\tilde{\omega}^2}{48}(1 + \frac{1}{12}\kappa)/(1 + \frac{1}{4}\kappa) \sim 1 + \frac{\tilde{\omega}^2}{48}\delta(\kappa) . \tag{86}$$

A particularly interesting feature is that the convective terms are coupled through the reaction for $\underset{\sim}{\Omega}_e$ is not diagonal except in the trivial case when $\underset{\sim}{\kappa}$ and $\underset{\sim}{D}$ are diagonal.

This work of Johns and DeGance [20] manipulates the equations with Taylorian virtuosity but leaves us rather in doubt as to what degree of approximation is attained. Their tie-back to Gill and Sankarasubramanian [29] is reassuring and since the scalar forms (86) show good agreement for $\kappa < 2$ and we have already raised doubts as to the value of a Taylor model for $\kappa > 0.1$, it would seem to be wanton to ask how their other method might extend to the multicomponent case. However, letting curiosity get the better of discretion, we will proceed, lit on our way by their assurance that "we can generalize the procedure to vector convective diffusion equations". [8 p. 191]. These we have in eqns. (10) - (13) and the equations for the moments (33) - (36) can be taken over as vector equations immediately. In particular, (lowering the case for the vector)

$$(\underset{\sim}{V}_o)_\rho = \underset{\sim}{D}\, L\, \underset{\sim}{v}_o$$

$$(\underset{\sim}{V}_o)_\rho = 0, \ \rho = 0; \quad \underset{\sim}{D}(\underset{\sim}{V}_o)_\rho + \underset{\sim}{\kappa}\,\underset{\sim}{v}_o = 0, \ \rho = 1$$

Now $\underset{\sim}{D}$ is positive definite and so $\underset{\sim}{\Delta} = \underset{\sim}{D}^{-\frac{1}{2}}$ exists and the solution can be written as the superposition of terms like

$$e^{-\lambda^2\tau}\, J_o(\lambda\underset{\sim}{\Delta}\rho)\underset{\sim}{a}$$

where J_o is the Bessel function of a matrix whose properties are discussed in the Appendix.

Now if the boundary condition at $\rho = 1$ is to be satisfied then,

$$\{-\underset{\sim}{D}\underset{\sim}{\Delta}\lambda\, J_1(\lambda\underset{\sim}{\Delta}) + \underset{\sim}{K}\, J_o(\lambda\underset{\sim}{\Delta})\}\,\underset{\sim}{a} = 0 \tag{87}$$

and hence the determinantal equation

$$|\underset{\sim}{\lambda\Delta}\, J_1(\lambda\underset{\sim}{\Delta}) - \underset{\sim}{D}^{-1}\underset{\sim}{\kappa}\, J_0(\lambda\underset{\sim}{\Delta})| = 0 \quad . \tag{88}$$

If $J_0(\lambda\underset{\sim}{\Delta})$ is non-singular this can be written

$$|Z(\lambda\underset{\sim}{\Delta}) - \underset{\sim}{D}^{-1}\underset{\sim}{\kappa}| = 0 \tag{89}$$

where

$$Z(z) = zJ_1(z)/J_0(z) \quad . \tag{90}$$

Moreover if $\underset{\sim}{\Delta}$ is diagonalizable, say

$$\underset{\sim}{L}\underset{\sim}{\Delta}\underset{\sim}{L}^{-1} = \mathrm{diag}(\delta_1, \delta_2, \ldots) \tag{91}$$

then

$$\underset{\sim}{L}Z(\lambda\underset{\sim}{\Delta})\underset{\sim}{L}^{-1} = \mathrm{diag}\,\{Z(\lambda\delta_1), Z(\lambda\delta_2) \ldots\} = Y(\lambda\underset{\sim}{\Delta}) \tag{92}$$

and

$$|Y(\lambda\underset{\sim}{\Delta}) - \underset{\sim}{L}\underset{\sim}{D}^{-1}\underset{\sim}{\kappa}\underset{\sim}{L}^{-1}| = |Y(\lambda\underset{\sim}{\Delta}) - \underset{\sim}{D}^{*}\underset{\sim}{L}\underset{\sim}{\kappa}\underset{\sim}{L}^{-1}| = 0 \tag{93}$$

where

$$\underset{\sim}{D}^{*} = \mathrm{diag}[\delta_1^{-2}, \delta_2^{-2}, \ldots] \quad . \tag{94}$$

In general there is an R-fold denumerable infinity of values of λ satisfying eqn. (88) which can be ordered $0 \leq \lambda_1 \leq \lambda_2 \leq \lambda_3 \ldots$. Then

$$\underset{\sim}{v}_0 = \sum_{n=1}^{\infty} C_n\, e^{-\lambda_n^2\tau} J_0(\lambda_n\underset{\sim}{\Delta}\rho)\underset{\sim}{a}_n \tag{95}$$

where the $\underset{\sim}{a}_n$ are the solutions of eqn. (87), and c_n are scalars.

In an asymptotic sense the first R terms give the behavior of v_0. Thus if $\underset{\sim}{\Lambda}^2 = \mathrm{diag}(\lambda_1^2 \ldots \lambda_R^2)$

$$\underset{\sim}{w}_0 = \underset{\sim}{N}\, e^{-\underset{\sim}{\Lambda}^2\tau}\, \underset{\sim}{C} = \underset{\sim}{N}\, e^{-\Lambda^2\tau}\, \underset{\sim}{N}^{-1}\, \underset{\sim}{w}_0(0)$$

$$= e^{-\underset{\sim}{K}_e\tau}\, \underset{\sim}{w}_0(0)$$

where

$$\underset{\sim}{K}_e = \underset{\sim}{N}\, \underset{\sim}{\Lambda}^2\, \underset{\sim}{N}^{-1} \tag{96}$$

and

$$\underset{\sim}{N} = 2\underset{\sim}{\Lambda}^{-1}\underset{\sim}{\Delta}^{-1} J_1(\underset{\sim}{\Lambda}\underset{\sim}{\Delta})\underset{\sim}{A} \tag{97}$$

$$\underset{\sim}{A} = [\underset{\sim}{a}_1, \ldots, \underset{\sim}{a}_R] \tag{98}$$

Noting that this comparison is free of any reference to the initial conditions we will turn aside to see how the moments develop in general, relying on the consecutive reaction scheme A → B → C to give us some particularity.

7. Moments in the equivalent model

Moments may be generated from the equivalent model (72) as follows. Since we are already averaging over the cross-section we can jump from $\bar{u}$ to $\bar{w}_p$ by

$$\bar{\underset{\sim}{w}}_p(\tau) = \int_{-\infty}^{\infty} \xi^p \, \bar{\underset{\sim}{u}}(\xi,\tau)d\xi \tag{99}$$

Then

$$(\bar{\underset{\sim}{w}}_p)_\tau = - \underset{\sim}{K}_e \bar{\underset{\sim}{w}}_p + p\tilde{\omega}\underset{\sim}{\Omega}_e \bar{\underset{\sim}{w}}_{p-1} + p(p-1) \underset{\sim}{D}_e \bar{\underset{\sim}{w}}_{p-2} \tag{100}$$

so that

$$\bar{\underset{\sim}{w}}_p(\tau) = e^{-\underset{\sim}{K}_e\tau} \bar{\underset{\sim}{w}}_{po} + \int_0^\tau e^{-\underset{\sim}{K}_e(\tau-\sigma)} \left[p\tilde{\omega}\underset{\sim}{\Omega}_e \bar{\underset{\sim}{w}}_{p-1} + p(p-1) \underset{\sim}{D}_e \bar{\underset{\sim}{w}}_{p-2}\right]d\sigma . \tag{101}$$

Let us write

$$\bar{\underset{\sim}{w}}_p(\tau) = \sum_{q=0}^{p} \underset{\sim}{W}_{pq}(\tau)\bar{w}_{qo} \tag{102}$$

then

$$\underset{\sim}{W}_{pp} = e^{-\underset{\sim}{K}_e\tau}, \quad p = 0, 1, 2, .. \tag{103}$$

$$\underset{\sim}{W}_{p,p-1} = p\tilde{\omega} \int_0^\tau e^{-\underset{\sim}{K}_e(\tau-\sigma)} \underset{\sim}{\Omega}_e \, e^{-\underset{\sim}{K}_e\sigma} \, d\sigma \tag{104}$$

and for q < p-1

$$\underset{\sim}{W}_{p,q}(\bar{\varepsilon}) = p\tilde{\omega} \int_0^\tau e^{-\underset{\sim}{K}_e(\tau-\sigma)} \underset{\sim}{\Omega}_e \underset{\sim}{W}_{p-1,q}(\sigma)d\sigma + p(p-1)\int_0^\tau e^{-\underset{\sim}{K}_e(\tau-\sigma)} \underset{\sim}{D}_e \underset{\sim}{W}_{p-2,q}(\sigma)d\sigma \tag{105}$$

For a consecutive reaction scheme all the matrices are lower triangular, and in particular for A → B → C we write

$$\underset{\sim}{K}_e = \begin{bmatrix} \tilde{\alpha} & \cdot \\ -\tilde{\beta} & \tilde{\gamma} \end{bmatrix}, \quad \underset{\sim}{\Omega}_e = \begin{bmatrix} a & \cdot \\ -b & c \end{bmatrix}, \quad \underset{\sim}{D}_e = \begin{bmatrix} \tilde{a} & \cdot \\ \tilde{b} & \tilde{c} \end{bmatrix} \tag{106}$$

It will also be convenient to write

$$E_n(\tau) = \frac{a^n e^{-\tilde{\alpha}\tau} - c^n e^{-\tilde{\gamma}\tau}}{\tilde{\gamma}-\tilde{\alpha}} \quad , \quad E_n(\tau) = \frac{\tilde{a}^n e^{-\tilde{\alpha}\tau} - \tilde{c}^n e^{-\tilde{\gamma}\tau}}{\tilde{\gamma}-\tilde{\alpha}}$$

$$d = (c-a)/(\tilde{\gamma}-\tilde{\alpha}) \quad , \quad \tilde{d} = (\tilde{c}-\tilde{a})/(\tilde{\gamma}-\tilde{\alpha}) \tag{107}$$

Then

$$\underset{\sim}{W}_{oo} = \begin{bmatrix} e^{-\tilde{\alpha}\tau} & \cdot \\ \tilde{\beta}E_o(\tau) & e^{-\tilde{\gamma}\tau} \end{bmatrix} = \underset{\sim}{W}_{11} = \underset{\sim}{W}_{22} \tag{108}$$

$$\underset{\sim}{W}_{10} = \tilde{\omega} \begin{bmatrix} a\tau e^{-\tilde{\alpha}\tau} & \cdot \\ \beta E_1(\tau) + (\tilde{\beta}d-b)E_o(\tau) & c\tau e^{-\tilde{\gamma}\tau} \end{bmatrix} = \frac{1}{2} W_{21} \tag{109}$$

$$\underset{\sim}{W}_{20} = 2 \begin{bmatrix} a\tau e^{-\tilde{\alpha}\tau} & \cdot \\ \beta\tilde{E}_1(\tau) + (\tilde{\beta}\tilde{d}+\tilde{b})E_o(\tau) & \tilde{c}\tau e^{-\tilde{\gamma}\tau} \end{bmatrix}$$

$$+ \tilde{\omega}^2 \begin{bmatrix} a^2\tau^2 e^{-\tilde{\alpha}\tau} & \cdot \\ \tilde{\beta}\tau^2 E_2(\tau) + 2(\tilde{\beta}d-b)\{E_1(\tau)\tau + dE_o(\tau)\} & c^2\tau^2 e^{-\tilde{\gamma}\tau} \end{bmatrix} \tag{110}$$

If $b = \tilde{\beta}d$, $\underset{\sim}{W}_{10} = \tilde{\omega}\tau\underset{\sim}{\Omega}_e \underset{\sim}{W}_{oo}$.

Clearly a particularly important initial condition would be $\bar{\underset{\sim}{w}}_{oo} = [1,0]$, $\bar{\underset{\sim}{w}}_{po} = [0,0]$, $p > 0$, so that we would have

$$\bar{w}_{01} = e^{-\tilde{\alpha}\tau}, \ \bar{w}_{11} = \tilde{\omega}a\tau e^{-\tilde{\alpha}\tau}, \ \bar{w}_{21} = (2\tilde{a}\tau + \tilde{\omega}^2 a^2\tau^2)e^{-\tilde{\alpha}\tau}$$

$\bar{w}_{02} = \tilde{\beta}E_o(\tau)$, $\bar{w}_{12} = \tilde{\omega}\{\tilde{\beta}\tau E_1(\tau) + (\tilde{\beta}d-b)E_o\}$, etc. The interpretation in terms of mean and variance gives means of $\tilde{\omega}a\tau$ and $\tilde{\omega}(E_1/E_o)\tau + \tilde{\omega}(d-b/\tilde{\beta})$. If $\tilde{\alpha} \ll \tilde{\gamma}$ the latter moves with the same speed as the former; if $\tilde{\alpha} \gg \tilde{\gamma}$ it moves with speed $\tilde{\omega}$ c. The variance of the first reaction is, as expected, $2\tilde{a}\tau$; that of the second is

$$2(\tilde{E}_1/E_o)\tau + 2\{\tilde{d}+(\tilde{b}/\tilde{\beta})\} + \tilde{\omega}^2\{d^2-(b/\tilde{\beta})^2\}$$
$$+ \tilde{\omega}^2\tau^2 \ (E_2/E_o - E_1^2)/E_o^2 \ .$$

The last term though involving τ^2 is small. It is clear that comparison of the matrices $\underset{\sim}{W}_{pq}$ is what is needed so let us turn to the consecutive reaction scheme in detail.

8. Consecutive irreversible reactions

Consider the reaction scheme* $A \to B \to C$ and take

$$\underset{\sim}{D} = \begin{bmatrix} 1 & . \\ . & \delta^{-2} \end{bmatrix} \qquad \underset{\sim}{\kappa} = \begin{bmatrix} \kappa & . \\ -\kappa & \kappa' \end{bmatrix} \tag{111}$$

so that $\Delta = \mathrm{diag}(1,\delta)$. Then

$$|\lambda\underset{\sim}{\Delta}\, J_1(\lambda\underset{\sim}{\Delta}) - \underset{\sim}{D}^{-1}\underset{\sim}{K}\, J_o(\lambda\underset{\sim}{\Delta})| =$$

$$\{\lambda\delta J_1(\lambda\delta) - K'\delta^2 J_o(\lambda\delta)\}\ \{\lambda J_1(\lambda) - KJ_o(\lambda)\} = 0 \tag{112}$$

so that we have two series of roots

$$\lambda_n = L_n(\kappa)\ ,\quad \lambda_n' = L_n(\kappa'\delta^2)/\delta \tag{113}$$

If κ and $\kappa'\delta^2$ are small

$$\lambda_1^2 = 2\kappa - \frac{1}{2}\kappa^2 + ..,\ \lambda_1^2 = 2\kappa' - \frac{1}{2}\kappa'^2\delta^2 + \ldots,$$

$$\lambda_n \doteqdot J_{1,n-1},\quad \lambda_n' \doteqdot J_{1,n-1}/\delta$$

If κ and $\kappa'\delta^2$ are large

$$\lambda_n = j_{o,n}\ (1 - \frac{1}{\kappa}..)\ ,\ \lambda_n' = \frac{j_{o,n}}{\delta}\ (1 - \frac{1}{\kappa'\delta^2}\ ..\ )$$

If κ and κ' are fixed and δ varies, the λ_n are fixed while the λ_n' decrease as δ increases. Thus, as we see in Fig. 6, there

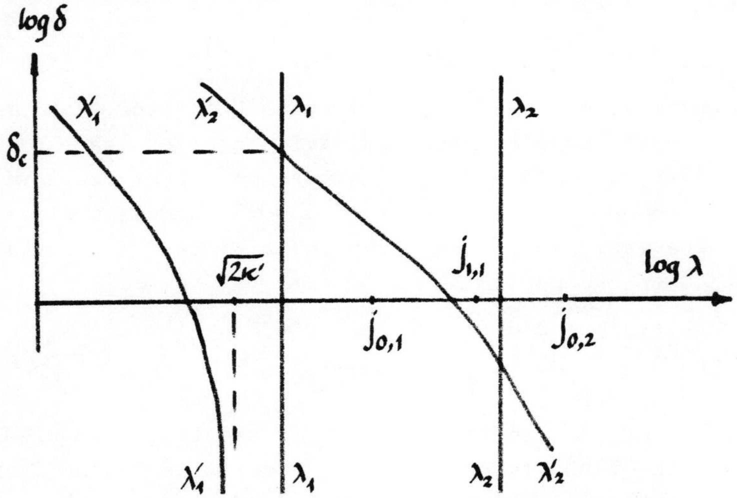

Fig. 6. Behavior of lowest roots of eq. 112.

*Hudson [18] considered this case for plug flow.

will be a critical δ_c above which the two least eigenvalues are λ_1' and λ_2' rather than λ_1 and λ_1'. Now eqns. (87) are

$$\begin{bmatrix} -\lambda J_1(\lambda) + \kappa J_0(\lambda) & 0 \\ -\kappa J_0(\lambda) & -\lambda J_1(\lambda\delta)/\delta + \kappa' J_0(\lambda\delta) \end{bmatrix} \begin{bmatrix} a_1 \\ a_2 \end{bmatrix} = 0 \tag{114}$$

Thus when $\lambda = \lambda_n = L_n(\kappa)$

$$\frac{a_{1n}}{a_{2n}} = -\frac{J_1(\lambda_n\delta)}{\delta J_1(\lambda_n)} + \frac{\kappa' J_0(\lambda_n\delta)}{\kappa J_0(\lambda_n)} = \frac{1}{e_n} \tag{115}$$

whereas, when $\lambda = \lambda_n' = L_n(\kappa'\delta^2)/\delta$, $a_{1n} = 0$.

We thus have

$$v_{01}(\rho,\tau) = \sum_{n=1}^{\infty} a_{1n}\, e^{-\lambda_n^2\tau} J_0(\lambda_n\rho) \tag{116}$$

$$v_{02}(\rho,\tau) = \sum_{n=1}^{\infty} a_{1n}\, e_n\, e^{-\lambda_n^2\tau} J_0(\lambda_n\delta\rho) + \sum_{n=1}^{\infty} a_{2n}\, e^{-\lambda_n'^2\tau} J_0(\lambda_n'\delta\rho) \tag{117}$$

The coefficients are determined by

$$a_{1n} = \frac{2}{J_0^2(\lambda_n)+J_1^2(\lambda_n)} \int_0^1 \rho\, v_{01}(\rho,0)\, J_0(\lambda_n\rho)d\rho \tag{118}$$

and

$$a_{2n} = \frac{2}{J_0^2(\lambda_n')+J_1^2(\lambda_n')} \int_0^1 \rho\{v_{02}(\rho,0) - \sum_{n=1}^{\infty} a_{1m}e_n\, J_0(\lambda_m\delta\rho)\}\, J_0(\lambda_n'\delta\rho)d\rho \tag{119}$$

For the average, $\underset{\sim}{w}_0 = 2\int_0^1 \rho\, \underset{\sim}{v}_0(\rho)d\rho$, we have

$$w_{01} = \Sigma\, a_{1n}[2J_1(\lambda_n)/\lambda_n]\, e^{-\lambda_n^2\tau} = \Sigma\, A_{1n}\, e^{-\lambda_n^2\tau}$$

$$w_{02} = \Sigma\, a_{1n}\, e_n\, [2J_1(\lambda_n\delta)/\lambda_n\delta]\, e^{-\lambda_n^2\tau} + \Sigma\, a_{2n}\, [2J_1(\lambda_n'\delta)/\lambda_n'\delta]\, e^{-\lambda_n^2\tau}$$

$$= \Sigma\, A_{2n}\, e^{-\lambda_n^2\tau} + \Sigma\, A_{2n}'\, e^{-\lambda_n'^2\tau}$$

If λ_1 and λ_1' dominate, i.e. $\delta < \delta_c$,

$$w_{01} = A_{11} e^{-\lambda_1^2 \tau}$$
$$w_{02} = A_{11} B_1 e^{-\lambda_1^2 \tau} + A_{21}' \, e^{-\lambda_1'^2 \tau} \tag{120}$$

where

$$B_1 = e_1 \frac{J_1(\lambda_1 \delta)}{\delta J_1(\lambda_1)} = \frac{\lambda_1 J_1(\lambda_1 \delta)}{\kappa' \delta J_0(\lambda_1 \delta) - \lambda_1 J_1(\lambda_1 \delta)} , \tag{121}$$

and B_1 is finite for $\delta < \delta_c$. But this can be written

$$\underset{\sim}{w}_0(\tau) = \underset{\sim}{N} \, e^{-\underset{\sim}{\Lambda}^2} \, \underset{\sim}{N}^{-1} \, \underset{\sim}{w}_\infty \tag{122}$$

where

$$\underset{\sim}{N} = \begin{bmatrix} 1 & \cdot \\ B_1 & 1 \end{bmatrix} \qquad \underset{\sim}{\Lambda} = \text{diag } (\lambda_1, \lambda_1') \tag{123}$$

Hence

$$\underset{\sim}{K}_e = \underset{\sim}{N} \underset{\sim}{\Lambda}^2 \underset{\sim}{N}^{-1} = \begin{bmatrix} \lambda_1^2 & \cdot \\ B_1(\lambda_1^2 - \lambda_1'^2) & \lambda_1'^2 \end{bmatrix} \tag{124}$$

If we ask how this compares with eqn. (83) we have to go to the limit of small κ and κ'. Then

$$\lambda_1^2 \doteqdot 2\kappa/(1 + \tfrac{1}{4}\kappa), \quad \lambda_1'^2 \doteqdot 2\kappa'/(1 + \tfrac{1}{4}\kappa'\delta^2)$$

and

$$B_1(\lambda_1^2 - \lambda_1'^2) = -2 \frac{\kappa}{1 + \frac{1}{4}\kappa} \; \frac{1}{1 + \frac{1}{4}\kappa'\delta^2}$$

But

$$\left[\underset{\sim}{I} + \tfrac{1}{4}\kappa \underset{\sim}{D}^{-1}\right]^{-1} = \begin{bmatrix} 1 + \frac{1}{4}\kappa & \cdot \\ -\frac{1}{4}\kappa & 1 + \frac{1}{4}\kappa'\delta^2 \end{bmatrix}^{-1} =$$

$$\begin{bmatrix} (1 + \frac{1}{4}\kappa)^{-1} & \cdot \\ \frac{1}{4}\kappa(1 + \frac{1}{4}\kappa)^{-1}(1 + \frac{1}{4}\kappa'\delta^2)^{-1} & (1 + \frac{1}{4}\kappa'\delta^2)^{-1} \end{bmatrix}$$

and

$$\underset{\sim}{K}_e = 2(\underset{\sim}{I} + \frac{1}{4}\underset{\sim}{\kappa}\underset{\sim}{D}^{-1})^{-1}\underset{\sim}{\kappa} =$$

$$\begin{bmatrix} 2\kappa/(1+\frac{1}{4}\kappa) & \\ -2\kappa/(1+\frac{1}{4}\kappa)(1+\frac{1}{4}\kappa'\delta^2) & 2\kappa'/(1+\frac{1}{4}\kappa'\delta^2) \end{bmatrix} \quad (125)$$

Thus there is agreement for small κ and κ' and eqn. (124) represents the extension of (125).

In taking the comparison further we will omit all save the dominant terms so

$$(\underset{\sim}{v}_p) = \underset{\sim}{D}L\underset{\sim}{v}_p + 2p\tilde{\omega}(1-\rho^2)\underset{\sim}{v}_{p-1} + p(p-1)\underset{\sim}{D}\,\underset{\sim}{v}_{p-2} \quad (126)$$

$$-\underset{\sim}{D}(\underset{\sim}{v}_p) + \underset{\sim}{\kappa}\,\underset{\sim}{v}_p = 0, \; \rho = 0 \quad (127)$$

simplify for $p = 1$ to

$$(v_{11})_\tau = Lv_{11} + 2\tilde{\omega}(1-\rho^2)a_{11}\, e^{-\lambda_1^2\tau} J_o(\lambda_1\rho) \quad (128)$$

$$(v_{12})_\tau = \delta^{-2}Lv_{12} + 2\tilde{\omega}(1-\rho^2)\,\{a_{11}e_1 e^{-\lambda_1^2\tau} J_o(\lambda_1\delta\rho) + + a_{21}\, e^{-\lambda_1'^2\tau} J_o(\lambda_1'\delta\rho)\} \quad (129)$$

With the first equation we know how to deal, for the solution

$$v_{11} = \tilde{\omega}\{b_{11} + \beta_1\tau + f_1(\rho)\}\, a_{11}\, e^{-\lambda_1^2\tau}\, J_o(\lambda_1\rho) \quad (130)$$

satisfies the boundary condition

$$v_{11\,\rho^*} - \kappa\, v_{11} = 0$$

if $f_1'(1) = 0$. But it satisfies eqn. (128) if

$$(\rho J_o^2(\lambda_1\rho)f_1')' = \rho\, J_o^2(\lambda_1\rho)\, \{\beta_1 - 2(1-\rho^2)\}$$

and β_1 is again given by eqn. (55). There is still an arbitrary constant in f_1 and if it is chosen so that the average of $f_1(\rho)\ J_o(\lambda_1\rho)$ is zero then

$$w_{11} = \tilde{\omega}\{b_{11} + \beta_1\tau\}\ A_{11}\ e^{-\lambda_1^2\tau} \tag{131}$$

$$= w_{11}(0)\ e^{-\lambda_1^2\tau} + \tilde{\omega}\beta_1\tau\ w_{01}(0)\ e^{-\lambda_1^2\tau}$$

The second equation is subject to the boundary condition

$$(v_{12}) + \delta^2\kappa\ v_{11} - \delta^2\kappa' v_{12} = 0 \text{ at } \rho = 1\ , \tag{132}$$

so that if we take a solution

$$v_{12} = \tilde{\omega}\{b_{11}^* + \beta_1^* + f_1^*(\rho)\}\ a_{11}e_1 e^{-\lambda_1^2\tau}\ J_o(\lambda_1\delta\rho)$$

$$+ \tilde{\omega}\{b_{12} + \beta_1'\tau + f_2(\rho)\}\ a_{21}\ e^{-\lambda_1'^2\tau}\ J_o(\lambda_1'\delta\rho) \tag{133}$$

we again have to choose β_1^* and β_1' to make $f_1^{*\prime}(1)$ and $f_2'(1)$ both vanish. Now

$$\int_o^1 \{\beta - 2(1-\rho^2)\}\rho\ J_o^2(\alpha\rho) = 0 \tag{135}$$

if

$$\beta = \frac{4}{3}\left\{ 1 + \frac{J_1(\alpha)\{J_1(\alpha) - \alpha J_o(\alpha)\}}{\alpha^2\{J_o^2(\alpha) + J_1^2(\alpha)\}} \right\} = \beta(\alpha)$$

so

$$\beta_1^* = \beta(\lambda_1\delta)\ , \quad \beta_2 = \beta(\lambda_1'\delta) \tag{136}$$

Note that, since $\lambda_1'\delta J_1(\lambda_1'\delta) = \kappa'\delta^2 J_o(\lambda_1'\delta)$,

$$\beta_1' = \frac{4}{3}\ \frac{\kappa'^2\delta^2 - \kappa'\delta^2\lambda_1'^2 + \lambda_1'^4 + \kappa'^2\delta^4\lambda_1'^2}{\lambda_1'^2(\lambda_1'^2 + \kappa'^2\delta^4)} \tag{137}$$

Again choosing the constants in f_1^* and f_2 so that the weighted average vanishes, we have

$$w_{11} = \tilde{\omega}\beta_1\tau\ A_{11}\ e^{-\lambda_1^2\tau} + B_{11}\ e^{-\lambda_1^2\tau}$$

$$w_{12} = \tilde{\omega}\beta_1^*\tau\ A_{11}B_1\ e^{-\lambda_1^2\tau} + \tilde{\omega}\beta_1'\ A_{21}'\ e^{-\lambda_1'^2\tau} + B_{21}e^{-\lambda_1^2\tau} +$$

$$B_{21}'\ e^{-\lambda_1'^2\tau} \tag{138}$$

where B_1 is given by Eqn. (21), the A's are as before and the B's are chosen to fit w_{110} and w_{120}. Thus the first terms correspond to the elements of W_{10},

$$\underset{\sim}{W}_{10} = \tilde{\omega}\tau \begin{bmatrix} \beta_1 e^{-\lambda_1^2\tau} & \cdot \\ \beta_1^* B_1 e^{-\lambda_1^2\tau} - \beta_1' e^{-\lambda_1'^2\tau} & \beta_1' e^{-\lambda_1'^2\tau} \end{bmatrix} \tag{139}$$

If this correspondence holds up we must have $b = \tilde{\beta}d$ to eliminate terms not proportional to τ and then $a = \beta_1$, $c = \beta_1'$. Now from the zeroth moment we know $\tilde{\alpha} = \lambda_1^2$, $\tilde{\gamma} = \lambda_1'^2$, and $\underset{-}{\beta} = B_1(\tilde{\gamma}-\tilde{\alpha})$, so $b = B_1(\beta_1'-\beta_1)$ and

$$\tilde{\omega}\tau\underset{\sim}{\Omega}_e \underset{\sim}{W}_\infty = \tilde{\omega}\tau \begin{bmatrix} \beta_1 e^{-\lambda_1^2\tau} & \cdot \\ B_1(\beta_1 e^{-\lambda_1^2\tau} - \beta_1' e^{-\lambda_1'^2\tau}) & \beta_1' e^{-\lambda_1'^2\tau} \end{bmatrix}$$

The off-diagonal term is not the same as in (139) and it would appear that

$$\underset{\sim}{\Omega}_e = \begin{bmatrix} \beta_1 & \cdot \\ B_1(\beta_1-\beta_1') & \beta_1' \end{bmatrix}$$

is not a valid equivalent. The nearest we can come to equivalence is

$$\underset{\sim}{W}_{10} = \tilde{\omega}\tau \begin{bmatrix} \beta_1 & \cdot \\ B_1(\beta_1^*-\beta_1') & \beta_1' \end{bmatrix} e^{-\underset{\sim}{K}_e\tau} \begin{bmatrix} 1 & \cdot \\ B_1-1 & 1 \end{bmatrix}$$

Conclusions

The classical non-reactive case of Taylor diffusion and the case of a single reactant are now well understood. Taylor's original treatment has been justified and a number of different arguments and calculations indicate that the asymptotic models are good for times longer than $a^2/2D$. For the reactive case this poses a limitation since asymptotics only make sense if not too much of the reaction has taken place e.g. if no more than 10% of the reactant is to be consumed k must be less than D/10a. The use of time-dependent effective coefficients is not very satisfactory since their behavior before they are constant depends on the initial conditions. Fife's asymptotic method is more revealing and there is a need for direct computation perhaps with an adaptation of some numerical or approximative technique [36].

The situation with multiple reactions is much less clear and the need for direct computation the greater. The asymptotic method is doubly suspect since the same set of exponentials does not necessarily dominate for each concentration. Johns and DeGance's Taylorian equivalent is probably a reasonable approximation for weak reactions but needs to be put to the proof in more general cases, for (ni fallor) the structure of the moments does not seem to bear the traditional comparisons.

Note added after the meeting of October 22

Rather than trying to incorporate the several points that came up in discussion, I will acknowledge my indebtedness to various correspondents and interlocutors by mentioning certain developments and suggestions that arose.

With regard to the difficulty raised by the fact that for certain values of the diffusivity ratio the dominant eigenvalues would all come from only part of the system, it was suggested that the limit should be taken seriously and an asymptotic analysis done which would explicitly take account of the extreme value of the diffusivity ratio. This seems to be an excellent suggestion and will be followed up.

A relevant paper of which I did not know at the time the present one was written is that by D. Huang and A. Varma on yield optimization noted below. It considers the plug flow and solves the equations by the same kinds of double Fourier Bessel series as arise here. A subsequent paper using the parabolic profile is forthcoming. [18]

However the work of Ramkrishna on normal operators [27,28] promises to have the most direct impact. He has shown that for the extended Graetz problem [that is the Graetz problem where longitudinal diffusivity or conductivity is taken into account] the non-selfadjoint problem can be split up into two parts. He has applied this both to the Neumann and Dirichlet boundary conditions, thus laying the foundation for the development to the Robin problem and multicomponent reactions which I hope to make.

References

1. Aris, R., Mathematical Modelling Techniques, Pitman, London, 1978.
2. Aris, R., On dispersion of a solute in a fluid flowing through a tube, Proc. Roy. Soc. A235, 67-77, 1956.
3. Brenner, H., A general theory of Taylor dispersion phenomena, 1979.
4. Carrier, G. F., On diffusive convection in tubes, Q. App. Math. 14, 108, 1956.
5. Chatwin, P. C., The approach to normality of the concentration distribution of a solute in a solvent flowing along a straight pipe, J. Fluid Mech. 43, 321-52, 1970.
6. Chatwin, P. C., The initial dispersion of contaminant in Poiseuille flow and the smoothing of the snout, J. Fluid Mech. 77, 593-602, 1976.
7. Chatwin, P. C., The initial development of longitudinal dispersion in straight tubes, J. Fluid Mech. 80, 33-48, 1977.
8. DeGance, A. E. and Johns, L. E., The theory of dispersion of chemically active solutes in a reactilinear flow field, Appl. Sci. Res. 34, 189-225, 1978.

9. DeGance, A. E. and Johns, L. E., On the dispersion coefficient for Poiseuille flow in a circular cylinder, Appl. Sci. Res. 34, 227-258, 1978.

10. Fife, P. C. and Nicholes, K. R. K., Dispersion in flow through small tubes, Proc. Roy. Soc. A344, 131-145, 1975.

11. Gill, W. N., A note on the solution of transient dispersion problems, Proc. Roy. Soc. A298, 335-339, 1967.

12. Gill, W. N., Ananthakrishnan, V., and Barduhn, A. J., Laminar dispersion in capillarits, AIChE Journal 11, 1063-72, 1965.

13. Gill, W. N. and Sankarasubramanian, R., Dispersion of non-uniformly distributed time-variable continuous sources in time-dependent flow, Proc. Roy. Soc. A327, 191-208, 1972.

14. Gill, W. N. and Sankarasubramanian, R., Dispersion from a prescribed concentration distribution in time variable flow, Proc. Roy. Soc. A329, 479-492, 1972.

15. Gill, W. N., and Sankarasubramanian, R., Dispersion of a non-uniform slug in time-dependent flow, Proc. Roy. Soc. A322, 101-117, 1971.

16. Gill, W. N., and Sankarasubramanian, R., Exact analysis of unsteady convective diffusion, Proc. Roy. Soc. A316, 341-350, 1970.

17. Horn, F. J. M., Calculation of dispersion coefficients by means of insonents, AIChE Journal 17, 613-620, 1971.

18. Huang, D. T-J. and Varma, A., Yield optimization in a tube-wall reactor, to be published in "Computer Applications to Chemical Engineering Process Design and Simulation", A.C.S. Symposium Series, 1980.

19. Hudson, J. L., Diffusion with consecutive heterogeneous reactions, A.I.Ch.E. J. 11, 943-45, 1965.

20. Johns, L. E., and DeGance, A. E., Dispersion approximations to the multicomponent convective diffusion equation for chemically active systems, Chem. Engng. Sci. 30, 1065-65, 1975.

21. Katz, S., Chemical reactions catalysed on a tube wall, Chem. Engng. Sci. 10, 202, 1959.

22. Lauwerier, H. A., The use of confluent hypergeometric functions in mathematical physics and the solution of an eigenvalue problem, Appl. Sci. Res. A2, 184-204, 1950.

23. Lighthill, J. M., Initial development of diffusion in Poiseuille flow, J. Inst. Math. Applics. 2, 97, 1966.

24. Lupa, A. J. and Dranoff, J. S., Chemical reaction on the wall of an annular reactor, Chem. Engng. Sci. 21, 861-866, 1966.

25. Pancharatnam, S. and Homsey, G. M., An asymptotic solution for tubular flow reactor with catalytic wall at high Peclet numbers, Chem. Engng. Sci. 27, 1337-40, 1972.

26. Ramkrishna, D., Papoutsakis, E. and Lim, H. C., The extended Graetz problem with Dirichlet wall boundary conditions, submitted to Applied Science Research, 1979.

27. Ramkrishna, D., Papoutsakis, E. and Lim, H. C., The extended Graetz problem with Neumann wall boundary conditions, submitted to AIChE Journal, 1979.

28. Ramkrishna, D., Normal operators in chemical engineering, Prod. 2nd Int. Cong. on Mathematical Modelling, St. Louis, 1979.

29. Sankarasubramanian, R., and Gill , W. N., Unsteady convective diffusion with interphase mass transfer, Proc. Roy. Soc. A333, 115-132 and A341, 407-8, 1973.

30. Smith, J. M., Models in ecology, Camb. Univ. Press, Cambridge, 1974.

31. Solomon, R. L. and Hudson, J. L., Heterogeneous and Homogeneous reactions in a tubular reactor, A.I.Ch.E. J. 13, 545-550, 1967.

32. Taylor, G. I., Dispersion of soluble matter in solvent flowing slowly through a tube, Proc. Roy. Soc. A219, 186-203, 1953.

33. Taylor, G. I., The dispersion of matter in turbulent flow through a pipe, Proc. Roy. Soc. A223, 446-465, 1954.

34. Taylor, G. I., Conditions under which dispersion of a

solute in a stream can be used to measure molecular diffusion, Proc. Roy. Soc. A225, 473-477, 1954.

35. Ulrichson, D. L. and Schmitz, R. A., Chemical reaction in the entrance length of a tubular reactor, IEC Fundls. 4, 2-7, 1965.

36. Yu, J. S., An approximate analysis of laminar dispersion in circular tubes, J. App. Mech. 98, 537, 1976.

37. Zeigler, B. P., Theory of modelling and simulation, John Wiley, New York, 1976.

Acknowledgement

Parts of this paper were stimulated, and to some extent supported, by grants for the study of monoliths (NSF GP43923) and chromatographic reactors (DOE: EY-76-5-02-2945). My chief debt however is to the Mathematics Research Center of the University of Wisconsin, where most of the writing was done during the summer of 1979 and to the Chemical Engineering Department of that university, where, as the Olaf Hougen Professor in the fall of that year, I greatly benefitted from discussions with members of its distinguished faculty. I am greatly indebted to Mr. Fred Sauer for computing the numbers that lie behind the figures.

Appendix. Matrix Bessel Functions

Just as the matrix exponential function can be defined so also can the Bessel function. If $\underset{\sim}{M}$ is an p x p matrix, so also is

$$J_n(\underset{\sim}{M}x) = (\tfrac{1}{2}\underset{\sim}{M}x)^n \sum_{k=0}^{\infty} (-)^k \frac{(\frac{1}{2}\underset{\sim}{M}x)^{2k}}{k!(k+n)!} \tag{A1}$$

When $\underset{\sim}{M}$ is diagonalizable say $\underset{\sim}{L}\underset{\sim}{M}\underset{\sim}{L}^{-1} = \mathrm{diag}(m_1,\ldots.m_p)$

$$J_n(\underset{\sim}{M}x) = L^{-1}\,\mathrm{diag}(J_n(m_1x),\ldots J_n(m_px))L. \tag{A2}$$

All powers of $\underset{\sim}{M}$ commute with $J_n(\underset{\sim}{M}x)$ so all the usual relations carry over; in particular

$$\frac{d}{dx} J_0(\underset{\sim}{M}x) = -\underset{\sim}{M}J_1(\underset{\sim}{M}x) = -J_1(\underset{\sim}{M}x)\underset{\sim}{M} \tag{A3}$$

and

$$\frac{d}{dx} x^{\pm n} J_n(\underset{\sim}{M}x) = x^{\pm n} J_{n\pm 1}(\underset{\sim}{M}x)\underset{\sim}{M} \tag{A4}$$

and

$$\frac{1}{x}\frac{d}{dx}\left(x \frac{d}{dx} J_o(\underset{\sim}{M}x)\right) = - \underset{\sim}{M}^2 J_o(\underset{\sim}{M}x) \tag{A5}$$

Department of Chemical
Engineering and Materials
Science
University of Minnesota
Minneapolis, MN 55455

Model Reduction of Chemically Reacting Systems

E. Dieter Gilles and Willi Ruppel

1. INTRODUCTION.

By using the example of the fixed-bed reactor, the problem of model reduction of chemically reacting systems will be treated in this paper. When designing a chemical reactor, one should try to use all available information. This means that the design is to be based on a set of balance equations, which should be as detailed as possible.

In dealing with the control of the fixed-bed reactor, one also needs a mathematical model in order to analyse the dynamic behavior, on the basis of which the controller can be designed. At first one tends to also make use of the detailed reactor model, and to treat the control problems by applying the theory of distributed parameter systems. This also includes the design of observers and filters for state estimation.

However, two serious problems arise when the reactor behavior is characterized by a strongly non-linear dependency of the reaction rate:

a) For non-linear systems with distributed parameters, there are no theoretical methods available which allow one to analyse the dynamic behavior and to design appropriate control techniques.
b) The computer time necessary for implementation of non-linear observers or filters based on the detailed reactor model is in general very high, and often not available for this

ISBN 0-12-669550-4

purpose.
In all attempts to overcome these difficulties, a model reduction was at first carried out with the goal of developing an appropriate control concept for the fixed-bed reactor. For this purpose, the partial differential equations of the reactor were approximated by a finite set of ordinary differential equations.

Today, essentially two different procedures can be distinguished, which are applied in order to achieve this model reduction.

a) The formalistic method

By means of the method of weighted residuals, the local temperature and concentration profiles are approximated by finite series of the form

$$x = \sum_{i=1}^{N} c_i(t) \cdot \varphi_i(z) \quad .$$

By introducing such a series into the corresponding partial differential equation, N ordinary differential equations are obtained for the $c_i(t)$. In order to apply the well-known state space methods, the reactor equations are linearized for given steady-state operating conditions. Adaptive-type controllers are used to surmount the problem of parametric sensitivity which is due to the nonlinear reaction rate dependency |1-4|. One drawback of this technique arises in the limitation of the number N. For practical reasons, N should be as small as possible. However, in this case steep temperature profiles, which occur for strongly exothermic reactions, cannot be approximated to a satisfactory accuracy. A further disadvantage of this formalistic procedure must be seen in the fact that it hardly allows for a profound insight into the physical nature of the process. Such an insight could be of great help, especially for solving control problems.

b) The heuristic method

By simulation of the detailed model, or just through experiments, those reactor variables being of particular interest for the solution of control problems are determined in their dynamic behavior. Examples of such quantities include the value or the location of the maximal temperature. The dependency of these quantities with respect to the input variables is described approximately by linear transfer functions. By this means, a reduced linear model is obtained for limited operating conditions of the reactor [5-7].

In the following, we shall show that a detailed analysis of certain nonlinear phenomena of the fixed-bed reactor can lead us to a reduced model of the system. It is our objective to approximate the reactor behavior, and this not only for the immediate vicinity of the steady-state operating conditions. The model reduction is to be achieved for a certain class of temperature- and reaction-rate profiles. Although these profiles can be characterized by only a small number of state variables, they enable us to describe the reactor behavior in a rather wide range of operation. This procedure of model reduction based on physical considerations is also helpful in the design of appropriate control concepts, as it implies a better understanding of the nonlinear phenomena in chemical reactors.

2. MODEL OF THE FIXED-BED REACTOR.

A large number of papers published in recent years deal with problems of the mathematical modelling of fixed-bed reactors [8-11].

Fig. 1 shows the fixed-bed reactor. The gaseous components react on the active surface of the catalyst, where the heat is released in the presence of an exothermic reaction. A rather simple model, which enables us to describe the reactor behavior in the case of a single reaction

$$A \rightarrow \text{products},$$

comprises a mass- and an energy balance equation:

$$u \frac{\partial c}{\partial z} = r(c,T)\ (1-\varepsilon) \quad ; \quad c(t,0) = c_o(t) \tag{1}$$

$$(\rho c_p)^* \frac{\partial T}{\partial t} + u\,\rho c_p \frac{\partial T}{\partial z} - \bar{\lambda} \frac{\partial^2 T}{\partial z^2} + \frac{2\alpha}{R}(T-T_W) = (-\Delta h_R) r(c,T)(1-\varepsilon) \tag{2}$$

$$T(t,0) = T_o(t) \quad ; \quad \left(\frac{\partial T}{\partial z}\right)_\ell = 0 \ .$$

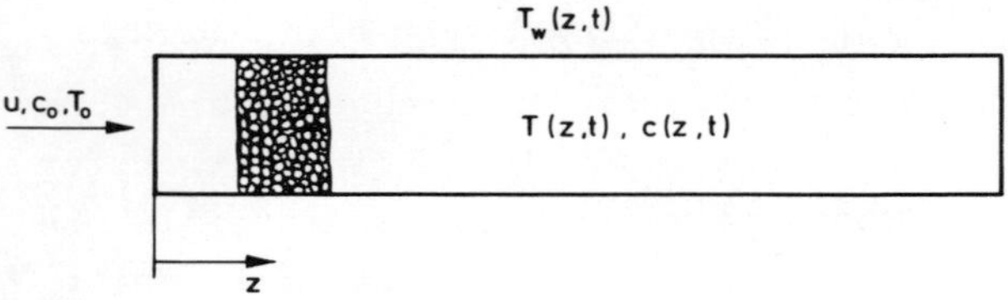

Fig. 1. Fixed-bed reactor.

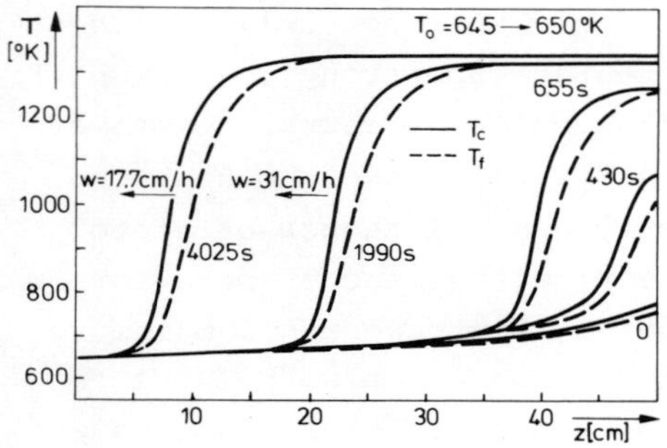

Fig. 2. Process of ignition, short-time increase of T_O.

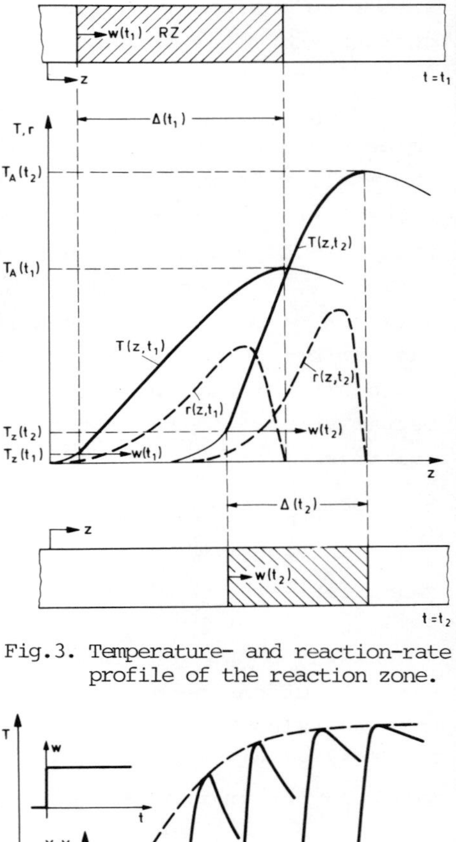

Fig. 3. Temperature- and reaction-rate profile of the reaction zone.

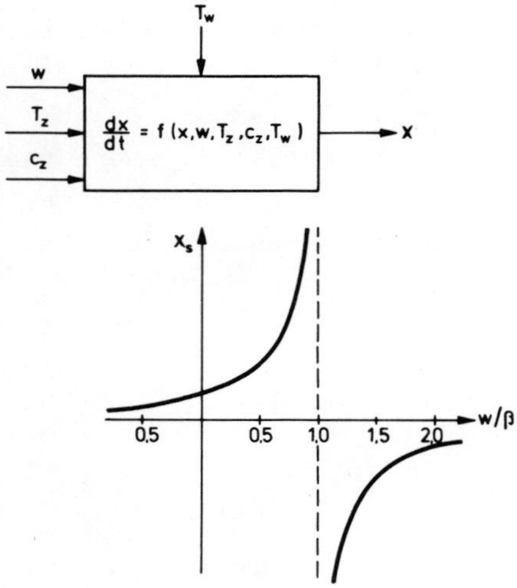

Fig. 4. Block diagram and steady-state behavior of the reaction zone.

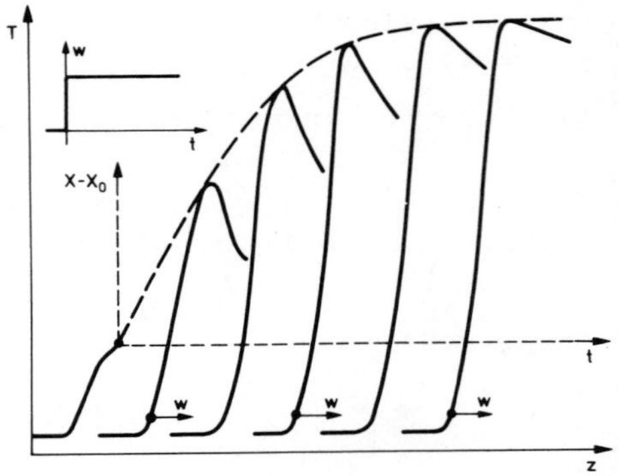

Fig. 5. Qualitative transient behavior of the reaction zone.

The state variables are the concentration c of the component A and the temperature T. The heat capacity of the catalyst is about thousand times larger than that of the fluid. This difference causes a temperature disturbance to propagate within the reactor at a velocity which is about a thousand times slower than that of a concentration disturbance. It is just this difference in the velocity of the convective heat and mass transport which must be considered a key to understanding some phenomena of the reactor behavior.

In light of our objective -- namely, model reduction of the nonlinear reactor equations -- we must first consider in greater detail the moving reaction zone as a nonlinear phenomenon of the fixed-bed reactor. We are going to demonstrate that this appearance may be seen as a prototype of an active transport process. Furthermore, it will be emphasized that a moving reaction zone is characterized by a certain feature, which we will call "form stability". This property proves to be useful for the considerable reduction of the nonlinear reactor equations. The first section of the following considerations deals with a rather qualitative analysis of the reactor behavior in order to find out the principal structure underlying the reactor behavior. The results are then used in the last section to obtain a reduced model allowing for a quantitative description of the reactor.

3. MOVING REACTION ZONE.

A strongly exothermic reaction in a fixed-bed reactor can result in a small reaction zone, that propagates within the reactor at a well-defined moving velocity. Such a reaction zone represents an ignited state of reaction. Within the zone, the reaction proceeds to a complete consumption of the reactants. The direction in which the reaction zone moves can be either up- or downstream along the reactor.

From the thermodynamic point of view, a moving reaction zone must be viewed as a dissipative structure. Basically, such a structure appears after the system has passed a stability boundary. A moving reaction zone arises when the reactor proceeds from an extinguished to an ignited state. Such a process of ignition is shown in Fig. 2 for a short time increase of the inlet temperature.

Again from the thermodynamic point of view, a reaction zone corresponds to a state of high order. Now it can be seen that the system tends to maintain such a state also against changes in its enviroment. This kind of self-stabilisation results from the dissipative processes occurring within the reaction zone. Furthermore, one can deduce from simulation results, that as a consequence of self-stabilisation, the behavior of the reaction zone is characterized by a property that we will call"form stability".This means that the system responds to the boundary conditions only by changing the moving velocity and the scaling factors of the dissipative structure. The principal shape of the temperature profile, however, remains nearly unchanged within the reaction zone.

Let us now consider Fig. 3, where the temperature- and the reaction-rate profile of the moving reaction zone are plotted. The starting point of the zone, which covers the length Δ, is assumed to move at the velocity w in the flow-direction. As a result of "form stability" we can deduce that the principal shape of the local temperature profile remains unchanged in time. The only variables which will be changing during a dynamic process are the length Δ and the temperature increase

$$x = T_A - T_Z$$

within the reaction zone. The left hand boundary value T_Z is called the "ignition temperature". All further considerations will be simplified if the balance equations are transformed into a coordinate system (z',t), the origin of which is coupled with the starting point of the reaction zone. In this new coordinate system the reactor equations read:

$$u \frac{\partial c}{\partial z'} = - r(c,T)(1-\varepsilon) \quad ; \quad c(z',0) = c_Z \tag{3}$$

$$\frac{\partial T}{\partial t} + (\dot{\beta}-w) \frac{\partial T}{\partial z'} = \frac{(-\Delta h_R)}{(\rho c_p)^*}(1-\varepsilon) r(c,T) + \frac{2\alpha}{R(\rho c_p)^*}(T_W - T) \tag{4}$$

$$+ \frac{\bar{\lambda}}{(\rho c_p)^*} \frac{\partial^2 T}{\partial z'^2} \quad ; \quad T(z'=0) = T_Z \; ; \; \left(\frac{\partial T}{\partial z'}\right)_{z'=\Delta} = 0 \; .$$

Now we assume that the local temperature profile within the reaction zone can be approximately described by:

$$T(z',t) = T_Z(t) + x(t) \sin \frac{\pi}{2} \frac{z'}{\Delta(t)} \; . \tag{5}$$

Using this relationship, a global energy balance of the reaction zone can be obtained by integrating equ. (4) over the length Δ. This balance equation reads:

$$\frac{dx}{dt} + \left\{\frac{\pi}{2\Delta}(\beta - w) + \frac{\pi^2}{4\Delta^2}\frac{\bar{\lambda}}{(\rho c_p)^*} + \frac{2\alpha}{R(\rho c_p)^*} + \left(1 - \frac{\pi}{2}\right)\frac{d}{dt}\ln\Delta\right\} x \quad (6)$$

$$= \frac{\pi}{2\Delta} q_R + \frac{\alpha\pi}{R(\rho c_p)^*}(T_W - T_Z) - \frac{\pi}{2}\frac{dT_Z}{dt} \quad .$$

In setting up this equation, it was presupposed that the reactants were completely exhausted within the reaction zone, so that

$$q_R = \frac{(-\Delta h_R)}{(\rho c_p)^*} u \, c_Z \quad . \qquad (7)$$

Using the conditional equations which are valid for the location of the maximal reaction rate, the length Δ can be expressed as a nonlinear function of x and T_Z:

$$\Delta = f(c_Z, T_Z, x) \quad . \qquad (8)$$

An increase of the temperature in the reaction zone raises the reaction rate, and this means that the reaction zone in general is reduced in its length Δ. From these considerations, it follows that the reaction zone can be approximately described by its temperature increase x, with x as the only state variable of this reactor part.

A very interesting property of the moving reaction zone can be seen by looking at the stationary state solution of equ. (6). Assuming adiabatic operating conditions of the reactor and neglecting the influence of heat conduction, we obtain for the temperature increase:

$$x_s = \frac{q_R/\beta}{1 - \frac{w}{\beta}} \quad . \qquad (9)$$

In this relationship, q_R/β is the adiabatic temperature increase which appears when the reaction zone is fixed. Fig. 4 shows x_s as a function of $\frac{w}{\beta}$. If the reaction zone moves in flow direction, the temperature rise increases in comparison with the fixed zone. The temperature rise even tends to infinity when the moving velocity w just equals the velocity β of the convective heat transport in the bed. Under this condition the convective heat transport out of the reaction zone vanishes. This means that the heat generated through the reaction remains at its full extension within the reaction zone, which consequently must lead to an infinitely high temperature increase. If the reaction zone moves opposite to the flow direction, the temperature increase x_s is lower than that in a fixed zone. This can be traced to the influence of the cold catalyst, which now enters the reaction zone from the left. The negative sign of x_s for $w > \beta$ means that the temperature increase in the reaction zone takes place in the negative z'-direction. This results from the convective heat transport conveying the heat in the same direction out of the zone.

If the moving velocity w is changed by a stepwise increase starting from a fixed zone, a nonlinear transient behavior of x is obtained, as qualitatively shown in Fig. 5. Such transients can be measured if a catalyst poison is used in order to cause a catalyst deactivating process, proceeding in the flow direction at a constant velocity w. The measuring results obtained by Butt and Price [12] show that for a small increase of w, x reaches its quasi-stationary value after the reaction zone has passed only a small area of the catalyst bed (Fig. 6). However, in the case of a fast deactivation process, a very pronounced transient behavior of x comes into appearance. Fig. 7 shows some measuring results obtained by Blaum, where a pilot plant reactor for the oxidation of CO on a NiO-catalyst was used [13, 14].

Thus far, the moving velocity of the reaction zone was considered a quantity which, for example, was predetermined by a proceeding catalyst deactivation process. The simulation results in Fig. 8,though, show that moving zones can also appear when the boundary conditions are constant in time. This phenomenon was first measured by Padberg and Wicke [15, 16]. Now, the moving velocity w must be taken as an internal quantity primarily determined by the heat conduction within the bed [17].

In order to illustrate the mechanism of propagation, let us imagine the reaction zone as bounded on its left hand side by an ignition zone, which also moves at the velocity w (Fig. 9). The breadth δ of the ignition zone can be viewed

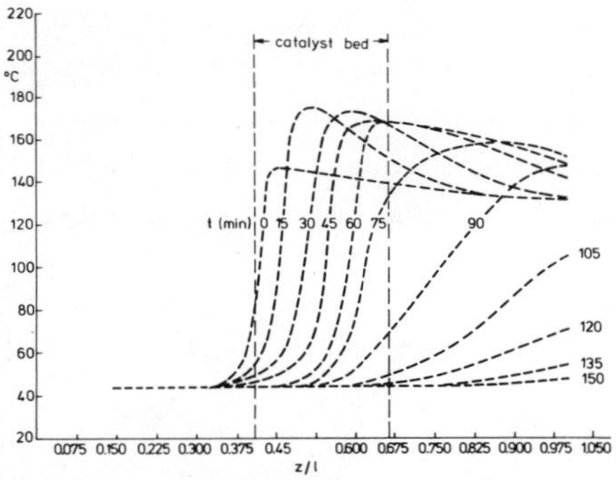

Fig.6. Moving reaction zone in case of catalyst poisoning.

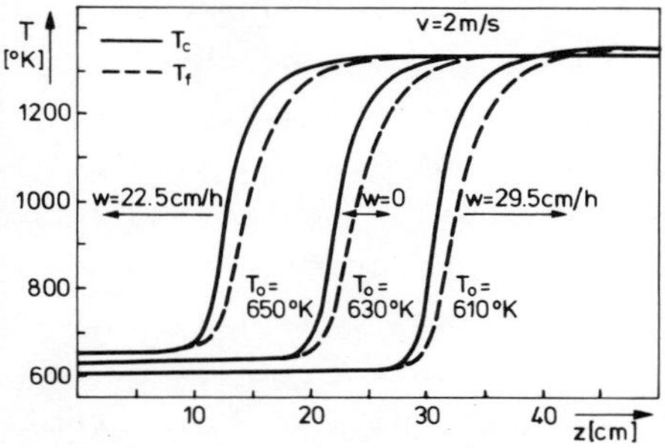

Fig.8. Moving reaction zone as a result of heat conduction.

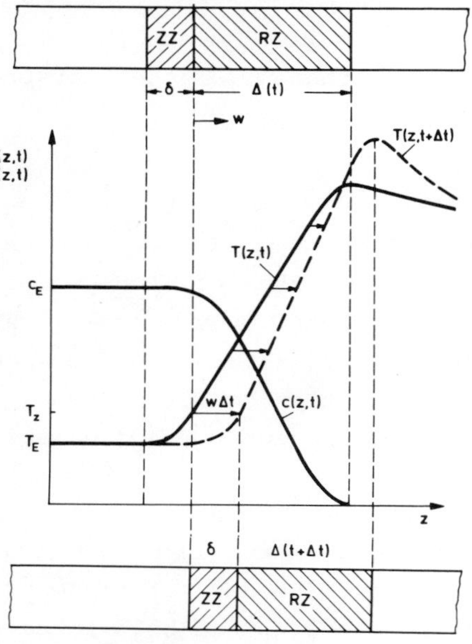

Fig.9. Moving reaction zone (RZ) and ignition zone (ZZ).

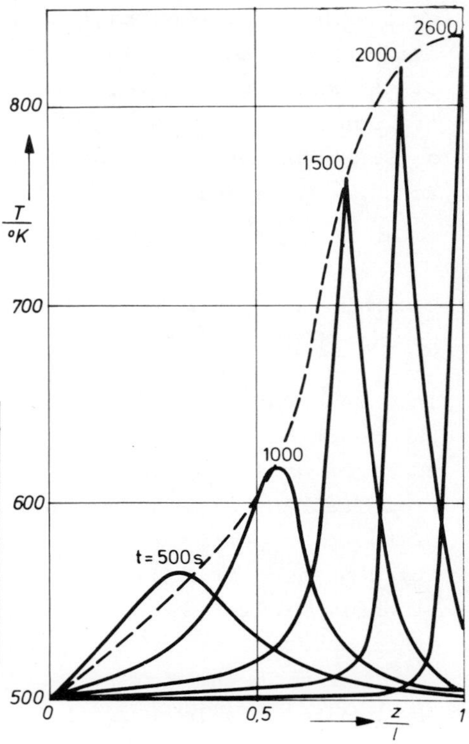

Fig.7. Moving reaction zone, fast catalyst deactivation process.

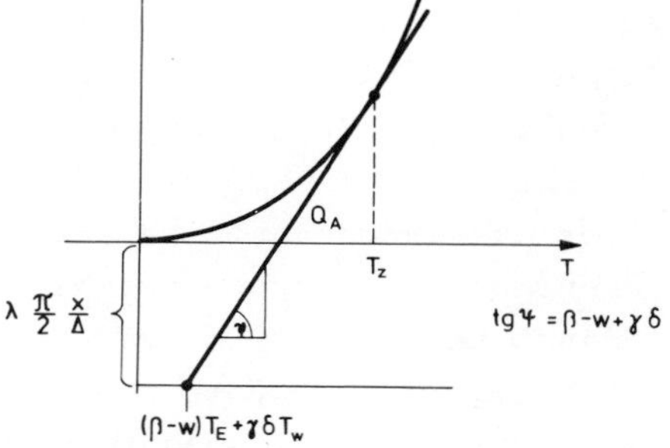

Fig.10. Energy balance of the ignition zone.

as the depth of penetration valid for the heat current generated by heat conduction. In relation to the dynamic behavior of the reaction zone, the processes within the ignition zone are assumed to always be in a quasi-stationary state. Assuming a CSTR-behavior, the global energy balance equation of this zone reads as follows:

$$(\beta-w)(T_Z-T_E)=q_R\frac{(1-\varepsilon)}{u\ c_o}\ r(T_Z,c_Z)\delta+\frac{2\alpha\ \ \delta}{R(\rho c_p)^*}(T_W-T_Z) \qquad (10)$$

$$+\frac{\bar{\lambda}}{(\rho c_p)^*}\ \ \frac{\pi}{2\Delta}\ x\quad .$$

In this equation, the amount of heat which is transferred from the reaction zone through heat conduction must be considered to be of great importance. This heat current leads to a spatial feedback within the reactor, and hence is the very origin of the propagating reaction zone. Obviously, the moving velocity w must be of such a quantity that the ignition zone permanently finds itself just at its stability boundary -- in other words, just before it passes over to an ignited state. Together with the global energy balance equation, this stability condition allows us to determine both the moving velocity w and the ignition temperature T_Z. The postulation that the ignition zone is found at the stability limit means that in Fig. 10, the heat removal line is the tangent of the heat production curve. This yields:

$$w = \beta + \frac{2\alpha}{R(\rho c_p)^*} - \frac{q_R(1-\varepsilon)}{u\ c_Z}\ \frac{\partial r}{\partial T_Z} \qquad (11)$$

Fig. 11 shows the block-diagram of both the reaction and the ignition zone. Within the reaction zone, the reaction proceeds to a complete consumption of the reactants. Only a small fraction of the generated heat is transported through heat conduction from the reaction zone to the ignition zone. Together with the input variables T_E and c_E of the ignition zone, this heat current determines the moving velocity w at which this phenomenon propagates within the reactor. The state variables adjust themselves in such a way as to permanently maintain a stability boundary in the ignition zone. Because of these properties, the propagating reaction zone can be held to be a prototype of an active transport process.

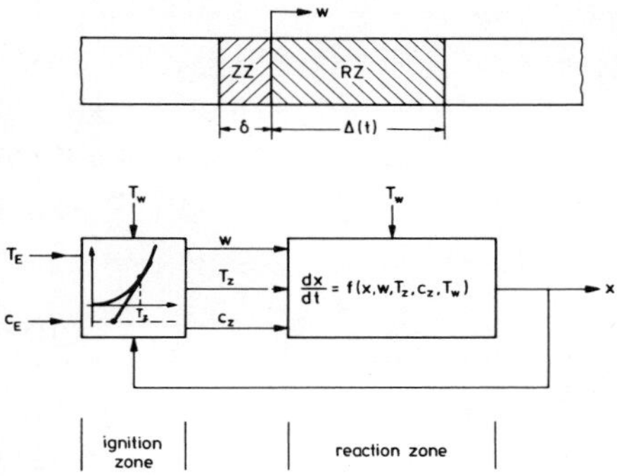

Fig.11. Block diagramm of the reaction- and the ignition zone.

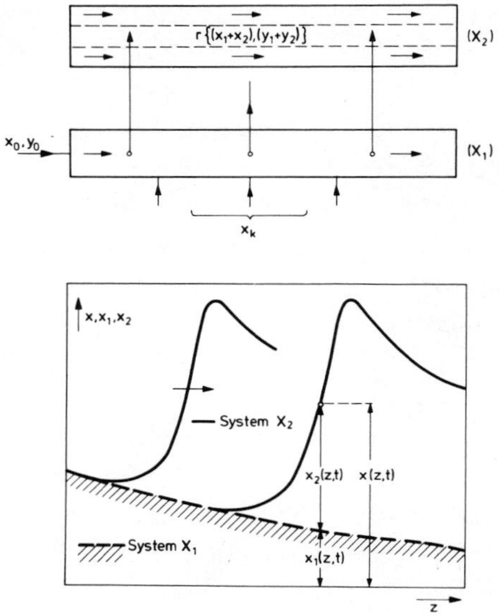

Fig.13. Subsystems S_1 and S_2 of the fixed-bed reactor.

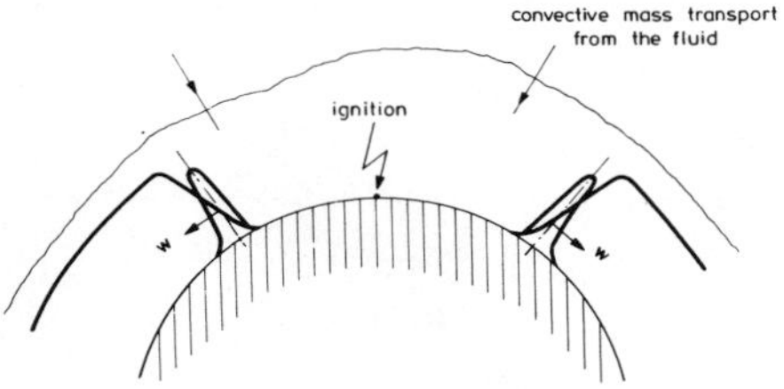

Fig.12. Moving reaction zone on a catalyst pellet.

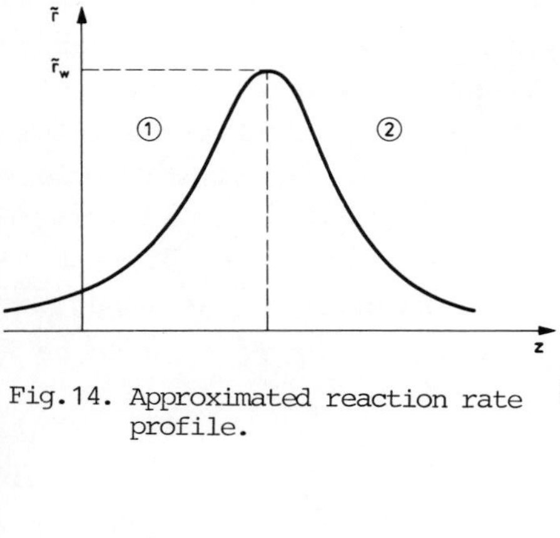

Fig.14. Approximated reaction rate profile.

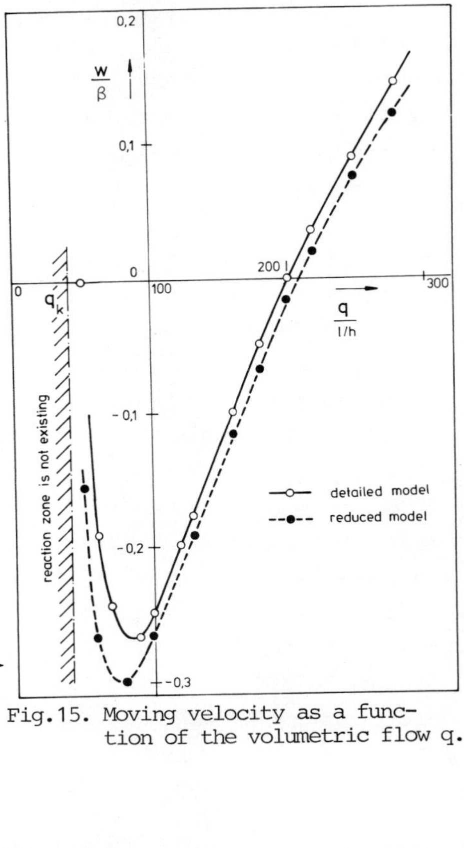

Fig.15. Moving velocity as a function of the volumetric flow q.

Through the example of the moving reaction zone, we perceive that a distributed parameter system can exhibit an active transport process only if it is provided with both an internal source and an internal feedback mechanism. In the form of the heat of reaction and the heat current through heat conduction, the fixed-bed reactor disposes of both prerequisites. Similarly, we can imagine that under isothermal conditions, a moving reaction zone can appear as a result of a purely kinetic mechanism. However, it must be presupposed that the reaction is autocatalytic. Such a purely kinetic-type reaction zone is in its principal structure identical to the thermal reaction zone in a fixed-bed reactor. The quantity x then represents the mass of a certain component produced in the reaction zone. A small amount of this mass is transported to the ignition zone through diffusion where it exerts an activating influence upon the reaction rate. By means of an appropriate moving velocity w, the ignition zone maintains a stability boundary.

Thus far we have considered examples in which the convective mass transport to the reaction zone and the propagation of this zone take place along the same coordinate. It is also conceivable that a reaction zone propagates transversally to the convective mass transport. Such types of active transport processes may be encountered on the surface of a catalyst pellet (Fig. 12). Let us imagine a short-time disturbance of stochastic nature which, on the catalyst surface, leads to an ignition of the reaction and to a formation of a reaction zone. Starting from the location of ignition, the reaction zone propagates along the surface in all directions, consuming on its way the stored reactants. Finally, the reaction disappears if the flow of reactants coming from the fluid phase is not strong enough. After extinction, the surface can be loaded again with reactants from the fluid, in order to prepare for a further process of ignition. In this way, a relaxation-type oscillation, characteristic of a distributed parameter system, might occur. It can be expected that the frequency of such an oscillation will vary according to the changing operating conditions of the pellet.

4. REDUCED MODEL OF THE FIXED-BED REACTOR.

Based on the previous considerations, we are now going to develop a reduced model which allows for a quantitative description of the fixed-bed reactor. The balance equations of this reactor type are now given in a normalized form:

$$\frac{\partial y}{\partial z} = - \text{Da}\ r(x,y) \quad ; \quad y(0,t) = y_o \tag{12}$$

$$\frac{\partial x}{\partial t} + \frac{\partial x}{\partial z} - \varepsilon \frac{\partial^2 x}{\partial z^2} + B(x-x_k) = \text{Da}\ r(x,y) \tag{13}$$

$$x(0,t) = x_o \quad \left(\frac{\partial x}{\partial z}\right)_{z=1} = 0$$

$$r = k_o \cdot e^{-\gamma/x} \cdot y \quad . \tag{14}$$

The reactor behavior can be interpreted as the result of the superposition of two subsystems S_1 and S_2. In order to achieve this splitting, we postulate that with

$$x = x_1 + x_2 \quad ; \quad y = y_1 + y_2 \ ,$$

the following relations are valid:

$$\frac{\partial x_1}{\partial t} + \frac{\partial x_1}{\partial z} - \varepsilon \frac{\partial^2 x_1}{\partial z^2} + B(x_1-x_k) = 0$$

$$S_1: \quad x_1(0) = x_o \quad ; \quad \left(\frac{\partial x_1}{\partial z}\right)_{z=1} = 0 \tag{15}$$

$$y_1 = y_o$$

$$\frac{\partial x_2}{\partial t} + \frac{\partial x_2}{\partial z} - \varepsilon \frac{\partial^2 x_2}{\partial z^2} + B\ x_2 = \text{Da}\ r(x,y)$$

$$S_2: \quad x_2(0) = 0 \quad ; \quad \left(\frac{\partial x_2}{\partial z}\right)_{z=1} = 0 \tag{16}$$

$$\frac{\partial y_2}{\partial z} = - \text{Da}\ r(x,y) \quad ; \quad y_2(0) = 0 \quad .$$

The advantage of this splitting must be seen in the fact that subsystem S_1 comprises all input variables affecting the reactor behavior either through the boundary conditions (x_o, y_o), or through the source function (x_k). The subsystem S_1 is linear and hence can be solved analytically. The influence of heat conduction may usually be disregarded in this part of the system, so that x_1 can be considered a pure flow system of which the solution is easily obtained.

The subsystem S_2 represents the autonomous part of the reactor. This part of the system is primarily determined by the internal sources through the chemical reaction which exhibits a strongly nonlinear dependency upon the state variables of the reactor. Fig. 13 illustrates the splitting of the system. There is only a unidirectional influence of S_1 upon S_2, and this takes place via the reaction rate. The flow system S_1 can be interpreted as a basis upon which the reaction zone of subsystem S_2 moves up- or downstream.

Our further considerations can be confined to the subsystem S_2. In order to achieve a model reduction for S_2, we again make use of the "form stability" property which characterizes the behavior of the developed reaction zone. However, we now do not start with an approximation of the temperature profile, but instead approximate the local profile of the reaction rate. For this purpose, the space domain of the reactor in Fig. 14 is formally extended to infinity on both sides and split into the two sections:

(1) the ignition domain: $z' \in (-\infty, 0]$

and

(2) the reaction domain: $z' \in [0, \infty)$.

We approximate the reaction rate profile by $\tilde{r}$, which is given as follows:

$$\tilde{r}^{(1)}(z,t) = r_z\, e^{\beta_1 (z-z_z)} \qquad z \leq z_z \tag{17}$$

$$\tilde{r}^{(2)}(z,t) = r_z\{1+(\beta_1+\beta_2)(z-z_z)\}e^{-\beta_2(z-z_z)} \qquad z \geq z_z .$$

The coordinate z_z is called the ignition point. Its moving velocity w will be determined later by the evaluation of the stability criterion. Due to the form stability, $\tilde{r}$ may be expected to be a valid approximation in both the steady and the unsteady state. By introducing

$$z' = z - z_z ,$$

a moving coordinate system of which the origin is coupled to the ignition point z_z, we obtain for the ignition domain:

$$\frac{\partial x_2^{(1)}}{\partial t} + (1-w)\frac{\partial x_2^{(1)}}{\partial z'} - \varepsilon\, \frac{\partial^2 x_2^{(1)}}{\partial z'^2} + B\, x_2^{(1)} = Da\, \tilde{r}^{(1)} \tag{18a}$$

$$x_2^{(1)}(-z_z') = 0 \; ; \quad x_2^{(1)}(0) = x_{2z} \; ; \quad \left(\frac{\partial x_2^{(1)}}{\partial z'}\right)_{z'=0} = q_{2z} \tag{18b}$$

$$\frac{\partial y_2^{(1)}}{\partial z'} = - \mathrm{Da}\, \tilde{r}^{(1)} \quad ; \quad y_2^{(1)}(-z_z) = 0 \tag{19}$$

The quantity x_{2z} represents the contribution of subsystem S_2 to the ignition temperature x_z. The corresponding contribution to the temperature gradient q_z at the ignition point is given by q_{2z}.

The ignition domain can be viewed as the formally extended domain of the ignition zone. In order to develop a simple approximate solution for this domain, it is advisable to split the function $x_2^{(1)}$ into a particular integral $\xi^{(1)}$ comprising the influence of the reaction, and into a homogeneous part $\varphi^{(1)}$. Thus,

$$x_2^{(1)} = \xi^{(1)} + \varphi^{(1)} \qquad z' \in (-\infty, 0] \tag{20}$$

The particular integral is assumed to obey the following differential equation:

$$\frac{\partial \xi^{(1)}}{\partial t} + (1-w)\frac{\partial \xi}{\partial z'} - \varepsilon \frac{\partial^2 \xi^{(1)}}{\partial z'^2} + B\, \xi^{(1)} = \mathrm{Da}\, r_z e^{\beta_1 z'} \tag{21}$$

With

$$\lim_{z' \to \infty} \xi^{(1)}(z',t) = 0$$

$\xi^{(1)}$ is chosen in such a way as to be identical with the solution of equ. (21) for a system which is extended to infinity on both sides. Thus, we obtain

$$\xi^{(1)}(z',t) = \xi_z^{(1)}\, e^{\beta_1 z'}, \tag{22}$$

where $\xi_z^{(1)}$ is determined by the differential equation:

$$\frac{d\xi_z^{(1)}}{dt} + \{(1-w)\beta_1 - \varepsilon \beta_1^2 + B\}\, \xi_z^{(1)} = \mathrm{Da}\, r_z \, . \tag{23}$$

The postulation that equ. (18) remains valid with this particular choice of $\xi^{(1)}$ leads us to the following conditional equations for the homogeneous part of $x_2^{(1)}$:

$$\frac{\partial \varphi^{(1)}}{\partial t} + (1-w)\frac{\partial \varphi^{(1)}}{\partial z'} - \varepsilon \frac{\partial^2 \varphi^{(1)}}{\partial z'^2} + B\, \varphi^{(1)} = 0 \tag{24}$$

$$\varphi^{(1)}(-z_z) = - \xi_z^{(1)}\, e^{-\beta_1 z_z}$$

$$\varphi^{(1)}(0) = x_{2z} - \xi_z^{(1)} \quad ; \quad \left(\frac{\partial\varphi^{(1)}}{\partial z'}\right)_{z'=0} = q_{2z}^{(2)} - \beta_1 \xi_z^{(1)} .$$

It is now advisable to also break up $\varphi^{(1)}$ into two parts, so that

$$\varphi^{(1)}(z',t) = \varphi_1^{(1)}(z',t) + \varphi_2^{(1)}(z',t) \tag{25}$$

The part $\varphi_1^{(1)}$ serves to satisfy the boundary condition of $x_2^{(1)}$ at the front of the reactor ($z' = -z_z$), while $\varphi_2^{(1)}$ allows us to connect the two spatial domains at $z' = 0$ in an appropriate manner. This means that $\varphi_2^{(1)}$ comprises the retroaction of the reaction domain upon the ignition domain.

In order to approximate $\varphi_1^{(1)}$, we write:

$$\varphi_1^{(1)}(z',t) = \varphi_{1z}^{(1)} \, e^{-\lambda_{11} z'} . \tag{26}$$

Introducing this into (24), we find:

$$\frac{d\varphi_{1z}^{(1)}}{dt} - \{(1-w)\lambda_{11} + \varepsilon\,\lambda_{11}^2 + B\}\,\varphi_{1z}^{(1)} = 0 . \tag{27}$$

The time dependent exponential λ_{11} serves to meet the boundary condition valid for $\varphi_1^{(1)}$ at $z' = -z_z$. This yields

$$\lambda_{11} = \frac{1}{z_z} \ln \left| \frac{\xi_z^{(1)}}{\varphi_{1z}^{(1)}} \right| - \beta_1 . \tag{28}$$

To obtain this result, it was presupposed that $\varphi_2^{(1)}(-z_z) \approx 0$. This assumption is possible whenever the penetration depth δ of the heat current through heat conduction is small in comparison with the coordinate of the ignition point. This means that

$$z_z >> \delta = \frac{1}{\frac{1-w}{2\varepsilon} + \sqrt{\frac{(1-w)^2}{4\varepsilon^2} + \frac{B}{\varepsilon}}} . \tag{29}$$

The second part of $\varphi^{(1)}$ is specified correspondingly:

$$\varphi_2^{(1)}(z',t) = \varphi_{2z}^{(1)} \, e^{\lambda_{12} z'} . \tag{30}$$

The amplitude $\varphi_{2z}^{(1)}$ of this approximate solution also satisfies a differential equation:

$$\frac{d\varphi_{2z}^{(1)}}{dt} + \{(1-w)\lambda_{12} - \varepsilon\,\lambda_{12}^2 + B\}\,\varphi_{2z}^{(1)} = 0 . \tag{31}$$

The connecting conditions of the two spatial domains at z'= O enable us to determine λ_{12}:

$$\lambda_{12} = \frac{q_{2z} + \lambda_{11}\varphi_{1z}^{(1)} - \beta_1\xi_z^{(1)}}{\varphi_{2z}^{(1)}} \quad . \qquad (32)$$

The completed splitting of the approximate solution of $x_2^{(1)}$ proves to be favourable, as it allows us to separate the dynamic behavior into a slow and a fast part. When the reaction zone is established, the particular integral $\xi^{(1)}$ may usually be considered a fast process in contrast to the part $\varphi_1^{(1)}$ of the homogeneous solution $\varphi^{(1)}$. This can be traced to the influence of the chemical reaction considerably accelerating the process described by $\xi^{(1)}$. One can recognize this effect by a detailed analysis of the eigenvalues of equ. (23) and (27). If the penetration depth δ is small, $\varphi_2^{(1)}$ can then also be numbered among the fast processes.

We now simplify our further considerations about the ignition domain by assuming that the fast dynamic processes are always found to be in a quasi-stationary state. We then obtain from equ. (31):

$$\lambda_{12} = \lambda_{12}^{o} = \frac{1}{\delta} \quad . \qquad (33)$$

When the boundary conditions at z'= O are placed in the quasi-stationary form of equ. (23), an algebraic relationship can be derived for x_{2z}:

$$\varepsilon\{\lambda_{12}^{o}x_{2z} - q_{2z}\} = \text{Da}\; r_z \frac{1}{\lambda_{11}^{o}+\beta_1} \{1 - \frac{\lambda_{12}^{o}+\lambda_{11}}{\lambda_{12}^{o}-\beta_1} e^{-(\beta_1+\lambda_{11})z_z}\} \qquad (34)$$

where

$$r_z = \frac{k_o\, e^{-\gamma/x_z}\,(y_o + \frac{\text{Da}\; r_o}{\beta_1})}{1 + \frac{\text{Da}}{\beta_1} k_o\, e^{-\gamma/x_z}} \qquad (35)$$

Equ. (34) corresponds to the energy balance of the ignition zone given by equ. (10). However, we should recognize that (34) does not comprise the energy balance of the linear subsystem S_1. A further difference exists, as (34) incorporates in form of the bracket term on its right-hand side the boundary condition valid for x_2 at the front of the reactor. In the energy balance of the ignition domain λ_{11} must be considered a time dependent variable.

The quantity $\varepsilon\, q_{2z}$ represents the heat current through heat conduction which flows at z'= 0 into the ignition domain. According to our considerations in the previous chapter, this heat current may be held to be a slowly-changing input variable of the ignition domain.

In order to determine β_1, we postulate that

$$\left(\frac{\partial \tilde{r}}{\partial z'}\right)_{z'=0} = \left(\frac{\partial r}{\partial z'}\right)_{z'=0} . \tag{36}$$

This yields

$$\beta_1 = \frac{\gamma}{x_z^2} \{q_{2z} + \left(\frac{\partial x_1}{\partial z'}\right)_{z'=0}\} - \text{Da}\, k_o\, e^{-\gamma/x_z} . \tag{37}$$

Now, we again make use of our previous considerations of the ignition zone in order to specify the moving velocity w. The right-hand side of equ. (34) can be interpreted as the heat production curve, while the terms on the left define the heat removal line. We now proceed from the assumption that the ignition section is found to be at its stability boundary. This means, in other words, that the heat removal line of this section is the tangent to the heat production curve. This condition corresponds to equ. (11) and reads:

$$\varepsilon\, \lambda_{12}^o = \text{Da} \frac{\partial}{\partial x_{2z}} \left[\frac{r_z}{\lambda_{11}^o + \beta_1} \{1 - \frac{\lambda_{12}^o + \lambda_{11}}{\lambda_{12}^o - \beta_1} \left(\frac{r_o}{r_z}\right)^{1+\frac{\lambda_{11}}{\beta_1}} \} \right] \tag{38}$$

In connection with the energy balance (34) the stability condition allows us to determine both w and x_{2z}, provided that the temperature gradient q_{2z} is known. Consequently, our final goal is to derive an equation for q_{2z}. This can be done by an analysis of the reactor behavior within the reaction domain.

Using the approximation (17) of the reaction rate the behavior of the subsystem S_2 within the reaction domain can be described as follows:

$$\frac{\partial x_2^{(2)}}{\partial t} + (1-w) \frac{\partial x_2^{(2)}}{\partial z'} - \varepsilon \frac{\partial^2 x_2^{(2)}}{\partial z'^2} + B\, x_2^{(2)} = \text{Da}\, \tilde{r}^{(2)} \tag{39a}$$

$$x_2^{(2)}(0) = x_{2z} \quad ; \quad \left(\frac{\partial x_2}{\partial z'}\right)_{z'=0} = q_{2z} \quad ; \quad \left(\frac{\partial x_2}{\partial z'}\right)_{z'=1-z_z} = 0 \quad (39b)$$

Within the scope of this paper we are going to confine ourselves to the quasi-stationary solution of this equation. In addition, it is presupposed that the reactor is long enough to be able to disregard the retroaction of the boundary condition at the reactor end. The solution of equ. (39) then reads:

$$x_2^{(2)}(z') = \xi_z^{(2)}(1+z')e^{-\beta_2 z'} + (x_{2z} - \xi_z^{(2)})e^{-\lambda_{21}^{o} z'} \quad (40)$$

where

$$\xi_z^{(2)} = \frac{Da\ r_z}{\{-(1-w)\beta_2 - \varepsilon\ \beta_2^2 + B\}}\left\{1 - \frac{\{1-w+2\varepsilon\ \beta_2\}(\beta_1+\beta_2)}{\{-(1-w)\beta_2 - \varepsilon\ \beta_2^2 + B\}}\right\} \quad (41)$$

and

$$\alpha = \frac{Da\ r_z}{\xi_z^{(2)}} \quad \frac{\beta_1 + \beta_2}{\{-(1-w)\beta_2 - \varepsilon\ \beta_2^{(2)} + B\}} \quad (42)$$

Using the normalizing condition of the reaction rate

$$y_o = Da \int_0^\infty \tilde{r}(z,t)\ dz, \quad (43)$$

β_2 can be expressed as follows:

$$\beta_2 = Da\ r_z\ \beta_1 \frac{1 + \sqrt{\frac{\beta_1\ y_o}{Da\ r_z} + e^{-\beta_1\ z_z}}}{y_o - Da\ r_z\ (1 - e^{-\beta_1\ z_z})} \quad (44)$$

The eigenvalue λ_{21}^{o} can be given as follows:

$$\lambda_{21}^{o} = -\frac{1-w}{2\varepsilon} + \sqrt{\frac{(1-w)^2}{4\varepsilon^2} + \frac{B}{\varepsilon}} \quad (45)$$

In connection with the algebraic relationships (41) to (45) equ. (40) can be seen as the integrated energy balance of the reaction domain. With $z' = \Delta$, it corresponds to the quasi-stationary energy balance (6) of the reaction zone. The length Δ can be easily obtained, when the end of the reaction zone is specified as the location of the maximal value of x_2.

We are now able to calculate q_{2z} as the decisive input variable of the ignition domain:

$$q_{2z} = (\alpha - \beta_2 + \lambda_{21})\xi_z^{(2)} - \lambda_{21}^{o}\ x_{2z} \quad (46)$$

Using the quasi-stationary energy balance of the reaction domain, the dynamic behavior of the reduced model is dominated by the state equation of the ignition point:

$$\frac{dz_z}{dt} = w \quad . \tag{47}$$

The reduced model of the fixed-bed reactor served to calculate the moving velocity w in a laboratory reactor used for the methanisation of CO_2 |19|. Fig. 15 shows w as a function of the volumetric flow. This function nearly coincides with corresponding results obtained by a numerical solution of the partial differential equations.

Presently, studies are being made to appropriately incorporate the dynamic behavior of the reaction domain into the reduced model, so that we can drop the assumption of quasi-stationarity.

CONCLUSION.

We started with a detailed analysis of the behavior of the moving reaction zone in a fixed-bed reactor. This appearance, which represents a spatially ordered dissipative structure, may be considered a prototype of an active transport process. It is characterized by the "form stability" property. This property is the key to a considerable reduction of the nonlinear reactor equations. The reduced model may be considered valid for quite a wide range of operating conditions. One may even expect that this model can serve as an ideal base for further simplifications. Thus, there is a good chance of finally arriving at a system of rather simple equations which allow us to describe the main features of the reactor at a sufficient accuracy. Studies are being conducted on the quantitative accuracy of the model with respect to various kinds of the dynamic excitation of the reactor. The results will soon be published.

It appears that the method of model reduction based on the form stability of a dissipative structure is not limited to chemical reacting systems. We have also studied the dynamic behavior of an extractive distillation column used in a large-scale water refinement process. We were able to show that the behavior of this column also is primarily determined by the motion of a dissipative structure, which in this case is a narrow zone of intensive mass exchange between both phases [19]. The form stability of the structure could also be used to obtain a considerably reduced model.

REFERENCES

1. Georgakis, G., Aris, R., and Amundson, N.R., Studies in the Control of Tubular Reactors I, II, III. Chem.Eng.Sci. 32 (1977), 1359.
2. Jutan, A., Tremblay, J.P., MacGregor, J.F., and J.D. Wright, Multivariable Computer Control of a Butane Hydrogenolysis Reactor I, II, III, AIChE Journal 23 (1977), 732.
3. Sørensen, J.P., Experimental Investigation of the Optimal Control of a Fixed-Bed Reactor. Chem. Eng. Sci. 32 (1977), 763.
4. Aström, K.J., Self-Tuning Control of a Fixed-Bed Chemical Reactor System. Lund Institute of Technology (1978) Report LKTFD2/(TFRT-3151).
5. Eigenberger, G., Modellbildung und Rechnersimulation als Werkzeug der sicheren Reaktionsführung. Chem.-Ing.-Tech.51 (1979), 1105.
6. Eigenberger, G., Zur Dynamik und Regelung von Rohrreaktoren mit stark exothermer Reaktion. Diss. Universität Stuttgart (1973).
7. Bär, W., and F. Stein, Ein einfaches Modell zur Regelung der Brennzone eines katalytischen Festbettreaktors. Chem.-Ing.-Tech. 51 (1979), 651.
8. Schmitz, R.A., Multiplicity, Stability and Sensitivity of States in Chemically Reacting Systems - A Review. Advan. Chem.Ser. (1975), 148.
9. Gilles, E.D., Reactor Models. Proceeding of the 4th Int. 6th Europ.Symp. on Chem. React. Eng., Heidelberg (1976), 257.
10. Ray, W.H., Bifurcation Phenomena in Chemically Reacting Systems. Proceedings Advanced Seminar on Applications of Bifurcation Theory, Math.Res.Center, University of Wisconsin (1976).

11. Aris, R., The Mathematical Theorie of Diffusion and Reaction in Permeable Catalysis. Vol. I and II. Clarendon Press, Oxford (1975).
12. Butt, J.B. and T.H. Price, Catalyst Poisoning and Fixed-Bed Reactor Dynamics II, Chem.Eng.Sci. 32, 393, (1977).
13. Blaum, E., Zur Dynamik des katalytischen Festbettreaktors bei Katalysatordesaktivierung. Chem.Eng.Sci. 29, 2263, (1974).
14. Gilles, E.D., Wandernde Brennzonen infolge Alterung des Katalysators, RT 6, 191, (1977).
15. Wicke, E. and G. Padberg, Einfluß von Stoff- und Wärmetransport bei Reaktionen gasförmig/fest am Beispiel katalytischer Brennzonen in adiabatischer Kontaktschicht, Chem.-Ing.-Tech. 40, 1033, (1968).
16. Padberg G., and E. Wicke, Stabiles und instabiles Verhalten eines adiabatischen Rohrreaktors am Beispiel der katalytischen CO-Oxidation, Chem.Eng.Sci., 22, 1035, (1967).
17. Gilles, E.D., Quasistationäres Verhalten von wandernden Brennzonen, Chem.Eng.Sci. 29, 1211, (1974).
18. Ruppel, W., Eine mathematische Beschreibung wandernder Reaktionszonen in Schüttschichten, Diss. Universität Stuttgart (1980).
19. Gilles, E.D., Retzbach, B., and F. Silberberger, Modeling, Simulation and Control of an Extractive Distillations Column, ACS Symposium Series, Washington 1979 (to be published).

Institut für Systemdynamik
und Regelungstechnik
Universität Stuttgart
Pfaffenwaldring 9
D-7000 Stuttgart 80

Chemical Oscillations, Multiple Equilibria, and Reaction Network Structure

Martin Feinberg

1. INTRODUCTION

Despite remarkable progress in chemical reactor theory during the past thirty years, it must be said that our understanding of reactors with complex chemistry remains rather primitive. At the very time that reactors of this kind are commanding our attention to a greater extent, the value of intuition derived from experience with simple reactors is coming into serious question. The dilemma has its roots in the fact that increased complexity in the underlying chemistry gives rise to qualitatively new physical phenomena that systems of lesser intricacy cannot exhibit.

Biochemists have understood this for some time. Their observations, both experimental and theoretical, that isothermal reaction systems of sufficient complexity can exhibit bistability and sustained composition oscillations have led them to conjecture that dynamic behavior of this sort might underlie the workings of biological clocks and switches. Although chemical engineers have known for decades that non-isothermal continuous stirred tank reactors with simple chemistry can display instability, it is only recently that experimental observations of multiple steady states, sustained composition oscillations, and even "chaotic" composition dynamics have been reported for isothermal continuous stirred tank reactors with complex chemistry, (See, for example, the work described by Roger Schmitz within this volume.) And, at

ISBN 0-12-669550-4

the same time, there has developed a parallel literature
concerning observations of sustained isothermal composition
oscillations in the oxidation of hydrogen or carbon monoxide
on metal catalysts.

The growing awareness of pathology in complex reaction
systems carries with it an element of pessimism. If, as seems
inevitable, reactors with complex chemistry are to play an
increasingly prominent role in the lives of chemists and
engineers, how are we to deal with these in any systematic
manner? Even if we know the chemistry in full detail, how are
we to know that a complex reactor will not display unexpected
pathology? Can we, at the very least, know on the basis of
the underlying chemistry which isothermal reactors share the
intrinsic stability of simple reactors and which warrant more
cautious study? In short, can there be a general theory of
complex reactors that is both powerful and easily applied?

At first glance, prospects would hardly seem bright for
precisely the same reason our understanding of complex
reaction systems remains so rudimentary: the differential
equations that govern these are generally large in number, are
highly coupled, and are nonlinear. Worse still, the general
complexion of the governing system of differential equations
changes markedly with the underlying chemistry.

A simple example will make concrete the difficulties that
must be addressed. Although we will eventually deal with
reactors that exchange matter with their surroundings, it will
be instructive to begin thinking about a closed well-stirred
reactor of unit volume maintained at constant temperature.
Suppose that the reactor contains five species, say
A_1, A_2,...,A_5, and that the chemical changes among these are
deemed to be modelled by the reaction network

$$A_1 \underset{\beta}{\overset{\alpha}{\rightleftarrows}} 2A_2$$

(1.1)

$$A_1 + A_3 \underset{\xi}{\overset{\gamma}{\rightleftarrows}} A_4$$

$$A_1 + A_3 \overset{\varepsilon}{\searrow} \quad \swarrow \delta$$
$$A_2 + A_5$$

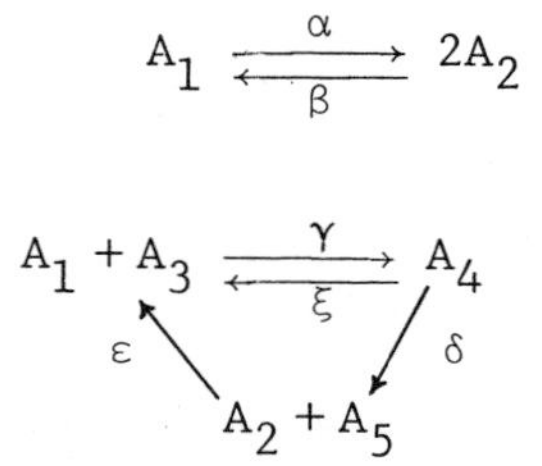

where the individual reaction rates are taken to be of mass action type with (positive) rate constants denoted by Greek letters beside the corresponding reaction arrows. Thus, if $c_1, c_2, \ldots, c_5$ denote the instantaneous molar concentrations of the five species, the reaction $A_1 \longrightarrow 2A_2$ will have occurrence rate αc_1, the reaction $2A_2 \longrightarrow A_1$ will have occurrence rate $\beta(c_2)^2$, $A_1 + A_3 \longrightarrow A_4$ will have occurrence rate $\gamma c_1 c_3$, and so on. Taking account of the contribution of individual reactions to the net production of each species, we can write the following system of ordinary differential equations to describe the temporal evolution of the reactor composition:

$$\begin{aligned}
\dot{c}_1 &= -\alpha c_1 + \beta(c_2)^2 - \gamma c_1 c_3 + \varepsilon c_2 c_5 + \xi c_4 \\
\dot{c}_2 &= 2\alpha c_1 - 2\beta(c_2)^2 + \delta c_4 - \varepsilon c_2 c_5 \\
\dot{c}_3 &= -\gamma c_1 c_3 + \varepsilon c_2 c_5 + \xi c_4 \\
\dot{c}_4 &= \gamma c_1 c_3 - (\delta + \xi) c_4 \\
\dot{c}_5 &= \delta c_4 - \varepsilon c_2 c_5 \quad .
\end{aligned} \tag{1.2}$$

Despite the simplicity of the reaction network from which it derives, the system (1.2) does not lend itself to easy analysis in the face of questions like these: Does the system (1.2) admit a positive equilibrium (in each stoichiometric compatibility class[†])? Can there be more than one positive equilibrium (within a stoichiometric compatibility class)? Can the positive equilibria be unstable? Are there cyclic solutions?

[†]The notion of stoichiometric compatibility class will be introduced in Section 3. Roughly speaking these are sets of reactor compositions with the following property: A composition trajectory that begins within a stoichiometric compatibility class must remain within that stoichiometric compatibility class for all subsequent times.

These are not simple questions*, and the answers might well depend on values of the rate constants $\alpha, \beta, \ldots, \xi$. Even if we could provide answers for every set of positive rate constants, what would we have accomplished? We would have understood one reaction system in qualitative terms, at least with respect to certain issues. But there are thousands of reaction systems that might present themselves for analysis in one context or another. Each has its system of differential equations, and we would like to understand these as well.

How are we to proceed? Must we examine in an ad hoc fashion the intricate system of differential equations for each complex reactor that commands our attention? Or might we not initiate a bolder program that would, in some sense, enable us to examine all chemical systems at once with an eye toward classifying them according to the kinds of behavior that could be exhibited?

If our example suggests that such an undertaking should be tempered with a certain amount of pessimism, it also provides a source of optimism. The fact is that we were able to construct the differential equations (1.2) for our system merely by inspecting the reaction diagram (1.1). That is, the underlying reaction network completely determined the differential equation up to values of the rate constants. Were we dealing with a different reaction network, we would certainly have written a different system of differential equations, but nevertheless these would have derived in a prescribed manner from the network at hand.

---

*Since the reactor under consideration is closed to its environment with respect to mass transport, some readers might argue that, on "thermodynamic grounds," we should not expect instability. Such an argument, whatever its merits, would not constitute a proof that network (1.1), deemed to model what might be a more intricate chemistry, necessarily induces a system of differential equations with the stability properties such readers might expect. Indeed, the questions posed above should be viewed as pertaining to qualities of the model invoked, and it would seem that such questions are well worth asking. In any case, our focus on closed reactors is temporary and is merely intended to motivate what follows.

Indeed, if there is anything that lends chemical dynamics its coherency as a subject it is the fact that, however dramatically the dynamical equations might vary from one chemical system to another, the essential form of those equations derives from the structure of the underlying reaction network. This raises the possibility that qualitative properties of solutions to the induced differential equations might ultimately be tied to aspects of network structure in a precise way.

There is more to this than wishful thinking. There are already in existence theorems which draw such connections, and they are quite easy to use. In fact, one of these (the Zero Deficiency Theorem) enables us to answer the four questions posed earlier about the system (1.2) merely by inspecting the reaction diagram (1.1). The answers are, respectively: yes, no, no, and no; these answers hold true for all positive values of the rate constants $\alpha, \beta, \ldots, \xi$.

It is my intention in this lecture to discuss theorems that connect reaction network structure to properties of the induced dynamical equations, and I will focus exclusively upon homogeneous ("well-stirred") reactors maintained at fixed temperature. Some of what I will say amounts to a review of older results. In particular, the Zero Deficiency Theorem in its present form is based upon work by Fritz Horn, Roy Jackson and me prior to 1974. However, I will also report some recent results that are as yet unpublished.

My sole purpose here will be to state theorems and illustrate how they work in examples. I will not prove them or even try to hint at how the proofs go. A fairly elementary sketch of ideas behind the Zero Deficiency Theorem can be found in [3], while a more technical and comprehensive account of these and ideas behind more recent results will be available in [4]. I have listed a few pertinent journal articles in the bibliography, but I would suggest that interested readers first look at the surveys [3] or [4] to develop some sense of how the scattered journal pieces fit together. Readers who do wish to track down the original literature should certainly begin with the crucial article by Horn and Jackson [14].

2. REACTION NETWORKS, KINETICS, AND THE INDUCED DIFFERENTIAL EQUATIONS

The mathematical framework for our discussion is presented in this section. In §2.1 we offer a few definitions pertaining to reaction network structure. Kinetics for networks are introduced in §2.2, and in §2.3 we indicate how a network taken with its kinetics induces a system of ordinary differential equations. Some elementary connections between network structure and properties of the induced differential equations are described in §2.4.

2.1. Reaction Networks and their Structure.

Loosely speaking, a reaction network for a mixture is a set of chemical reactions deemed to reflect reasonably well those chemical events which lend the mixture its dynamic character. Thus, the network

$$A_1 + A_2 \longrightarrow A_3 \begin{array}{l} \nearrow A_4 + A_5 \\ \rightleftarrows 2A_6 \end{array} \tag{2.1}$$

$$A_1 + A_6 \rightleftarrows A_7$$

is intended to suggest that one molecule of species A_1 reacts with one molecule of A_2 to form one molecule of A_3; a molecule of A_3 decomposes to form two molecules of A_6 or, alternatively, a molecule of A_4 and a molecule of A_5; two molecules of A_6 can combine to form one of A_3; and so on.

We require a modest vocabulary with which reaction network structure might be discussed. Some of the terminology we are about to present will find no use until Section 5. Although it is generally a good idea to introduce new words just before they are brought into play, we will find some advantage in having our entire lexicon located in one place to be drawn upon as the need arises. The definitions offered here are more suggestive than precise*, and we rely heavily upon examples. Readers might wish to consult Table I (Section 5) for still more illustrations†.

The letter N is used to denote the number of species in a network. The vector space of "N-tuples" of real numbers is designated by $\mathbb{R}^N$. The standard basis for $\mathbb{R}^N$ is denoted by $\{\underset{\sim}{e}_1, \underset{\sim}{e}_2, \ldots, \underset{\sim}{e}_N\}$, where $\underset{\sim}{e}_L$ is the element of $\mathbb{R}^N$ with 1 in the Lth place and zero in all other places.

2.1.A. <u>The complexes of a network</u>. The <u>complexes</u> of a network are the entities that appear before or after the reaction arrows. Thus, the complexes of network (2.1) are $\{A_1 + A_2, A_3, A_4 + A_5, 2A_6, A_1 + A_6, A_7\}$. We reserve the symbol n for the <u>number of distinct complexes</u> in a network. For network (2.1), n = 6.

With each complex of an N species network we associate a <u>complex vector</u> in $\mathbb{R}^N$. This is best explained in terms of mechanism (2.1) for which N = 7. With the complex $A_1 + A_2$ we associate the vector $\underset{\sim}{e}_1 + \underset{\sim}{e}_2$ in $\mathbb{R}^7$; with the complex A_3 we associate the vector $\underset{\sim}{e}_3$; with $2A_6$ we associate $2\underset{\sim}{e}_6$; and so on.§ The set of complex vectors for a network will be denoted by $\underset{\sim}{y}_1, \underset{\sim}{y}_2, \ldots, \underset{\sim}{y}_n$. With this in mind, <u>we shall write</u> $i \rightarrow j$ <u>to indicate the reaction whereby the</u> i^{th} <u>complex reacts to the</u> j^{th} <u>complex</u>.

2.1.B. <u>The reaction vectors for a network</u>. With each reaction of an N species network there is associated a <u>reaction vector</u> in $\mathbb{R}^N$ formed by subtracting the "reactant" complex vector from the "product" complex vector. Thus, for the reaction $A_1 + A_2 \longrightarrow A_3$ in network (2.1) the corresponding reaction

---

*Readers interested in more formal definitions might consult [4], [5], or [15].

†The seemingly "peculiar" networks in Table I are explained in Section 3. These networks encode the physico-chemical makeup of certain open reactors.

§In certain networks of Table I the <u>zero complex</u> (0) appears. The complex vector corresponding to the zero complex is the zero vector of $\mathbb{R}^N$. The manner in which the zero complex arises will be explained in Section 3.

vector in $\mathbb{R}^7$ is $e_3 - (e_1 + e_2)$; for the reaction $2A_6 \longrightarrow A_3$, the reaction vector is $e_3 - 2e_6$. The full set of six reaction vectors for network (2.1) is shown below:

$$\{e_3-e_1-e_2,\ e_4+e_5-e_3,\ 2e_6-e_3,\ e_3-2e_6,\ e_7-e_1-e_6,\ e_1+e_6-e_7\} \tag{2.2}$$

2.1.C. <u>The stoichiometric subspace for a network</u>. The <u>stoichiometric subspace</u> of an N species network is the span* of its reaction vectors; clearly, the stoichiometric subspace is a linear subspace of $\mathbb{R}^N$. Thus, the stoichiometric subspace for network (2.1) is the span of those vectors of $\mathbb{R}^7$ displayed in (2.2). We reserve the symbol S to designate the stoichiometric subspace of a network.

We shall denote by s the <u>dimension of the stoichiometric subspace</u> for a network. Equivalently, s is the number of elements in the largest linearly independent set composed of reaction vectors for the network. For network (2.1) $s = 4$ since the four element set

$$\{e_3-e_1-e_2,\ e_4+e_5-e_3,\ 2e_6-e_3,\ e_1+e_6-e_7\}$$

is linearly independent, while any five element set of reaction vectors for network (2.1) is linearly dependent. Computation of s for a given network is readily effected using standard techniques of linear algebra.†

2.1.D. <u>The linkage classes of a network</u>. If a network is displayed such that each complex appears <u>no more than once</u>§, the manner in which the various complexes are interconnected by reaction arrows may be seen with greater clarity. (We call

*The span of a set of vectors is the linear subspace comprised of all possible linear combinations of that set.

†Remember that reaction vectors are N-tuples. These may be "listed" in an r×N matrix, where r is the number of reactions. The number s is precisely the rank of such a matrix, and the rank may be readily deduced by row reduction.

§There are, of course, other ways in which a network might be displayed. For example, all the reactions might be listed in some arbitrary order.

such a display a <u>standard</u> <u>reaction</u> <u>diagram</u>.) Network (2.1) is
displayed in such a fashion, and it is evident that its
standard reaction diagram is composed of two disconnected
"pieces":

$$A_1 + A_2 \longrightarrow A_3 \begin{array}{c} \nearrow A_4 + A_5 \\ \searrow \\ 2A_6 \end{array}$$

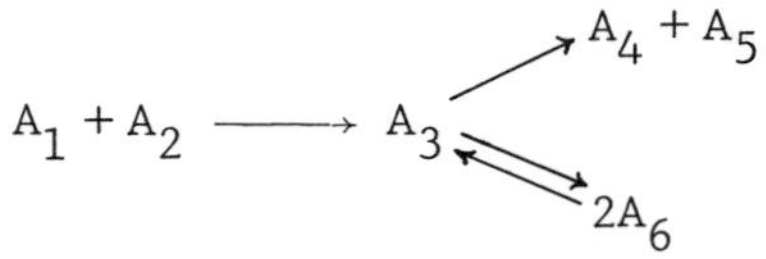

and

$$A_1 + A_6 \rightleftharpoons A_7 \ .$$

The standard reaction diagram for any network is easily constructed, and the "pieces" of which the diagram is composed
are readily discerned. The number of such pieces, designated
by the symbol ℓ, is called the <u>number</u> <u>of</u> <u>linkage</u> <u>classes</u> in
the network. Thus, for network (2.1) $\ell = 2$.

2.1.E. <u>The</u> <u>deficiency</u> <u>of</u> <u>a</u> <u>network</u>. Recall that n denotes
the number of complexes of a network, ℓ the number of its
linkage classes, and s the dimension of its stoichiometric
subspace. The number δ defined by

$$\delta = n - \ell - s$$

is called the <u>deficiency</u> of the network. It is not difficult
to show that δ cannot be negative. For network (2.1) $n = 6$,
$\ell = 2$, and $s = 4$ so that $\delta = 0$.

2.1.F. <u>Weakly</u> <u>reversible</u> <u>networks</u>. A network is <u>weakly</u>
<u>reversible</u> if its standard reaction diagram has the following
property: Whenever there exists a directed arrow pathway
(consisting of one or more arrows) "pointing" from one complex
to another, there also exists a directed arrow pathway
"pointing" from the second complex to the first. (Equivalently,
a network is weakly reversible if in the standard reaction
diagram each reaction arrow is contained in a directed
reaction cycle.) Thus, the network

$$\begin{array}{c} 2A_1 \\ \nearrow \quad \searrow \\ A_3 \rightleftharpoons A_2 \end{array} \qquad 2A_2 \rightleftharpoons A_4 \qquad (2.3)$$

is not weakly reversible: There is a directed arrow path connecting the complex A_3 to the complex $2A_1$, but there is no directed arrow path from $2A_1$ back to A_3. On the other hand, the network

(2.4)

is weakly reversible: As in (2.3) there is a directed arrow path connecting A_3 to $2A_1$, but in (2.4) there is also a directed arrow path connecting $2A_1$ back to A_3 (via A_2). By checking all other connected pairs of complexes, the reader may readily confirm that (2.4) satisfies the requirement for weak reversibility. Clearly, network (2.1) is not weakly reversible.

In the standard language of chemistry, a reversible network is one in which each reaction is accompanied by its "anti-reaction" (i.e., $i \to j$ implies that $j \to i$). Every reversible network is weakly reversible, but weak reversibility does not, of course, imply reversibility. For our purposes, weak reversibility is the more useful of the two ideas.

The reader is reminded that Table I (Section 5) provides further examples of networks along with specifications of n, ℓ, s and δ for each and an indication of whether each is weakly reversible.

2.2 Kinetics for Networks.

Consider for the moment a spatially homogeneous reactor in which the molar concentration of species A_L is denoted c_L. By a composition of species $\{A_1, A_2, \ldots, A_N\}$ in such a reactor we mean the element $\underset{\sim}{c} = [c_1, c_2, \ldots, c_N]$ in $\mathbb{R}^N$. Because molar concentrations are non-negative, it will prove useful to designate the set of vectors in $\mathbb{R}^N$ with strictly positive components by the symbol $\mathbb{P}^N$ and the set of vectors in $\mathbb{R}^N$ with non-negative components by $\bar{\mathbb{P}}^N$.

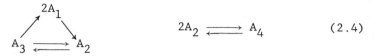

By a kinetics for a network with N species we mean an assignment to each reaction $i \to j$ of a rate function $f_{i \to j}(\cdot)$ defined on $\bar{\mathbb{P}}^N$ which takes non-negative values. The non-negative number $f_{i \to j}(\underset{\sim}{c})$ is the rate of reaction* $i \to j$ at composition $\underset{\sim}{c} \in \bar{\mathbb{P}}^N$. We shall require of a kinetics that, for each reaction $i \to j$ in the network,

(i) $f_{i \to j}(\cdot)$ is continuous

and

(ii) if $\underset{\sim}{y}_i$ is the vector corresponding to the ith complex, then $f_{i \to j}(\underset{\sim}{c}) > 0$ if and only if $c_L > 0$ for all $1 \leq L \leq N$ such that $y_{iL} > 0$.†

This last requirement amounts to an assertion that, in a homogeneous mixture, a reaction proceeds at a positive (i.e., non-zero) rate if and only if all its "reactants" are present in the mixture. For example, the reaction $A_1 + A_2 \longrightarrow A_3$ in network (2.1) is deemed to have a positive rate in a homogeneous mixture if and only if both A_1 and A_2 are present (i.e., if and only if $c_1 > 0$ and $c_2 > 0$).

The archetypal example is provided by the class of mass action kinetics: A kinetics for a network is mass action if, for each reaction $i \to j$ of the network, there exists a positive rate constant $k_{i \to j}$ such that

$$f_{i \to j}(\underset{\sim}{c}) \equiv k_{i \to j} \prod_{L=1}^{N} (c_L)^{y_{iL}} . \tag{2.5}$$

2.3. The Differential Equations Induced by a Reaction Network.

Consider an N-species reaction network with complex vectors $\{\underset{\sim}{y}_1, \underset{\sim}{y}_2, \ldots, \underset{\sim}{y}_n\} \subset \bar{\mathbb{P}}^N$, and suppose the network is endowed with a kinetics $\{f_{i \to j}(\cdot)\}$. By the (vector)

---

*More precisely, $f_{i \to j}(\underset{\sim}{c})$ is the rate of reaction $i \to j$ per unit volume (in bulk mixtures) or per unit area (in surface mixtures).

†Here y_{iL} is the Lth component of the complex vector $\underset{\sim}{y}_i$ - that is, the stoichiometric coefficient of species A_L in the ith complex. We tacitly assume that stoichiometric coefficients are non-negative in every complex. In standard mathematical language, (ii) has a more incisive statement: $f_{i \to j}(\underset{\sim}{c}) > 0$ if and only if support $\underset{\sim}{y}_i \subset$ support $\underset{\sim}{c}$.

differential equation induced by such a system we mean

$$\dot{\underset{\sim}{c}} = \sum_{i \to j} f_{i \to j}(\underset{\sim}{c})(\underset{\sim}{y}_j - \underset{\sim}{y}_i) \ , \tag{2.6}$$

where the sum is taken over all reactions of the network. Equation (2.6) is deemed to govern the evolution of composition in a spatially homogeneous mixture with dynamics generated by the reaction network under scrutiny. Note that the right side of (2.6) is formed by summing the reaction vectors for the network, each weighted by the corresponding reaction rate function.

The component equation of (2.6) corresponding to species A_L is

$$\dot{c}_L = \sum_{i \to j} f_{i \to j}(\underset{\sim}{c})(y_{jL} - y_{iL}) \ . \tag{2.7}$$

The number $y_{jL} - y_{iL}$ is the net number of molecules of A_L produced with every occurrence of the reaction whereby the ith complex reacts to the jth complex, and $f_{i \to j}(\underset{\sim}{c})$ is the occurrence rate of that reaction when the reactor composition is $\underset{\sim}{c}$. Thus, the right side of (2.7) reflects the additive contribution of all reactions to the net production rate of species A_L in a reactor of composition $\underset{\sim}{c}$.

By an <u>equilibrium</u> of (2.6) we mean an element $\underset{\sim}{c} \in \overline{\mathbb{P}}^N$ such that the right side of (2.6) is the zero vector. By a <u>positive equilibrium</u> we mean an equilibrium $\underset{\sim}{c} \in \mathbb{P}^N$ - that is, an equilibrium with strictly positive components.

When the kinetics is mass action (2.6) takes the form

$$\dot{\underset{\sim}{c}} = \sum_{i \to j} k_{i \to j} \prod_{L=1}^{N} (c_L)^{y_{iL}} (\underset{\sim}{y}_j - \underset{\sim}{y}_i) \ . \tag{2.8}$$

For all but the simplest networks (e.g., networks of unimolecular reactions) the vector equation (2.8) represents a system of N non-linear coupled scalar differential equations.

The mathematical complexity of such a system is apparent. To be sure we need only recall the rather intricate system (1.2) of differential equations induced by the relatively simple network (1.1) endowed with mass action kinetics. It is not difficult to confirm that the five scalar equations displayed in (1.2) are just the component equations of (2.8) written for network (1.1).

Equations (2.6) and (2.8) bring into sharper focus the close connection between reaction network structure and the shape of the induced differential equations. We can see that relationship in (2.6) even when the kinetics is arbitrary. When the kinetics is mass action the connection becomes substantially more intimate since the rate function for each reaction assumes a form dictated by its reactant complex.

2.4. An Elementary Property of Differential Equations Induced by Reaction Networks

One simple relationship between solutions to equation (2.6) and the structure of the underlying reaction network may be deduced without excessive difficulty. We draw that relationship in this section not because it is profound, but rather because it provides a setting within which certain questions might be posed more precisely.

Consider an N-species reaction network endowed with an arbitrary kinetics subject only to the constraints postulated in §2.2, and suppose that S is the stoichiometric subspace of the network (§2.1.C). For each $\underset{\sim}{c} \in \bar{\mathbb{P}}^N$ the right side of (2.6) is clearly a linear combination of the reaction vectors for the network and therefore lies in S. Thus, if $\underset{\sim}{c}(\cdot)$ is a solution to (2.6) it is necessary that $\dot{\underset{\sim}{c}}(t)$ lie in S for each t, and this suggests that solutions to (2.6) cannot wander through $\mathbb{R}^N$ in an arbitrary fashion. Rather, they are constrained by the nature of the stoichiometric subspace and, therefore, by the nature of the reaction vectors for the network.

In fact, it is not difficult to show that if $\underset{\sim}{c}$ and $\underset{\sim}{c}'$ are two compositions along a solution to (2.6) then $\underset{\sim}{c} - \underset{\sim}{c}'$ must lie in S. Thus, a composition $\underset{\sim}{c}$ can evolve to a composition $\underset{\sim}{c}'$ only if such a composition change is compatible with stoichiometric constraints imposed by the underlying reaction network.

Moreover, it can also be shown that a composition trajectory that originates in $\bar{\mathbb{P}}^N$ cannot leave $\bar{\mathbb{P}}^N$; the constraints imposed upon a kinetics in §2.2 taken together with equation (2.6) ensure that $\dot{c}_L \geq 0$ whenever $c_L = 0$.

Motivated by these considerations we say that two compositions $\underset{\sim}{c} \in \bar{\mathbb{P}}^N$ and $\underset{\sim}{c}' \in \bar{\mathbb{P}}^N$ are *stoichiometrically compatible* if $\underset{\sim}{c} - \underset{\sim}{c}' \in S$. Accordingly, the set of all possible compositions ($\bar{\mathbb{P}}^N$) can be partitioned into *stoichiometric compatibility classes*. Each of these is the intersection of a parallel* of S with $\bar{\mathbb{P}}^N$. A solution to (2.6) which originates within a stoichiometric compatibility class must remain within that stoichiometric compatibility class for all subsequent time.

Readers unfamiliar with these ideas might find helpful Figures 2.1 and 2.2. Figure 2.1 shows the reaction vectors, the stoichiometric subspace and stoichiometric compatibility classes for the two species network (2.9). The stoichiometric

$$2A_1 \rightleftarrows A_2 \tag{2.9}$$

subspace S is one-dimensional and contains the two reaction vectors $2\underset{\sim}{e}_1 - \underset{\sim}{e}_2$ and $\underset{\sim}{e}_2 - 2\underset{\sim}{e}_1$. The stoichiometric compatibility classes are those parts of parallels of S that lie in the non-negative orthant. In Figure 2.2 we show the stoichiometric subspace and a stoichiometric compatibility class for three species network (2.10). The stoichiometric subspace is two-

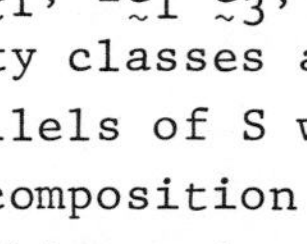

(2.10)

dimensional and contains the four reaction vectors (not shown) $\underset{\sim}{e}_2 - 2\underset{\sim}{e}_1$, $2\underset{\sim}{e}_1 - \underset{\sim}{e}_3$, $\underset{\sim}{e}_2 - \underset{\sim}{e}_3$, and $\underset{\sim}{e}_3 - \underset{\sim}{e}_2$. The stoichiometric compatibility classes are those triangles which are intersections of parallels of S with the non-negative orthant. In both figures the composition trajectories shown reside entirely within stoichiometric compatibility classes as required.

*A set X in $\mathbb{R}^N$ is said to be a parallel of S if there exists a vector $\underset{\sim}{a}$ in $\mathbb{R}^N$ such that X consists of all elements of the form $\underset{\sim}{a} + \underset{\sim}{y}$, where $\underset{\sim}{y} \in S$.

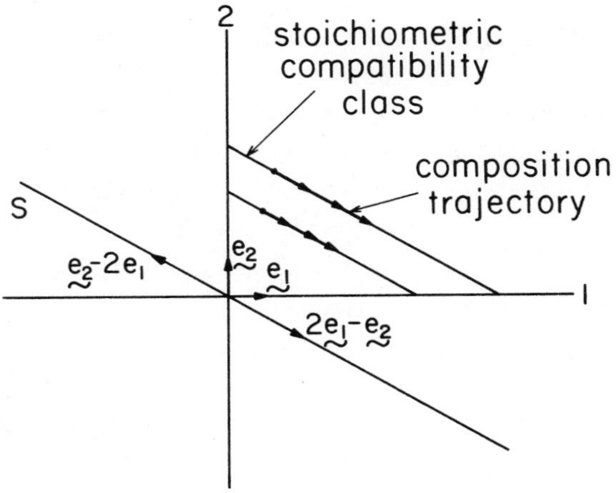

Figure 2.1. Stoichiometric Subspace and a Stoichiometric Compatibility Class for Network (2.9).

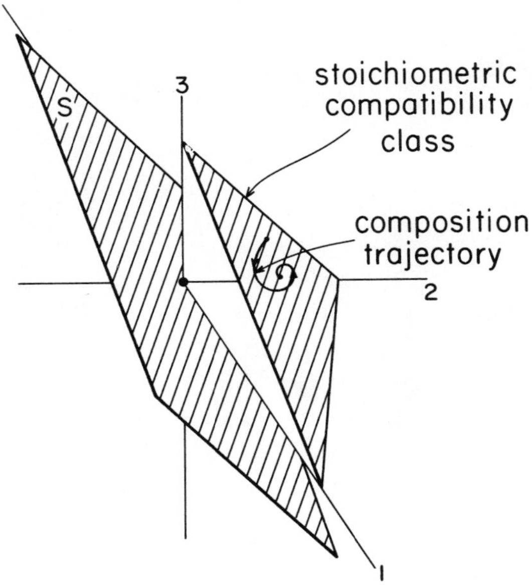

Figure 2.2. Stoichiometric Subspace and a Stoichiometric Compatibility Class for Network (2.10).

We introduce the notion of stoichiometric compatibility in anticipation of certain questions we shall wish to raise in a subsequent section. In particular, we shall be interested in the existence of multiple equilibria. To ask, without qualification, whether the differential equations induced by a network admit more than one equilibrium is to ask a question too broadly posed, for even very simple systems give rise to a wealth of equilibria. The question of real interest is whether those differential equations can admit multiple equilibria _within a stoichiometric compatibility class_ - whether there can exist two or more equilibria which are stoichiometrically compatible with the same initial composition. Similarly, when we speak of stability of an equilibrium _we shall always mean stability relative to initial conditions within the stoichiometric compatibility class containing that equilibrium_.

3. SYSTEMS EMBRACED BY THE MATHEMATICAL FRAMEWORK

It is the purpose of this section to indicate the variety of chemical systems embraced by the mathematical setting sketched in Section 2. At first glance it might seem that the framework described is suited only to spatially homogeneous isothermal reactors which are closed with respect to mass transport, for in systems of this kind composition changes result _only_ from the occurrence of chemical reactions. In fact, a large class of _open_ isothermal reactors (including all those of the "CSTR" variety) can be drawn into this same framework, provided that our conception of reaction network is suitably broadened to incorporate certain "pseudo-reactions" tailored to encode the infusion or effusion of those species supplied to or removed from the reactor under scrutiny.

We should make explicit the tactical importance of our ability to draw open systems into the framework of Section 2 by the means described. Suppose for the moment that the possibilities suggested in the Introduction could be realized, that there could be developed a decent theory of equation (2.8) which would, for example, indicate that the differential equations for networks of a certain large class cannot admit sustained oscillations for any set of rate constants. Suppose further that an open system could be modelled in terms of a reaction network (comprised in part of "pseudo-reactions") taken with mass action kinetics such that (2.8), when written for the model network and its kinetics, would yield precisely those differential equations one would write from first principles (i.e., species balances). In such a case, the aforementioned theory of equation (2.8) would connect qualitative properties of these equations with the structure of the model network; and the possibility of sustained composition oscillations in the system under scrutiny might, for example, be decided solely on the basis of model network structure. As we shall see, our capacity to subsume a large class of open systems within the framework of Section 2 renders any theory of equation (2.8) (or of (2.6)) generally applicable to those open systems in which chemical engineers and biologists have most interest.

A procedure for constructing model networks for open reactors is offered in [3] and [6]; it is not our intention to review that procedure here. In this section we merely offer several examples to suggest the variety of physical systems embraced by the theory and to indicate how each of these may be assigned a reaction network (with kinetics) such that the differential equations constructed according to the formalism of §2.3 are precisely those one would derive from first principles.

We begin with consideration of batch (or tubular) reactors and then turn to several reactors with less apparent connections to the mathematical setting of Section 2.

3.1. Example. Batch and Plug Flow Tubular Reactors. Consider a homogeneous (well stirred) batch reactor whose contents, a solution of species A_1, A_2, A_3, A_4, and A_5 are maintained at constant volume and temperature. These species are presumed to undergo those reactions shown in network (1.1), and the kinetics is presumed mass action with rate constants as indicated in (1.1). Species balances yield the system of five ordinary differential equations displayed in (1.2). These are, of course, the component equations of the vector differential equation (2.8) applied to the network (1.1) with the kinetics indicated. Thus, any general theory of equation (2.8) which serves to draw connections between dynamics and network structure will therefore serve to relate the nature of solutions to (1.2) with the structure of network (1.1). Because the differential equations of batch and steady ideal plug flow reactors are the same under the usual identification of time in the batch reactor with downstream distance in the plug flow reactor, the remarks made here apply to plug flow reactors as well.

In either case, it is important to recognize that a network upon which reactor differential equations are based might suggest but not coincide with the true chemistry. Indeed, complete knowledge of all the reactions in a complex system is rare; and, more often than not, the differential equations are constructed on the basis of a network believed to reflect the dominant features of the chemistry reasonably well. With this in mind, we shall not narrow our focus to preclude networks or kinetics incompatible with properties some might deem essential to the true chemistry (e.g., "microscopic reversibility" or "detailed balance"). On the contrary, we admit all networks (and kinetics) as potential models and ask what relationship these bear to properties of the differential equations they induce.

3.2. Example. Continuous Stirred Tank Reactors (CSTRs). Consider a continuous stirred tank reactor, the contents of which - a liquid mixture of species A_1 and A_2 - are maintained homogeneous, isothermal, and of fixed volume V. Feed of fixed composition is continuously supplied to the reactor at a constant volumetric flow rate g with molar concentrations of A_1 and A_2 in the feed equal to c_1^f and c_2^f, respectively. The contents of the reactor are continuously removed at volumetric flow rate g. In the reactor the only chemical reactions which occur are

$$2A_1 \underset{k'}{\overset{k}{\rightleftarrows}} A_2 \quad . \tag{3.1}$$

The kinetics is mass action with rate constants denoted as shown above. Species balances for A_1 and A_2 yield the following pair of differential equations for c_1 and c_2, the molar concentrations of A_1 and A_2 within the reactor:

$$\begin{aligned} \dot{c}_1 &= \frac{g}{V} c_1^f - \frac{g}{V} c_1 + 2k'c_2 - 2kc_1^2 \\ \dot{c}_2 &= \frac{g}{V} c_2^f - \frac{g}{V} c_2 + kc_1^2 - k'c_2 \,. \end{aligned} \tag{3.2}$$

Now consider the <u>model</u> reaction network shown below.

$$2A_1 \underset{k'}{\overset{k}{\rightleftarrows}} A_2 \underset{gc_2^f/V}{\overset{g/V}{\rightleftarrows}} 0 \underset{g/V}{\overset{gc_1^f/V}{\rightleftarrows}} A_1 \tag{3.3}$$

The entity "0" in this network is the <u>zero complex</u>; the stoichiometric coefficient of every species in the zero complex is zero, and the complex vector (in the sense of §2.1.A) corresponding to the zero complex is the zero vector of $\mathbb{R}^2$ (in this two species case). The pseudo-reactions $0 \to A_1$ and $0 \to A_2$ are adjoined to the reactions $2A_1 \rightleftarrows A_2$ to reflect the infusion of species A_1 and A_2, and the pseudo-reactions $A_1 \to 0$ and $A_2 \to 0$ are similarly adjoined to reflect their efflux. If we assign to network (3.3) a mass action kinetics with rate constants as indicated alongside the reaction arrows, then equation (2.8) written for this network with kinetics so assigned, yields a vector differential equation whose component equations are precisely those shown in (3.2).† Thus, any theory of equation (2.8) which draws connections between dynamics and network structure becomes applicable to this CSTR, so long as it is understood that the network of interest is that shown in (3.3) rather than the simpler (3.1).

It is not difficult to see that this example is readily generalized to arbitrary isothermal continuous stirred tank reactors. To obtain the required model network, one merely adjoins to the set of chemical reactions an "effluent reaction"

---

†Note that according to the mass action prescription (2.5), a reaction of the form $0 \to A$ with assigned rate constant α proceeds at a constant rate equal to α; since the stoichiometric coefficient of every species is zero in the zero complex, the "power product" in (2.5) takes the value 1.

$A_L \rightarrow 0$ for each species and a "feed reaction" $0 \rightarrow A_J$ for each species A_J in the feed. The true chemical reactions are assigned their usual rate constants; each effluent reaction is assigned the rate constant g/V, where g is the total volumetric flow rate of feed and V is the reactor volume; and each feed reaction $0 \rightarrow A_J$ is assigned the rate constant gc_J^f/V, where c_J^f is the molar concentration of A_J in the feed stream.

Interconnected systems of well-stirred cells can be afforded network descriptions in a similar manner [19].

3.3. Example: A Biological Reactor. Biochemists sometimes find it useful to consider homogeneous reactors in which the molar concentrations of certain species are deemed time-invariant despite their participation in chemical reactions. This may happen because these species are available in such large supply relative to other reaction participants that, on any practical time scale, the changes in their concentrations are imperceptible. Or, in more formal terms, the concentrations of these species within the reactor might be regulated by external means - e.g., by controlled infusion or effusion of these species at appropriate rates.

We offer an example of this kind, first studied by Edelstein [2] in consideration of the autocatalytic production of a species (A_1) which is subsequently degraded by an enzyme (A_2). A homogeneous reactor, whose contents are maintained at constant volume and temperature, contains a mixture of species A_1, A_2, A_3, B, and D. The following reactions occur on the interior of the reactor:

$$\begin{gathered} A_1 + B \underset{k_b}{\overset{k_a}{\rightleftarrows}} 2A_1 \\ A_1 + A_2 \underset{k_d}{\overset{k_c}{\rightleftarrows}} A_3 \underset{k_f}{\overset{k_e}{\rightleftarrows}} A_2 + D \end{gathered} \tag{3.4}$$

The kinetics is mass action with rate constants as designated in (3.4). The concentrations of B and D are maintained constant at values c_B^* and c_D^*. The governing differential equations, derived from first principles, for the concentrations of A_1, A_2, and A_3 are:

$$\dot{c}_1 = k_a c_B^* c_1 - k_b c_1^2 - k_c c_1 c_2 + k_d c_3$$
$$\dot{c}_2 = (k_d + k_e) c_3 - k_c c_1 c_2 - k_f c_D^* c_2 \quad (3.5)$$
$$\dot{c}_3 = k_c c_1 c_2 + k_f c_D^* c_2 - (k_d + k_e) c_3$$

Now we consider the <u>model</u> network shown below:

$$A_1 \underset{k_b}{\overset{k_a c_B^*}{\rightleftarrows}} 2A_1$$
$$A_1 + A_2 \underset{k_d}{\overset{k_c}{\rightleftarrows}} A_3 \underset{k_f c_D^*}{\overset{k_e}{\rightleftarrows}} A_2 \quad (3.6)$$

If we associate with this network a mass action kinetics with rate constants as indicated beside the arrows in (3.6) then equation (2.8), written for network (3.6) with its assigned kinetics, yields a vector differential equation whose component equations are precisely those shown in (3.5). Thus, any theory of equation (2.8) which draws connections between dynamics and network structure becomes applicable to the Edelstein system, so long as it is understood that the network of interest is that shown in (3.6) rather than in (3.4).

This example is readily generalized to other reactors of similar type. To obtain the appropriate model network, one merely "strips away" species with time-invariant concentrations from the original network, and one modifies certain rate constants in a manner suggested by the example.

3.4. <u>Example</u>: <u>Heterogeneous Catalysis</u>. In an attempt to understand recent experimental observations of sustained isothermal concentration oscillations on catalyst surfaces, Pikios and Luss [18] considered the following mechanism:

$$B + A_1 \underset{k_b}{\overset{k_a}{\rightleftarrows}} A_2$$

$$D + A_1 \underset{k_d}{\overset{k_c}{\rightleftarrows}} A_3 \tag{3.7}$$

$$A_2 + A_3 \xrightarrow[k_e]{} E + 2A_1 \ .$$

Here B and D represent two species present in a gaseous phase above a catalyst surface. A_1 represents an unoccupied catalyst site, while A_2 and A_3 represent species B and D bound to catalyst sites. Thus, the first two lines of (3.7) represent reversible binding of B and D to unoccupied sites. The third line represents a reaction whereby bound B and D react to form a species E which enters the gas phase, leaving behind two unoccupied catalyst sites.

Let us presume, at least for the moment, that the kinetics is mass action with rate constants as indicated in (3.7) and that, in formulating rate functions, volume concentrations are used for gas phase species while surface concentrations are used for bound species and unoccupied sites. Pikios and Luss treated gas phase concentrations of B and D (denoted here by c_B^* and c_D^*) as time invariant, presumably on the grounds that gaseous forms of these species are available in large excess immediately above the catalyst. In the absence of spatial inhomogeneities in concentrations along the catalytic surface, the differential equations for surface concentrations of A_1, A_2, and A_3 are

$$\begin{aligned} \dot{c}_1 &= -(k_a c_B^* + k_c c_D^*) c_1 + k_b c_2 + k_d c_3 + 2k_e c_2 c_3 \\ \dot{c}_2 &= k_a c_B^* c_1 - k_b c_2 - k_e c_2 c_3 \\ \dot{c}_3 &= k_c c_D^* c_1 - k_d c_3 - k_e c_2 c_3 \ . \end{aligned} \tag{3.8}$$

Now consider the model network

$$A_3 \underset{k_d}{\overset{k_c c_D^*}{\rightleftarrows}} A_1 \underset{k_b}{\overset{k_a c_B^*}{\rightleftarrows}} A_2$$

$$A_2 + A_3 \xrightarrow{k_e} 2A_1 \quad . \tag{3.9}$$

taken with mass action kinetics; the rate constants are indicated beside the corresponding reaction arrow. Equation (2.8), written for network (3.9) with its assigned kinetics, yields a vector differential equation with component equations precisely those shown in (3.8). Thus, any theory of equation (2.8) which draws connections between dynamics and network structure becomes applicable to this example so long as it is understood that the network of interest is (3.9) rather than (3.7).

It should be mentioned that Pikios and Luss did not endow the last reaction in (3.7) with a mass action rate function in the strict sense of §2.2, for they allowed k_e to depend upon surface concentrations of A_1, A_2, and A_3 in the following way*:

$$k_e(c_1, c_2, c_3) = \gamma \exp\{-\mu(\frac{c_3}{c_1+c_2+c_3})\} \quad . \tag{3.10}$$

Here $\gamma > 0$ and μ are catalyst parameters. Clearly, k_e becomes a rate constant in the sense of §2.2 only when $\mu = 0$.

The differential equations with which Pikios and Luss were concerned are readily derived by substituting (3.10) into (3.8). Strictly speaking, these would not be subsumed under a general theory of equation (2.8), for the Pikios-Luss kinetics is not mass action when $\mu \neq 0$. Nevertheless, their differential equations would be subsumed under broader theory developed for the more general equation (2.6) which, it will be recalled, is written for arbitrary kinetics. The network

*Pikios and Luss considered a situation in which the activation energy of the last step in (3.7) depends upon surface coverage.

in question would remain that shown in (3.9), but the kinetics would differ from the mass action case to the extent that k_e depends upon surface composition. Nevertheless, we shall have something to say about the relationship between network structure and the possibility of sustained oscillations for *arbitrary* kinetics in §5.

The examples considered in this section suggest that a large class of chemical systems - including important categories of open systems - are embraced by the mathematical framework of Section 2 *once we are prepared to consider networks with seemingly peculiar reactions* (e.g., $A \rightarrow 2A$, $0 \rightarrow A$, $A \rightarrow 0$) *and to endow these networks with kinetics which might violate principles* (e.g. detailed balance*) *believed appropriate to exact descriptions of closed systems*. With this in mind, we shall want any theory we might construct to subsume networks and kinetics unconstrained by tenets to which we might hold in a more narrowly focused effort. Clearly, the wider the class of networks and kinetics for which we can develop theory, the more widely applicable will such theory be.

4. SOME QUESTIONS

In the spirit of the preceding section we shall henceforth use the term "reaction network" in its broad sense: we shall understand that one or more reactions might in fact be "pseudo-reactions" (with assigned rate functions) incorporated in the network to reflect certain physico-chemical effects for a particular system under study. That is, we shall always understand the reaction network and kinetics for a system to be such that the appropriate differential equations for the system are induced by the network and its kinetics according

*It is worth noting explicitly that whatever the values of k_a, k_b, k_c, k_d, k_e, and k_f in the Edelstein system (Example 3.3), the set of rate constants for the *model* network (3.6) - $k_a c_B^*$, k_b, k_c, k_d, k_e, and $k_f c_D^*$ - may be *made* to violate the Wegscheider conditions for detailed balance by selecting the time invariant concentrations c_B^* and c_D^* appropriately. At least in principle, these may be maintained at desired values by means of controlled infusion or effusion of B and D.

to the formalism outlined in §2.3. With this in mind, we can begin to ask questions about the relationship between reaction network structure and properties of the induced differential equations.

Recall that one such elementary relationship was already articulated in §2.4. The network induces a partition of $\overline{\mathbb{P}}^N$ into stoichiometric compatibility classes, and composition trajectories that originate within a stoichiometric compatibility class must remain within that stoichiometric compatibility class for all subsequent times. We would, of course, like to know substantially more. We would like to know in qualitative terms what happens within these stoichiometric compatibility classes, and we would like to tie that qualitative behavior to reaction network structure.

Our objectives will be rather broad, and we should try to make clear what these are. We seek to classify reaction networks according to their capacity to induce differential equations which admit behavior of a specified kind. If we restrict our attention to networks endowed with mass action kinetics we will not ask, for example, whether the differential equations for a particular network taken with a specified set of rate constants admit sustained composition oscillations. Rather, we will ask if the network is such that the induced differential equations admit sustained oscillations for at least one set of rate constants. The network itself is our object of study, not the network endowed with a particular set of rate constants.

The balance of this section is devoted to the statement of some problems we would like to solve; the list is not intended to be exhaustive. Some of the problem statements are motivated by consideration of "play" networks that hardly reflect real chemistry. These involve only two or three species and were chosen so that equilibrium sets and phase portraits might be sketched and discussed easily.

All the problems are posed in the following way: Describe the class of networks which, when taken with mass action kinetics, are such that the induced differential equations have property X regardless of values of the rate constants. Clearly, a complete solution to a problem of this type gives a solution to a complementary problem: Describe the class of networks which, when taken with mass action kinetics, are such that the induced differential equations fail to have property X for some values of the rate constants. For example, if we could delineate the full class of networks that generate no periodic composition trajectories for any set of rate constants we would then know the complementary class of networks which have the capacity to generate periodic orbits for at least one set of rate constants.

4.1. The Existence of Positive Equilibria. Some reaction networks (e.g., $A_1 \rightarrow A_2$) have the property that, when taken with mass action kinetics, the induced differential equations admit no positive equilibria regardless of values that the rate constants might take. Whatever equilibria do exist are characterized by the "extinction" of one or more species. On the other hand, some networks (e.g., $A_1 \rightleftarrows A_2$) taken with mass action kinetics admit a positive equilibrium in each stoichiometric compatibility class for every set of rate constants.

Even if we grant that the existence or non-existence of positive equilibria for simple networks may be easy to decide, this is certainly not true of complicated networks; for ultimately one is confronted with a large system of polynomial equations in many variables (species concentrations) in which many parameters (rate constants) appear. Recall the relatively simple system (1.2).

We pose the following problem: Describe the class of networks which, when endowed with mass action kinetics, induce differential equations that admit a positive equilibrium within each stoichiometric compatibility class, regardless of values the rate constants might take.

4.2. The Uniqueness of Positive Equilibria. The Edelstein network (3.6), when taken with mass action kinetics, has the property that for some values of rate constants the induced differential equations admit multiple positive equilibria within certain stoichiometric compatibility classes. This is the case for the rate constants indicated in (4.1):

$$A_1 \underset{1}{\overset{8.5}{\rightleftarrows}} 2A_1$$

$$A_1 + A_2 \underset{1}{\overset{1}{\rightleftarrows}} A_3 \underset{0.2}{\overset{1}{\rightleftarrows}} A_2 \quad . \tag{4.1}$$

The locus of equilibrium compositions (excluding the origin) is sketched in Figure 4.1 along with two stoichiometric compatibility classes. These are parallels of the two-dimensional stoichiometric subspace (not shown) containing the six reaction vectors. (In particular, the stoichiometric subspace contains and is spanned by the vectors $\underset{\sim}{e}_1$ and $\underset{\sim}{e}_2-\underset{\sim}{e}_3$.) The lower stoichiometric compatibility class (the dashed rectangle) is pierced by the locus of equilibria in one point, while the higher stoichiometric compatibility class (the solid rectangle) is pierced in three points.

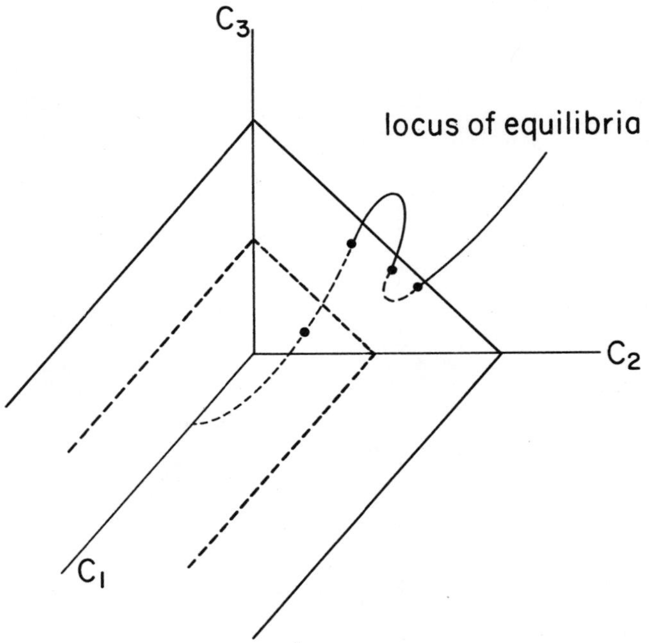

Fig. 4.1. A Sketch of the Equilibrium Set for the System (4.1).

Another network which has the capacity to admit multiple positive equilibria was studied by Horn and Jackson [14] and is shown in (4.2). The kinetics is presumed to be mass action with rate constants as indicated. The induced differential

$$\begin{array}{ccc} 3A_1 & \xrightarrow{1} & A_1 + 2A_2 \\ {\scriptstyle .1}\uparrow & & \downarrow{\scriptstyle .1} \\ 2A_1 + A_2 & \xleftarrow[1]{} & 3A_2 \end{array} \qquad (4.2)$$

equations admit three positive equilibria within each stoichiometric compatibility class (except for the trivial stoichiometric compatibility class which contains only the origin). The locus of equilibria is sketched in Figure 4.2 along with some composition trajectories. The stoichiometric subspace S is one-dimensional, and the stoichiometric compatibility classes are those parts of parallels of S which lie in the non-negative orthant.

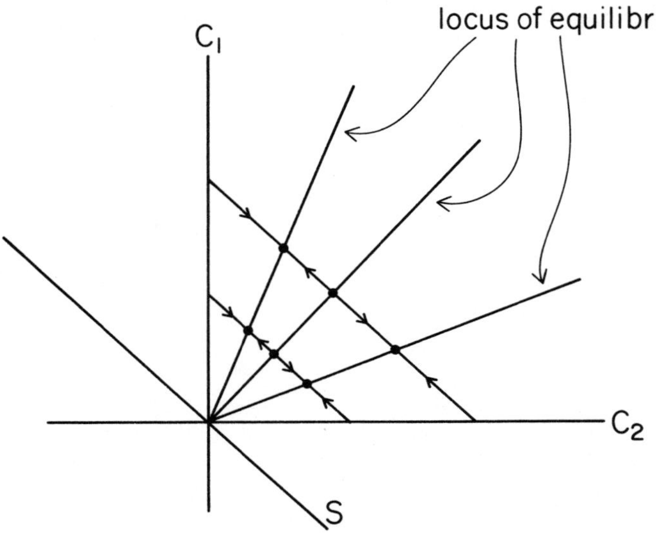

Fig. 4.2. A Sketch of the Equilibrium Set for System (4.2).

Still another example of a network which has the capacity to admit multiple positive equilibria derives from consideration of a simple (albeit unrealistic) continuous stirred tank reactor. (Recall §3.2.) Consider a continuous stirred tank reactor, the contents of which - a liquid mixture of A_1 and A_2 - are maintained homogeneous, isothermal, and of unit volume ($V = 1$). Feed of fixed composition is continuously supplied to the reactor at a constant volumetric flow rate $g = 13/12$ with molar concentrations of A_1 and A_2 in the feed equal to $c_1^f = 0.076$ and $c_2^f = 1.59$. The contents of the reactor are removed continuously at volumetric flow rate $g = 13/12$. In the reactor the only chemical reactions which occur are

$$2A_1 + A_2 \underset{3/10}{\overset{39/20}{\rightleftarrows}} 3A_1 \quad .$$

The kinetics is mass action with rate constants as shown.

In the sense of §3.2 the appropriate differential equations are induced by the network (4.3) taken with mass action kinetics with the indicated rate constants. Here there are two species, and the dimension of the stoichiometric subspace

$$\begin{gathered} A_1 \underset{\frac{13}{12}\text{x}.076}{\overset{13/12}{\rightleftarrows}} 0 \underset{13/12}{\overset{\frac{13}{12}\text{x}1.59}{\rightleftarrows}} A_2 \\ \\ 2A_1 + A_2 \underset{3/10}{\overset{39/20}{\rightleftarrows}} 3A_1 \end{gathered} \tag{4.3}$$

for network (4.3) is two. Thus, the stoichiometric subspace is all of $\mathbb{R}^2$, and there is but one stoichiometric compatibility class - the entire non-negative orthant. (That is, there are no constraints on composition trajectories imposed solely by "stoichiometry".)

The differential equations induced by the system (4.3) admit three positive equilibria. These are indicated in Figure 4.3 along with some composition trajectories.

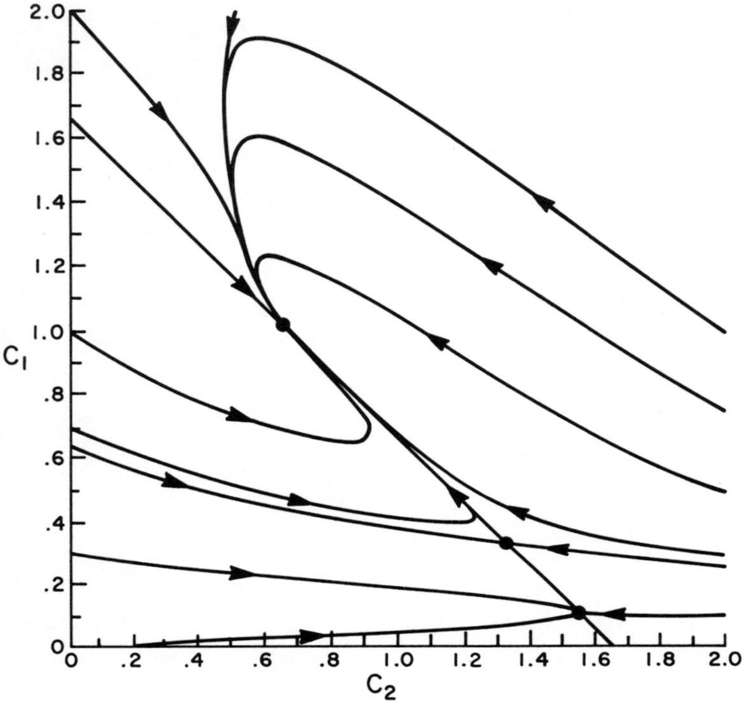

Fig. 4.3. The Three Equilibria and Some Composition Trajectories for System (4.3).

In contrast to networks (4.1)-(4.3) which, when taken with mass action kinetics, admit multiple positive equilibria for certain values of the rate constants, there exist networks which admit precisely one positive equilibrium in each stoichiometric compatibility class for <u>every</u> set of rate constants. We would like to distinguish between those networks which, when endowed with mass action kinetics, have the capacity to exhibit multiple positive equilibria and those which do not.

With this in mind, we pose the following problem: <u>Describe the class of networks which, when taken with mass action kinetics, induce differential equations that admit precisely one positive equilibrium within each stoichiometric compatibility class, regardless of values the rate constants might take</u>.

4.3. <u>The Stability of Positive Equilibria</u>. Each of the networks (4.1)-(4.3) not only admits multiple positive equilibria when the rate constants are as indicated, each also admits unstable (as well as stable) positive equilibria. Even among those networks that admit a unique positive equilibrium for <u>every</u> set of rate constants, there exist networks which have the property that for <u>certain</u> values of the rate constants the lone positive equilibrium is unstable. An example is provided by the "Brusselator" [11], shown here as network (4.4). (In the sense of §3.3, network (4.4) has been obtained from that considered in [11] by "stripping away" species deemed to have time-invariant concentration.) When (4.4) is given mass action kinetics with the rate constants indicated, the induced differential equations admit

$$0 \underset{1}{\overset{1}{\rightleftarrows}} A_1 \xrightarrow{3} A_2$$

$$2A_1 + A_2 \xrightarrow{1} 3A_1 \tag{4.4}$$

a unique positive equilibrium. That equilibrium is unstable and is enclosed within a stable limit cycle.

A sketch of the phase portrait is shown in Figure 4.4. Once again we have a two species network for which the stoichiometric subspace is two-dimensional and is therefore all of $\mathbb{R}^2$. There is but one stoichiometric compatibility class - the entire non-negative orthant of $\mathbb{R}^2$.

In contrast to networks (4.1)-(4.4) there exist networks which, when taken with mass action kinetics, induce differential equations that admit only asymptotically stable positive equilibria regardless of values the rate constants might take. We would like to be able to distinguish between those networks which have the capacity to generate unstable positive equilibria and those which do not.

Consequently, we pose the following problem: <u>Describe the class of networks which, when endowed with mass action kinetics, induce differential equations such that every positive equilibrium is asymptotically stable regardless of values the rate constants might take</u>. (Recall that when we speak of the stability of an equilibrium we shall always mean stability relative to initial conditions within the stoichiometric compatibility class containing that equilibrium.)

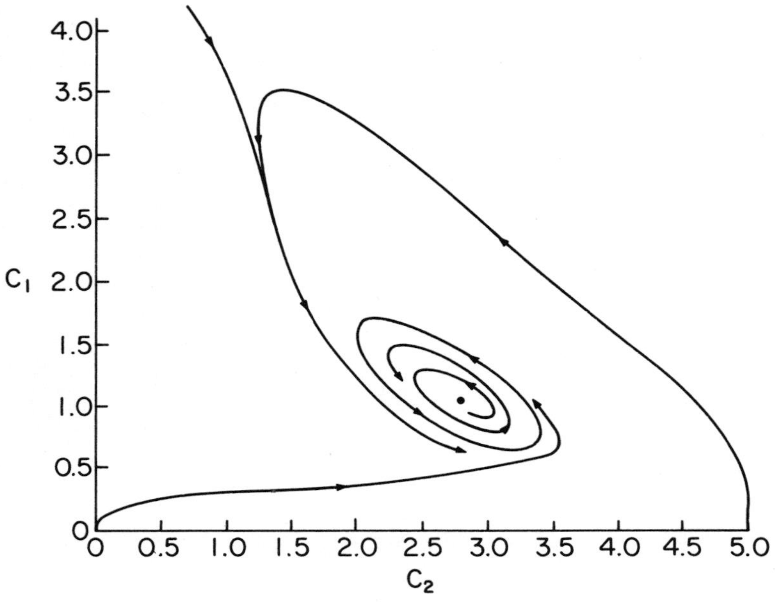

Fig. 4.4. Phase Portrait for System (4.4).

4.4. The Existence of Periodic Solutions. Network (4.4) taken with mass action kinetics has the property that for *some* values of the rate constants (in particular, for those shown) the induced differential equations admit periodic solutions. On the other hand, there exist networks which, when taken with mass action kinetics, fail to exhibit periodic solutions for *any* values of the rate constants. We would like to be able to distinguish between those networks which have the capacity to generate composition cycles and those which do not.

Therefore, we pose the following problem: Describe the class of networks which, when endowed with mass action kinetics, induce differential equations that admit no periodic solutions regardless of values the rate constants might take.

5. THE ZERO DEFICIENCY THEOREM.

For each of the problems posed in the preceding section we seek a characterization of the class of networks which have a particular property. We would like each such class to be described completely and, if possible, simply so that we might readily decide whether a given network has the capacity to admit behavior of one kind or another. But the questions posed are not easy ones, and complete answers are not likely to be forthcoming in the near future. In the absence of full resolution we would be content with partial answers of the kind that characterize large, if incomplete, classes of networks that might generate dynamics of specified types.

An advance in this direction came with proof of the Zero Deficiency Theorem by Horn, Jackson, and Feinberg. Its statement draws upon language introduced in Section 2, which the reader might wish to review. Recall in particular that for each reaction network we can calculate a non-negative integer called its deficiency and that we can inspect the network to decide whether it is weakly reversible. Networks of deficiency zero are quite common. But no matter how complicated these networks might be and no matter how intricate might be the induced differential equations, the theorem tells us that the resulting dynamics must have a special and, to some extent, dull quality.

Theorem 5.1 (The Zero Deficiency Theorem). For any reaction network of deficiency zero the following statements hold true:

a. If the network is not weakly reversible, then for arbitrary kinetics (mass action or otherwise) the induced differential equations cannot give rise to a positive equilibrium.
b. If the network is not weakly reversible, then for arbitrary kinetics (mass action or otherwise) the induced differential equations cannot give rise to a cyclic composition trajectory which passes through a state wherein all species concentrations are positive.
c. If the network is weakly reversible, then for mass action kinetics with any choice of (positive) rate constants, the induced differential equations give rise to precisely one positive equilibrium in each stoichiometric compatibility class; every positive equilibrium composition is asymtotically stable; and the induced differential equations admit no cyclic composition trajectories [except the trivial $\underset{\sim}{c}(\cdot) = \text{constant}$].

A sketch of the proof is offered in [3], and a more detailed account can be found in [4].

This theorem represents substantial progress toward resolution of the problems posed in §4 insofar as it characterizes the qualitative features of solutions to the differential equations induced by a rather broad class of networks. For the class of zero deficiency networks it settles completely - at least when the kinetics is mass action - virtually all issues concerning existence, uniqueness, and asymptotic stability of positive equilibria, and the theorem serves to preclude cyclic composition trajectories. Moreover, it tells us quite a bit even when the kinetics is of unspecified form. In terms more suggestive than precise, the theorem assures us that we can expect essentially "normal" behavior from reactors well modelled by zero deficiency networks, and it tells us that networks which are to serve as models for biochemical or catalytic systems that display certain kinds of "exotic" dynamics should have deficiencies in excess of zero.

In Table I we examine several networks to see what light the Zero Deficiency Theorem sheds and to confirm, at least for a few well-studied examples, that networks capable of generating "abnormal" dynamics do in fact have deficiencies of of one or more. Some of the networks shown (e.g., the "Brusselator, the "Oregonator") have been discussed extensively in the literature, where their appearance bears only a faint resemblance to their appearance in the table. Recall however that the networks with which we work have, in the sense of Example 3.3, been "stripped" of all species deemed to have time-invariant molar concentration. The differential equations induced by the various networks are not displayed, but it should be kept in mind that these are generally quite intricate. The narrative offered here is intended to supplement the commentary provided in Table I.

For networks 1 and 2 the Zero Deficiency Theorem merely confirms what is easy to decide by other means.

However, network 3 provides an example for which questions of the kind we have been asking cannot be settled easily on the basis of straightforward intuitive deduction. Can network 3, taken with mass action kinetics, give rise to a positive equilibrium? At first glance it might seem that the reaction $A_1 \rightarrow 2A_2$ would serve to deplete A_1, driving its concentration to zero in any equilibrium state. But so long as A_5 is present, the reaction $A_2 + A_5 \rightarrow A_1 + A_3$ serves to regenerate A_1, as does the reaction $A_4 \rightarrow A_1 + A_3$ when A_4 is present. Indeed, <u>all</u> species of network 3 are <u>both</u> produced and consumed by reactions of the network, and the existence or nonexistence of positive equilibria is not readily apparent. Can network 3, taken with mass action kinetics, give rise to a cyclic composition trajectory along which all species concentrations remain positive? This question is more difficult still. Yet the Zero Deficiency Theorem answers both questions, not only for mass action kinetics but for <u>arbitrary kinetics</u> (subject only

Table I

Network	# of complexes (n)	# of linkage classes (ℓ)	dimension of stoichiometric subspace (s)	deficiency ($\delta \equiv n-\ell-s$)	weakly reversible ?	Comments
1. $2A_1 \rightarrow A_2$	2	1	1	0	No	Zero Def. Thm., pts. (a) & (b), apply.
2. $2A_1 \rightleftarrows A_2$	2	1	1	0	Yes	Zero Def. Thm., pt. (c), applies.
3. $A_1 \rightarrow 2A_2$ $A_1 + A_3 \rightleftarrows A_4$ $A_4 \rightarrow A_2 + A_5$ $A_2 + A_5 \rightarrow A_1 + A_3$	5	2	3	0	No	Zero Def. Thm., pts. (a) & (b) apply: For any kinetics (mass action or otherwise) no positive equil., no sustained oscillations with all species concentrations positive.

Network	n	ℓ	s	$\delta = n - \ell - s$	weakly reversible ?	Comments
4. 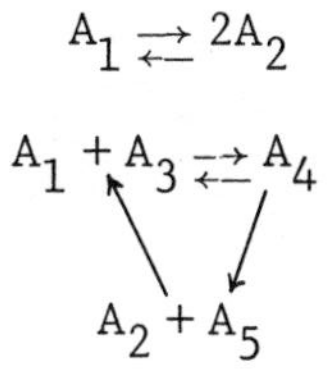	5	2	3	0	Yes	Zero Def. Thm., pt. (c), applies: For mass action kinetics with <u>any</u> set of rate constants there exists precisely one positive equilibrium in each stoichiometric compatibility class; all positive equilibria are asymptotically stable; no sustained oscillations.
5. 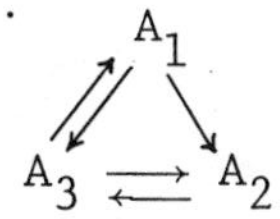	3	1	2	0	Yes	Same as 4.
6. 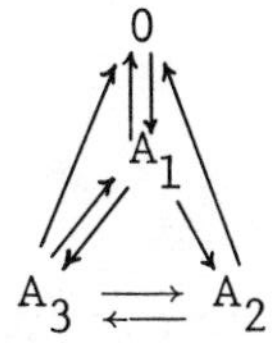	4	1	3	0	Yes	Same as 4. In the sense of §3.2, network 6 serves to model a CSTR with internal chemistry described by network 5. The reactor is fed A_1 ($0 \to A_1$), while A_1, A_2, and A_3 are removed in the effluent ($A_1 \to 0$, $A_2 \to 0$, $A_3 \to 0$).

Network	n	ℓ	s	$\delta = n - \ell - s$	weakly reversible ?	Comments
7. $A_1+A_3 \leftarrow A_4 \rightleftarrows 2A_2$ $A_2+A_3 \rightleftarrows A_5 \rightleftarrows A_1+A_4$	6	2	4	0	No	Same as 3.
8. $A_1+A_3 \rightleftarrows A_4 \rightleftarrows 2A_2$ $A_2+A_3 \rightleftarrows A_5 \rightleftarrows A_1+A_4$	6	2	4	0	Yes	Same as 4.
9. $A_1 \rightleftarrows 2A_1$ $A_1+A_2 \rightleftarrows A_3 \rightleftarrows A_2$	5	2	2	1	Yes	This is the Edelstein network (§3.3, §4.2). Taken with mass action kinetics (for certain rate constants) it gives rise to three positive equilibria within a stoichiometric compatibility class. One of these is unstable. Note that $\delta = 1$.

Network	n	ℓ	s	$\delta = n-\ell-s$	weakly reversible ?	Comments
10. $A_1 \rightleftarrows 2A_1$ $A_1+A_2 \rightleftarrows A_3 \rightleftarrows 2A_2$	5	2	3	0	Yes	Network 10 differs from the Edelstein network in only one complex. The Zero Def. Thm., part (c) applies: For <u>every</u> set of rate constants there is precisely one equilibrium, and that equilibrium is asymptotically stable. (There is but one stoichiometric compatibility class - the entire non-negative orthant of $\mathbb{R}^3$.)
11. $A_1 \longrightarrow 2A_1$ $A_1+A_2 \rightarrow 2A_2$ $A_2 \rightarrow 0$	6	3	2	1	No	This is the Lotka network [17]. Since $\delta=1$, Zero Def. Thm. does not apply. Taken with mass action kinetics, the Lotka network gives rise to sustained oscillations and a positive equilibrium. (Note contrast with pts. (a) & (b) of Zero Def. Thm.)

Network	n	ℓ	s	$\delta = n-\ell-s$	weakly reversible ?	Comments
12. $0 \longrightarrow A_1$ $A_1+A_2 \rightarrow A_3+A_4$ $A_3 \rightarrow A_2 \rightleftarrows A_4+A_5$ $A_4+A_6 \rightarrow A_7 \rightarrow A_6$	10	4	5	1	No	This is Higgins' model for glycolysis [13]. Taken with mass action kinetics, network 12 gives rise to sustained oscillations and a positive equilibrium for some sets of rate constants. Note that $\delta = 1$.
13. $0 \rightleftarrows A_1 \rightarrow A_2$ $2A_1 + A_2 \rightarrow 3A_1$	5	2	2	1	No	This is the "Brusselator" (§4.3). Taken with mass action kinetics, network 13 gives rise to an unstable positive equilibrium surrounded by a stable limit cycle for some sets of rate constants. Note that $\delta = 1$.

Network	n	ℓ	s	$\delta = n-\ell-s$	weakly reversible ?	Comments
14. $A_2 \to A_1 \to 2A_1 + A_3$ $A_1 + A_2 \to 0 \leftarrow 2A_1$ $A_3 \to 2A_2$	8	3	3	2	No	This is the Field-Noyes [10] "Oregonator" model for the Belousov-Zhabotinskii system. (Here the F.-N. "stoichiometric parameter" f has been set to two.) Taken with mass action kinetics the network gives rise to an unstable equilibrium and periodic orbits for certain rate constants [12,16]. Note that $\delta = 2$.
15. $A_3 \rightleftarrows A_1 \rightleftarrows A_2$ $A_3 + A_2 \to 2A_1$	5	2	2	1	No	This is the Pikios-Luss model for oscillations in heterogeneous catalysis. Taken with Pikios-Luss kinetics (§3.4) the network gives rise to sustained oscillations (for certain rate parameters) and a positive equilibrium. Note that $\delta = 1$.

Network	n	ℓ	s	$\delta = n-\ell-s$	weakly reversible ?	Comments
16. $A_3 \underset{\rightarrow}{\leftarrow} A_1 \underset{\leftarrow}{\rightarrow} 2A_2$ $A_3 + A_2 \rightarrow 2A_1$	5	2	3	0	No	Network 16 differs from Pikios-Luss network in only one complex. Because $\delta = 0$, Zero Def. Thm., pts. (a) & (b), preclude positive equilibria or sustained oscillations (in which $c_1 > 0$, $c_2 > 0$, $c_3 > 0$) for any kinetics – in contrast to behavior of Pikios-Luss model.
17. $3A_1 \rightarrow A_1 + A_2$ $2A_1 + A_2 \leftarrow 3A_2$	4	1	1	2	Yes	This is the network studied by Horn & Jackson (§4.2). For certain rate constants (in the context of mass action kinetics) there exist three positive equilibria in each stoichiometric compatibility class, one of which is unstable. Note that $\delta = 2$.

to the weak requirements of §2.2): The differential equations induced by network 3 _admit no positive equilibria or composition cycles of the type described_.

Network 4 is precisely the one with which we began our discussion in the Introduction. It differs from network 3 only insofar as the "arrow structure" is slightly different: the reaction $2A_2 \rightarrow A_1$ has been added. Since network 4 is weakly reversible, part (c) of the Zero Deficiency Theorem applies. If the network is endowed with mass action kinetics, then not only do we know that there exist positive equilibria, we know that there exists precisely one such equilibrium in each stoichiometric compatibility class and that all positive equilibria are asymptotically stable. Moreoever, we know that the differential equations induced by the network cannot give rise to cyclic composition trajectories. These statements hold true for _any_ set of rate constants. Lest it be supposed that results of this kind are readily deducible from an ad hoc study of the differential equations induced by network 4, the reader is encouraged to glance once again at (1.2), where these differential equations are displayed in full detail. Qualitative properties of their solutions are hardly apparent.

The system of linear differential equations induced by network 5 (taken with mass action kinetics) is amenable to solution. Still, the properties of solutions ensured by part (c) of the Zero Deficiency Theorem for _all_ sets of rate constants are not _immediately_ apparent. In fact, _all_ networks with only unimolecular complexes, however intricate their arrow structure, are of deficiency zero so that all such networks are embraced by the Zero Deficiency Theorem.

Networks 6, 7, and 8 provide further examples of networks which have zero deficiency.

Network 9 is the Edelstein model for autocatalytic production of a species followed by its enzymatic degradation. (Recall Example 3.3 and §4.2.) Taken with mass action kinetics, network 9 has the capacity (for some sets of rate constants) to admit three positive equilibria within a stoichiometric compatibility class, one of which is unstable. The Zero Deficiency Theorem requires that for a weakly reversible network (taken with mass action kinetics) to admit multiple positive equilibria or unstable positive equilibria its deficiency must exceed zero. In fact, the weakly reversible network 9 has deficiency one.

Network 10 differs from network 9 in only one complex; the complex A_2 in network 9 is replaced by the complex $2A_2$ in network 10. Taken with mass action kinetics, network 10 induces differential equations which are "more nonlinear" than those for network 9 - the rate of reaction $2A_2 \rightarrow A_3$ is proportional to $(c_2)^2$, while the rate of reaction $A_2 \rightarrow A_3$ is proportional to c_2. With this in mind, we might suppose that network 10 has the capacity to generate dynamics at least as exotic as those generated by network 9. Yet, this is not the case. Unlike network 9, network 10 has deficiency <u>zero</u>. Therefore, for <u>every</u> set of rate constants network 10 must give rise to <u>precisely one</u> positive equilibrium in each stoichiometric compatibility class, and that equilibrium is asymptotically stable. Thus, the Zero Deficiency Theorem serves to distinguish between a network which gives rise to "abnormal" dynamics and a very similar one which cannot. Moreover, the information the theorem offers runs contrary to what crude intuitive reasoning would suggest.

Networks 11-14, when taken with mass action kinetics, all have the capacity to admit positive equilibria despite the fact that none is weakly reversible, and each has the capacity to generate sustained oscillations wherein all species concentrations remain positive. As required by the Zero Deficiency Theorem each network has a deficiency in <u>excess</u> of zero.

Network 15, which encodes the Pikios-Luss model for sustained oscillations in heterogeneous catalysis, gives rise to positive equilibria and cyclic composition trajectories (wherein all species concentrations remain positive) for certain values of parameters in the Pikios-Luss kinetics. (Recall §3.4.) As required by either part (a) or part (b) of the Zero Deficiency Theorem, the deficiency of the Pikios-Luss network <u>exceeds</u> zero.

On the other hand network 16, which differs from the Pikios-Luss network in only one complex, has zero deficiency. Thus, it cannot give rise to positive equilibria or to cyclic composition trajectories (wherein all species concentrations remain positive) for <u>any</u> kinetics.

Network 17, when taken with mass action kinetics, gives rise to multiple positive equilibria for certain sets of rate constants. Part (c) of the Zero Deficiency Theorem tells us that any weakly reversible network which, when taken with mass action kinetics, gives rise to multiple equilibria must have a deficiency greater than zero. In fact, the deficiency of network 17 is two.

Before closing this section we should take note of what the Zero Deficiency Theorem, part (c), does <u>not</u> say. When a weakly reversible zero deficiency network is endowed with mass action kinetics we know that within each stoichiometric compatibility class there will be precisely one <u>positive</u> equilibrium and that that equilibrium is asymptotically stable. However, there is as yet no proof that every composition trajectory beginning on the (relative) interior of a particular stoichiometric compatibility class will ultimately approach as a limit the positive equilibrum contained within that stoichiometric compatibility class. The theorem merely ensures that trajectories which begin sufficiently close to the positive equilibrium will approach it as a limit with increasing time.

It is conceivable, for example, that a composition trajectory originating in the relative interior of a stoichiometric compatibility class (i.e., at a composition for which all species concentrations are positive) might ultimately evolve toward an equilibrium on the boundary of that stoichiometric compatibility class (i.e., toward an equilibrium characterized by the extinction of one or more species). This is certainly possible for networks that are not weakly reversible (e.g., $A_1 \rightarrow A_2$). On the other hand, it is unknown whether behavior of this kind can obtain for a weakly reversible network (of <u>any</u> deficiency) taken with mass action kinetics.

There is, however, something which can be said for weakly reversible zero deficiency networks when the kinetics is mass action: Consider a compact positively invariant set contained within the relative interior of a stoichiometric compatibility class. Then that set contains the unique positive equilibrium for the stoichiometric compatibility class, and all solutions originating within the invariant set will approach this equilibrium as a limit with increasing time.

6. NETWORKS OF POSITIVE DEFICIENCY

Everything said thus far amounts to a review of older work: the Zero Deficiency Theorem as stated here was known - if not widely known - by the end of 1973. We turn now to some new and as yet unpublished results. The next theorem we shall state is reminiscent of part (c) of the Zero Deficiency Theorem. It is somewhat less generous in the scope of conclusions it draws; but, on the other hand, it draws those conclusions for a wider class of networks.

Theorem 6.1. Consider a weakly reversible network with ℓ linkage classes. Let δ denote the deficiency of the network; let δ_i $(i=1,2,\ldots,\ell)$ denote the deficiency of the ith linkage class (viewed as a network unto itself); and suppose that these numbers satisfy the following conditions:

(i) $\delta_i \leq 1,\ i = 1,2,\ldots,\ell.$

(ii)* $\sum_{i=1}^{\ell} \delta_i = \delta$.

If the network is endowed with mass action kinetics, then, for every set of rate constants, the induced differential equations admit precisely one positive equilibrium in each stoichiometric compatibility class.

---

*Condition (ii) is equivalent to the requirement that the stoichiometric subspace for the network be the direct sum of the stoichiometric subspaces generated by the linkage classes separately.

A sketch of parts of the proof will be available in
[4], and it is expected that a full proof will be published
in the Archive for Rational Mechanics and Analysis [8].

Theorem 6.1 addresses issues concerning the existence
and uniqueness of positive equilibria; but, in contrast to
part (c) of the Zero Deficiency Theorem, it says nothing about
the stability of positive equilibria or about the possibility
of periodic composition trajectories. However, the Zero
Deficiency Theorem subsumes those and only those networks for
which condition (ii) is satisfied and for which condition (i)
is replaced by the stronger requirement: $\delta_i = 0$, $i = 1, 2, \ldots, \ell$.

We illustrate the use of Theorem 6.1 in the following
example:

Example: Consider network (6.1):

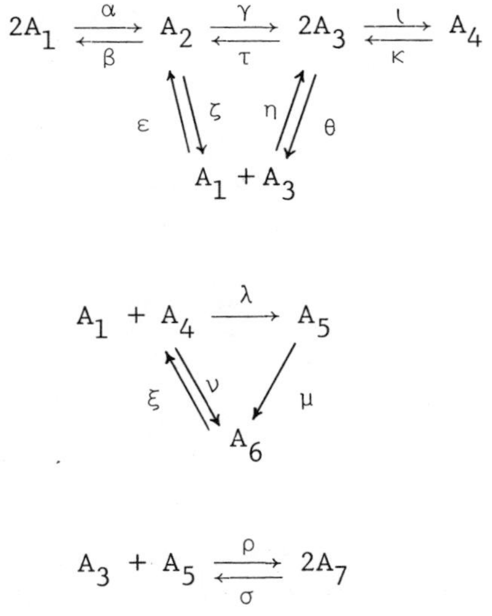

$$2A_1 \underset{\beta}{\overset{\alpha}{\rightleftarrows}} A_2 \underset{\tau}{\overset{\gamma}{\rightleftarrows}} 2A_3 \underset{\kappa}{\overset{\iota}{\rightleftarrows}} A_4$$

$$\varepsilon \Updownarrow \zeta \quad \eta \nearrow\!\!\!\!\swarrow \theta$$

$$A_1 + A_3$$

$$A_1 + A_4 \overset{\lambda}{\longrightarrow} A_5 \qquad\qquad (6.1)$$

$$\xi \Updownarrow \nu \quad \swarrow \mu$$

$$A_6$$

$$A_3 + A_5 \underset{\sigma}{\overset{\rho}{\rightleftarrows}} 2A_7$$

Taken with mass action kinetics - rate constants are indicated
by Greek letters - the network induces the system of differential equations displayed as (6.2).

$$\begin{aligned}
\dot{c}_1 &= -2\alpha c_1^2 + (2\beta+\zeta)c_2 - (\varepsilon+\eta)c_1c_3 + \theta c_3^2 - (\lambda+\nu)c_1c_4 + \xi c_6 \\
\dot{c}_2 &= \alpha c_1^2 - (\beta+\gamma+\zeta)c_2 + \varepsilon c_1c_3 + \tau c_3^2 \\
\dot{c}_3 &= (2\gamma+\zeta)c_2 + (\eta-\varepsilon)c_1c_3 - (\theta+2\iota)c_3^2 + 2\kappa c_4 - \rho c_3c_5 + \sigma c_7^2 \\
\dot{c}_4 &= \iota c_3^2 - \kappa c_4 - (\lambda+\nu)c_1c_4 + \xi c_6 \\
\dot{c}_5 &= \lambda c_1c_4 - \mu c_5 - \rho c_3c_5 + \sigma c_7^2 \\
\dot{c}_6 &= \nu c_1c_4 + \mu c_5 - \xi c_6 \\
\dot{c}_7 &= 2\rho c_3c_5 - 2\sigma c_7^2
\end{aligned} \tag{6.2}$$

Can this system of differential equations yield multiple positive equilibria in a stoichiometric compatibility class for some choice of the sixteen rate constants? An attempt to resolve the question on an ad hoc basis would require study of an intractable system of polynomial equations in seven variables in which sixteen unspecified parameters appear. Yet, the theorem assures us that the answer is no. The deficiency of the entire network is readily calculated to be one, while the deficiencies of the three linkage classes (from top to bottom) are $\delta_1 = 1$, $\delta_2 = 0$, and $\delta_3 = 0$. The network is weakly reversible and therefore falls within the class described by the theorem.

Theorem 6.1 has an immediate corollary. It results from the observation that a weakly reversible network which is of deficiency zero or one and which consists of a single linkage class satisfies the hypothesis of the theorem trivially.

Corollary 6.2. Consider a weakly reversible network consisting of a single linkage class, and suppose that the network is endowed with mass action kinetics. Then the induced differential equations can admit multiple positive equilibria within a stoichiometric compatibility class only if the network is of deficiency two or more.

In our next example Corollary 6.2 provides what, at first glance, might appear to be a surprising conclusion.

Example: Recall that the Edelstein network, discussed in §3.3, §4.2 and rewritten here as network (6.3), when taken with mass action kinetics has the capacity to generate multiple positive equilibria within a stoichiometric compatibility class. (Recall Figure 4.1.) These are not precluded by Theorem 6.1. Although condition (i) is satisfied, condition (ii) is not: the deficiency of the network is one ($n = 6$, $\ell = 2$, $s = 2$), and the deficiency of each linkage class is zero.

$$\begin{array}{c} A_1 \rightleftarrows 2A_1 \\ \\ A_1 + A_2 \rightleftarrows A_3 \rightleftarrows A_2 \end{array} \tag{6.3}$$

Now suppose that the Edelstein network is augmented to permit reversible transition between A_1 and A_3 as shown in (6.4) and suppose again that the kinetics is mass action.

$$\begin{array}{ccccc} & & A_1 & \rightleftarrows & 2A_1 \\ & & \updownarrow & & \\ A_1 + A_2 & \rightleftarrows & A_3 & \rightleftarrows & A_2 \end{array} \tag{6.4}$$

The augmented network is also of deficiency one ($n=5, \ell=1, s=3$) but now consists of a single linkage class. Thus, Corollary 6.2 asserts that, for whatever values the eight rate constants might take, there is precisely one equilibrium in the positive orthant† of $\mathbb{R}^3$. In particular, we note that however small might be the rate constants for the reactions added, the augmented Edelstein network no longer has the capacity to generate multiple positive equilibria. In fact, when the Edelstein network is "perturbed" as indicated, the entire locus of positive equilibria (Figure 4.1) collapses: all but one of the points on that locus become "disequilibrated" even when the rate constants for the two additional reactions are arbitrarily small. The positive equilibrium that persists under perturbation is determined by the ratio of those rate constants.

For some time after Theorem 6.1 was proved it seemed reasonable to conjecture that networks which satisfy its hypothesis would enjoy all the properties described in part (c) of the Zero Deficiency Theorem. That conjecture is false. My colleague Paul Berner found that, for certain values of the eight rate constants, the single positive equilibrium generated by network (6.4) is unstable and that the induced differential equations admit periodic solutions.

* * * * * * * * * *

Before closing this section we show by means of counter-examples that Theorem 6.1 becomes false if any of the three constraints on network structure - weak reversibility, condition (i) and condition (ii) - is dropped from its hypothesis.

It is not difficult to see that the absence of weak reversibility might preclude the existence of positive equilibria for certain networks which otherwise satisfy all the requirements imposed by the theorem. What is less apparent, however, is that the absence of weak reversibility might impinge upon the uniqueness of positive equilibria when they

---

†Because $s=3$ there is but one stoichiometric compatibility class - the entire non-negative orthant of $\mathbb{R}^3$.

do exist. An example is provided by network (6.5) taken with mass action kinetics. When the two rate constants are unequal there are no positive equilibria. When the two rate constants

$$2A_1 \longleftarrow A_1 + A_2 \longrightarrow 2A_2 \qquad (6.5)$$

are equal the induced differential equations reduce to $\dot{c}_1 \equiv 0$, $\dot{c}_2 \equiv 0$. In this case every composition is an equilibrium, and there are an infinity of positive equilibria within each stoichiometric compatibility class (apart from the trivial one consisting only of the origin of $\mathbb{R}^2$).* Network (6.5) is of deficiency one ($n = 3$, $\ell = 1$, $s = 1$), and it satisfies both conditions (i) and (ii) of Theorem 6.1.

That condition (i) cannot be dropped from the hypothesis of Theorem 6.1 follows from consideration of network 17 in Table 1. Its deficiency is two, and there is but one linkage class. The network is weakly reversible and satisfies condition (ii) trivially but fails to satisfy condition (i). As was illustrated in Figure 4.2 the network, when taken with mass action kinetics, has the capacity to generate multiple positive equilibria within a stoichiometric compatibility class.

That condition (ii) cannot be dropped from the hypothesis of Theorem 6.1 follows immediately from consideration of the Edetlstein network (6.3). It is weakly reversible; and, since the deficiency of each linkage class is zero, condition (i) is satisfied. On the other hand the network is of deficiency one so that condition (ii) fails to hold. Taken with mass action kinetics, the network has the capacity to generate multiple positive equilibria within a stoichiometric compatibility class.

---

*Although uniqueness of positive equilibria cannot be ensured when weak reversibility is dropped from the hypothesis of Theorem 6.1, uniqueness is ensured (when such equilibria exist) provided the weak reversibility condition is replaced by the less stringent requirement that each linkage class contain precisely one terminal strong linkage class. (See Def. 10 of [7].) There are two terminal strong linkage classes in network (6.5) - $\{2A_1\}$ and $\{2A_2\}$.

7. NETWORKS OF DEFICIENCY ONE: SPECIAL THEORY

In this section we shall focus upon the special problem of deciding whether a weakly reversible network of deficiency one, when taken with mass action kinetics, has the capacity to generate multiple positive equilibria within a stoichiometric compatibility class (for at least one set of rate constants). Although Theorem 6.1 answers this question in the negative for certain deficiency one networks, it provides no information for others. In fact, it is not difficult to see that, among the weakly reversible networks of deficiency one, Theorem 6.1 fails to subsume precisely those which are composed of two or more linkage classes and which have the property that each linkage class is of deficiency zero.

The remaining class of weakly reversible deficiency one networks is neither small nor unimportant. Consequently, we are compelled to examine this class with some care and develop theory appropriate to it. In order that we might develop a sense of what demands must be placed on such a theory, it will be useful to consider three weakly reversible deficiency one networks, each composed of two linkage classes of deficiency zero. These are not chosen to reflect real chemistry but, rather, to indicate difficulties a complete deficiency one theory must surmount. The networks are shown in Table II along with commentary; each is presumed to be endowed with mass action kinetics.

Of the three networks displayed only the second has the capacity to exhibit multiple positive equilibria. Yet the second network differs from the first only to the extent that A_1 has been added to both sides of the reactions $A_1 + A_2 \rightleftarrows 2A_1$. And the second network differs from the third only insofar as the complexes 0 and A_1 are permuted.

The table demonstrates that, among the class of weakly reversible deficiency one networks that are not subsumed by Theorem 6.1, some have the capacity to admit multiple positive equilibria while others do not. Even more, the table indicates that any theory purporting to distinguish between these must be of considerable delicacy if it is to discriminate between networks so similar as the three exhibited. To distinguish

TABLE II

Reaction Network	Comments
1. $A_1 \rightleftarrows 0 \rightleftarrows A_2$ $A_1 + A_2 \rightleftarrows 2A_1$	This network gives rise to precisely <u>one</u> positive equilibrium regardless of values that the six rate constants take. In the sense of §3.2, the network describes a continous stirred tank reactor in which the "internal chemistry" is $A_1 + A_2 \rightleftarrows 2A_1$ and both A_1 and A_2 are present in the feed.
2. $A_1 \rightleftarrows 0 \rightleftarrows A_2$ $2A_1 + A_2 \rightleftarrows 3A_1$	For <u>some</u> values of the rate constants this network gives rise to <u>three</u> positive equilibria. The network describes a continuous stirred tank reactor in which the "internal chemistry" is $2A_1 + A_2 \rightleftarrows 3A_1$ and both A_1 and A_2 are in the feed. (Recall §4.2.)
3. $0 \rightleftarrows A_1 \rightleftarrows A_2$ $2A_1 + A_2 \rightleftarrows 3A_1$	For <u>all</u> choices of rate constants this network gives rise to precisely <u>one</u> positive equilibrium. However, for some values of the rate constants, the induced differential equations admit a stable limit cycle. This is the reversible Brusselator. (Recall §4.3.)

between networks 1 and 2 such a theory would have to be sensitive to seemingly minor changes in the complexes, and to distinguish between networks 2 and 3 the theory would have to be sensitive to the precise manner in which complexes are linked by reaction arrows.

Theorems that speak _solely_ in terms of deficiency could not do the job. Stoichiometry influences deficiency only insofar as it influences the _difference_ in stoichiometric coefficients of the species across reaction arrows; in this respect, networks 1 and 2 are identical. Moreover, the precise manner in which complexes are linked by reaction arrows influences deficiency only to the extent that the complexes are partitioned into linkage classes; in this respect, networks 2 and 3 are identical. Thus, any theorems that sort the remaining weakly reversible deficiency one networks according to their capacity to exhibit multiple positive equilibria must respect the fine details of network structure to a far greater extent than do Theorems 5.1 and 6.1.

Nevertheless, we would like any additional theorems to share with deficiency-type theorems the remarkable ease with which they resolve otherwise intractable questions concerning intricate systems of non-linear equations in several variables and several unspecified parameters. Can there be developed such a _practical_ deficiency one theory, a theory which is easily used but which also relates more delicate aspects of network structure to the admissibility of multiple positive equilibria?

Recent research suggests that the answer is _yes_. Because it is still under development a full description of the theory would be premature. However, we can describe in some detail how the theory works for _certain_ networks of deficiency one. In particular, we shall confine our attention to _reversible_ (§2.1.F) deficiency one networks which have the following properties:

1. There are _two_ linkage classes, and each linkage class (viewed as a network unto itself) is of deficiency zero.

2. Each linkage class is <u>tree-like</u>: We call a linkage class tree-like if the removal of the reaction arrows ($\rightleftarrows$) between any pair of adjacent complexes results in the "disconnection" of the linkage class. For example, the linkage class

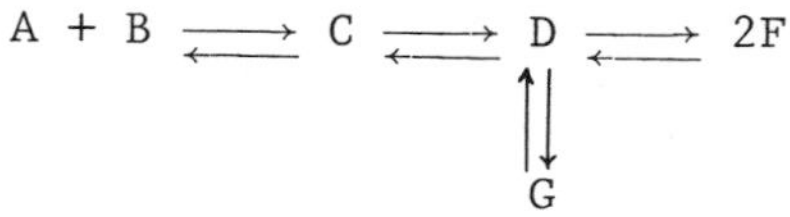

(7.1)

is tree-like, but the linkage class

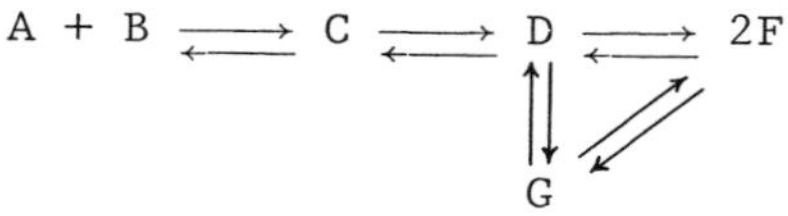

(7.2)

is not tree-like. In (7.1) removal of the reaction arrows between any pair of adjacent complexes "disconnects" the linkage class. On the other hand, removal of the reaction arrows between complexes G and 2F in (7.2) leaves the linkage class connected.

Note that the three networks shown in Table II fall within the class of networks under consideration.

We shall describe a simple algorithm tailored to this circumscribed class which helps decide whether a particular network, when taken with mass action kinetics, has the capacity to generate multiple positive equilibria (for at least one set of rate constants). Before we outline the detailed workings of the algorithm, we shall first provide an overview of its operation.

Suppose the network under study contains N species, say $A_1, A_2, \ldots, A_N$. With these species we associate N "variables", $\mu_1, \mu_2, \ldots, \mu_N$. Now with each complex of the network we construct a linear form in these variables in which the coefficient of μ_L is just the stoichiometric coefficient of A_L in that complex. (For example, the complex $2A_1 + A_2$ results in the linear form $2\mu_1 + \mu_2$.) If there are n complexes in the

TABLE III

Network	Linear Systems Produced by Algorithm	Comment
1. $A_1 \rightleftarrows 0 \rightleftarrows A_2$ $A_1 + A_2 \rightleftarrows 2A_1$	(a) $\mu_1 > 0 > \mu_2 > 2\mu_1 > \mu_1 + \mu_2$ (b) $\mu_2 > 0 > \mu_1 > \mu_1 + \mu_2 > 2\mu_1$	Neither system (a) nor (b) admits a solution. Hence, no set of rate constants for network 1 will give rise to multiple positive equilibria.
2. $A_1 \rightleftarrows 0 \rightleftarrows A_2$ $2A_1 + A_2 \rightleftarrows 3A_1$	(a) $\mu_1 > 0 > \mu_2 > 3\mu_1 > 2\mu_1 + \mu_2$ (b) $\mu_2 > 0 > \mu_1 > 2\mu_1 + \mu_2 > 3\mu_1$	System (a) admits no solution, but system (b) does: $\mu_1 = -2$, $\mu_2 = 1$.
3. $0 \rightleftarrows A_1 \rightleftarrows A_2$ $2A_1 + A_2 \rightleftarrows 3A_1$	(a) $0 = \mu_1 > \mu_2 > 3\mu_1 > 2\mu_1 + \mu_2$ (b) $\mu_2 > \mu_1 = 0 > 2\mu_1 + \mu_2 > 3\mu_1$	Neither system (a) nor (b) admits a solution. Hence, no set of rate constants for network 3 will give rise to multiple equilibria.

network there will be n such linear forms. For example, network 1 of Table II (for which n=5, N = 2) results in a set of five linear forms in two variables:

$$\{\mu_1,\ 0,\ \mu_2,\ \mu_1+\mu_2,\ 2\mu_1\}. \tag{7.3}$$

The algorithm will indicate how, for the particular network under study, these form should be joined by equality or inequality signs to produce two systems of linear inequalities or equations in the variables $\mu_1, \mu_2, \ldots, \mu_N$. For each of the networks shown in Table II, the resulting pair of linear systems produced by the algorithm is shown in Table III.

If neither linear system admits a solution then there can exist no set of rate constants for which the dynamical equations induced by the network admit multiple positive equilibria within a stoichiometric compatibility class, no matter how complex those equations might be. We shall postpone for the moment discussion of inferences to be drawn when solutions do exist.

Having described the algorithm in broad terms, we are in a position to outline its workings in detail. In fact, it remains only to indicate how the linear forms should be joined by equality or inequality signs to produce the pair of linear systems we would inspect for solution. Before we describe the steps one takes toward this end, we should say a few words about language and notation.

For terminological purposes, we shall find it convenient to confuse a complex (e.g., $A_1 + A_2$) with its representation (in the sense of §2.1.A) as a vector in $\mathbb{R}^N$ (e.g., $e_1 + e_2$). Thus, we will not hesitate to speak of "the complex y_i" or of "the reaction pair $y_i \rightleftarrows y_j$". Of the two linkage classes in a network under study, we shall arbitrarily designate one to be the first linkage class and denote by k the number of its complexes. The complexes of the first linkage class will be designated $\{y_1, y_2, \ldots, y_k\}$, and complexes of the second linkage class will be designated $\{y_{k+1}, \ldots, y_n\}$.

Example. We shall illustrate the steps of the algorithm by executing each for network 1 of Table II. With this in mind, we designate as the first linkage class of network 1 the one containing three complexes ($k = 3$) and designate those complexes as follows:

$$y_1 = e_1 = [1,0], \quad y_2 = 0 = [0,0], \quad y_3 = e_2 = [0,1]. \tag{7.4.a}$$

The complexes of the second linkage class are designated

$$y_4 = e_1 + e_2 = [1,1], \quad y_5 = 2e_1 = [2,0]. \tag{7.4.b}$$

Finally, we take note of the fact that the linear form corresponding to complex $y_i \in \mathbb{R}^N$ can be written $y_i \cdot \mu$, where $\mu = [\mu_1, \mu_2, \ldots, \mu_N] \in \mathbb{R}^N$ and "$\cdot$" denotes the standard scalar product in $\mathbb{R}^N$. Thus, the linear forms corresponding to the n complexes will be denoted

$$\{y_1 \cdot \mu, \; y_2 \cdot \mu, \ldots, y_n \cdot \mu\}.$$

Example. For network 1 we have

$$\begin{aligned} y_1 \cdot \mu &= [1,0] \cdot [\mu_1, \mu_2] = \mu_1 \\ y_2 \cdot \mu &= [0,0] \cdot [\mu_1, \mu_2] = 0 \\ y_3 \cdot \mu &= [0,1] \cdot [\mu_1, \mu_2] = \mu_2 \\ y_4 \cdot \mu &= [1,1] \cdot [\mu_1, \mu_2] = \mu_1 + \mu_2 \\ y_5 \cdot \mu &= [2,0] \cdot [\mu_1, \mu_2] = 2\mu_1 \end{aligned} \tag{7.5}$$

We are now in a position to describe the steps of the algorithm.

Step One. Find a set of numbers $\{g_i\}_{i=1,2,...,n}$, not all zero, such that

$$\sum_{i=1}^{n} g_i \underset{\sim}{y}_i = \underset{\sim}{0}, \quad \sum_{i=1}^{k} g_i = 0, \text{ and } \sum_{i=k+1}^{n} g_i = 0 . \qquad (7.6)$$

A network of deficiency one will always admit such sets of numbers, and all such sets will be identical up to multiplication by a constant.

Example. For network 1 of Table II (with complexes designated as in (7.4)) we shall take $g_1 = 1$, $g_2 = 0$, $g_3 = -1$, $g_4 = 1$, and $g_5 = -1$. These numbers satisfy (7.6) since $g_1 + g_2 + g_3 = 0$, $g_4 + g_5 = 0$, and

$1(\underset{\sim}{e}_1) + 0(\underset{\sim}{0}) + (-1)(\underset{\sim}{e}_2) + (1)\ (\underset{\sim}{e}_1 + \underset{\sim}{e}_2) + (-1)(2\underset{\sim}{e}_1) = \underset{\sim}{0}.$

Step Two. Now focus upon the first linkage class. Because each linkage class is tree-like, removal of any "reversible pair" of reaction arrows ($\rightleftarrows$) will disconnect the first linkage class into two "pieces". For each reaction pair, say $\underset{\sim}{y}_p \rightleftarrows \underset{\sim}{y}_q$, do the following: Disconnect the linkage class by removing the reaction arrows connecting $\underset{\sim}{y}_p$ to $\underset{\sim}{y}_q$. One of the resulting pieces will contain $\underset{\sim}{y}_p$, while the other will contain $\underset{\sim}{y}_q$. Choose one of these pieces, say that containing $\underset{\sim}{y}_p$, and sum all the g_i (determined in Step One) corresponding to complexes in this piece. If the sum is positive, write $\underset{\sim}{y}_p \cdot \underset{\sim}{\mu} > \underset{\sim}{y}_q \cdot \underset{\sim}{\mu}$; if the sum is negative, write $\underset{\sim}{y}_p \cdot \underset{\sim}{\mu} < \underset{\sim}{y}_q \cdot \underset{\sim}{\mu}$; and if the sum is zero, write $\underset{\sim}{y}_p \cdot \underset{\sim}{\mu} = \underset{\sim}{y}_q \cdot \underset{\sim}{\mu}$. Thus, there will result an inequality or an equation for every reaction pair in the first linkage class.

Example. The first linkage class in network 2 is $y_1 \rightleftarrows y_2 \rightleftarrows y_3$. Removal of the reaction arrows connecting y_1 to y_2 results in the disconnection of the linkage class into two pieces: y_1 and $y_2 \rightleftarrows y_3$. The first of these contains only the complex y_1, and $g_1 = 1 > 0$. Thus, we write $y_1 \cdot \mu > y_2 \cdot \mu$.* Removal of the reaction arrows joining y_2 to y_3 results in the disconnection of the first linkage classes into two pieces: $y_1 \rightleftarrows y_2$ and y_3. Summing the g_i for complexes in the piece containing y_2, we obtain $g_1 + g_2 = 1 > 0$. Thus, we write $y_2 \cdot \mu > y_3 \cdot \mu$. Step One therefore results in the following pair of inequalities:

$$y_1 \cdot \mu > y_2 \cdot \mu > y_3 \cdot \mu \quad . \tag{7.7}$$

Step Three. Repeat Step Two for the second linkage class, and then reverse all inequalities obtained (for the second linkage class).

Example. The second linkage class of network 2 consists only of the reaction pair $y_4 \rightleftarrows y_5$. Removal of the reaction arrows results in a disconnection into two pieces: y_4 and y_5. The first of these contains only y_4, and $g_4 = 1 > 0$. Thus, we write $y_4 \cdot \mu > y_5 \cdot \mu$. As required by Step Three we now reverse this inequality to obtain

$$y_5 \cdot \mu > y_4 \cdot \mu \quad . \tag{7.8}$$

*Had we chosen instead to sum the g_i in the "piece" containing y_2, we would have obtained the same result: since $g_2 + g_3 = -1$, we would have written $y_2 \cdot \mu < y_1 \cdot \mu$. That the choice of "pieces" is immaterial follows from (7.6).

Step Four. For every complex y_i in the first linkage class and every complex y_j in the second linkage class, write

$$y_i \cdot \mu > y_j \cdot \mu \ .$$

Example. For network 1 Step Four results in the six inequalities

$$\begin{array}{lll} y_1 \cdot \mu > y_4 \cdot \mu \ , & y_2 \cdot \mu > y_4 \cdot \mu \ , & y_3 \cdot \mu > y_4 \cdot \mu \\ y_1 \cdot \mu > y_5 \cdot \mu \ , & y_2 \cdot \mu > y_5 \cdot \mu \ , & y_3 \cdot \mu > y_5 \cdot \mu \ . \end{array} \tag{7.9}$$

Step Five. To obtain the first inequality system* for the network, combine the results obtained in Steps Two, Three, and Four.

Example. Taken with inequalities (7.7) and (7.8), the single inequality $y_3 \cdot \mu > y_5 \cdot \mu$ obtained in Step Four renders redundant the remaining five inequalities shown in (7.9). Thus, Step Five results in the following string of inequalities:

$$y_1 \cdot \mu > y_2 \cdot \mu > y_3 \cdot \mu > y_5 \cdot \mu > y_4 \cdot \mu \ . \tag{7.10}$$

Drawing upon (7.5) we may rewrite this string in terms of the variables μ_1 and μ_2:

$$\mu_1 > 0 > \mu_2 > 2\mu_1 > \mu_1 + \mu_2 \ . \tag{7.11}$$

Thus, we have obtained the first linear system shown for network 1 in Table III.

---

*We use the term "inequality system" loosely; certain relations among the linear forms might be equalities.

Step Six. To obtain the second inequality system for the network rewrite all the results of Steps Two and Three with inequalities reversed, and combine these with inequalities produced in Step Four.

Example. Rewriting the results of Steps Two and Three with inequality signs reversed, we obtain

$$y_3 \cdot \mu > y_2 \cdot \mu > y_1 \cdot \mu \quad \text{and} \quad y_4 \cdot \mu > y_5 \cdot \mu \ . \tag{7.12}$$

Taken with (7.12), the single inequality $y_1 \cdot \mu > y_4 \cdot \mu$ obtained in Step Four renders redundant the remaining five inequalities shown in (7.9). Thus, the results of Step Six reduce to

$$y_3 \cdot \mu > y_2 \cdot \mu > y_1 \cdot \mu > y_4 \cdot \mu > y_5 \cdot \mu \ . \tag{7.13}$$

Rewriting (7.13) in terms of μ_1 and μ_2, we obtain

$$\mu_2 > 0 > \mu_1 > \mu_1 + \mu_2 > 2\mu_1 \ . \tag{7.14}$$

This is the second linear system of inequalities shown for network 1 in Table III.

Step Seven. Examine both the first and second inequality systems produced in Steps Five and Six to determine whether either admits a solution. That is, determine whether there can exist a set of numbers $\mu_1, \mu_2, \ldots, \mu_N$ compatible with either inequality system. If neither admits a solution then the network under study, when taken with mass action kinetics, induces differential equations that admit at most one positive equilibrium within each stoichiometric compatibility class. This statement holds true regardless of values the rate constants might take.

Example. Inspection of the inequality systems (7.11) and (7.14) produced by the algorithm for network 1 tells us immediately that neither admits a solution. Therefore, network 1, taken with mass action kinetics, does not have the capacity to generate multiple positive equilibria within a stoichiometric compatibility class. In fact, there is but one stoichiometric compatibility class for network 1 - the entire non-negative orthant of $\mathbb{R}^2$ - so that there can be but one positive equilibrium whatever may be the values of the rate constants.

So ends the algorithm.

The reader might wish to confirm that the inequality systems produced by the algorithm for networks 2 and 3 of Table II are (up to reversal of all inequality signs*) those shown in Table III. Because neither inequality system for network 3 admits a solution that network cannot, when taken with mass action kinetics, give rise to multiple positive equilibria. On the other hand, one of the inequality systems for network 2 does admit a solution. In such a case we have as yet made no pronouncements which enable us, on the basis of the algorithm, to affirm or deny the capacity of the network to generate multiple positive equilibria.

Consequently, we must ask what inferences might be drawn when an inequality system produced by the algorithm does in fact admit solution. Fortunately, there is quite a bit we can say once we have a small amount of language at our disposal.

Consider an N-species network with stoichiometric subspace S. A vector $\underset{\sim}{\mu} = [\mu_1, \mu_2, \ldots, \mu_N] \in \mathbb{R}^N$ is sign-compatible with S if the components of $\underset{\sim}{\mu}$ agree in sign with the

*Depending upon certain arbitrary choices made in the execution of the algorithm (e.g., designation of the first and second linkage classes) the sense of all inequalities obtained might be exactly opposite to those displayed in Table III. Reversal of all inequality signs within a particular inequality system will not, of course, affect the existence of a solution.

corresponding components of some vector contained in S - that is, if there exists $\underset{\sim}{\sigma} = [\sigma_1, \sigma_2, \ldots, \sigma_N] \in S$ such that, for each $1 \leq L \leq N$, μ_L is positive if σ_L is positive, μ_L is negative if σ_L is negative, and $\mu_L = 0$ if $\sigma_L = 0$.

We have suggested that the existence of a solution to either inequality system generated by the algorithm is a matter of some importance. But the decisive issue is whether either admits a solution that is sign-compatible with the stoichiometric subspace for the network at hand.

Suppose that either inequality system generated by the algorithm admits a solution $\underset{\sim}{\mu} = [\mu_1, \mu_2, \ldots, \mu_N]$ that is sign-compatible with the stoichiometric subspace for the network under study. Then, if the network is endowed with mass action kinetics, there will exist a set of rate constants such that the induced differential equations admit multiple positive equilibria within a stoichiometric compatibility class. If neither inequality system admits such a solution then, for every set of rate constants, the induced differential equations will admit at most one positive equilibrium within each stoichiometric compatibility class.

Among the class of networks to which we have confined our attention there are networks which have particularly pleasant properties. Consider, for example, those networks for which the dimension of the stoichiometric subspace is identical to the number of species $(s = N)$. (All the networks shown in Table II are of this kind.) For such networks the stoichiometric subspace coincides with $\mathbb{R}^N$ so that every vector of $\mathbb{R}^N$ is sign-compatible with the stoichiometric subspace. Moreoever, there is but one stoichiometric compatibility class - the entire non-negative orthant of $\mathbb{R}^N$. Thus, we can make the following assertion:

Consider a network (of the class under consideration) for which the dimension of the stoichiometric subspace is identical to the number of species $(s = N)$, and suppose the network is endowed with mass action kinetics. There exists a set of rate constants for which the induced differential equations admit multiple positive equilibria if and only if either inequality system produced by the algorithm admits solution.

Example. What is new here is the assertion that, if s = N, the existence of a solution to either inequality system is not only a necessary condition but also a sufficient condition for the network, when taken with mass action kinetics, to have the capacity to generate multiple positive equilibria. With this in mind we can assert that network 2 of Tables II and III does indeed have the capacity to generate multiple positive equilibria for at least certain values of the rate constants. In fact, such a set of rate constants was exhibited in §4.2. Recall in particular Figure 4.3.

Only slightly less pleasant are networks for which the dimension of the stoichiometric subspace is one less than the number of species (s = N-1). (The Edelstein network (6.3) is of this kind.) In such a case, the stoichiometric subspace admits a one-dimensional orthogonal complement. That is, there exists a non-zero vector $\underset{\sim}{m} \in \mathbb{R}^N$ orthogonal to every reaction vector for the network - i.e., $\underset{\sim}{m} \cdot (\underset{\sim}{y}_j - \underset{\sim}{y}_i) = 0$ whenever i → j - and every vector having this property is a scalar multiple of $\underset{\sim}{m}$. Because specification of any such vector $\underset{\sim}{m}$ serves to specify the stoichiometric subspace completely, it is not surprising that sign-compatibility of a vector μ with the stoichiometric subspace can be expressed solely in terms of the relationship between components of μ and $\underset{\sim}{m}$. It is this idea that provides the basis for the following statement:

Consider a network (of the class under consideration) for which s = N-1, and let $\underset{\sim}{m} = [m_1, m_2, \ldots, m_N]$ be any non-zero vector of $\mathbb{R}^N$ orthogonal to every reaction vector for the network. Suppose that either inequality system generated by the algorithm admits a solution $\mu_1, \mu_2, \ldots, \mu_N$ such that the set

$$\{\mu_1 m_1, \ \mu_2 m_2, \ \ldots, \ \mu_N m_N\}$$

contains within it a positive number and a negative number or else consists entirely of zeros. Then, if the network is taken with mass action kinetics, there exists a set of rate constants such that the induced differential equations admit multiple positive equilibria within a stoichiometric compatibility class. If neither inequality system admits such a solution then, for every set of rate constants, the induced differential equations admit within each stoichiometric compatibility class at most one positive equilibrium.

Example. The Edelstein network

$$A_1 \rightleftarrows 2A_1$$

$$A_1 + A_2 \rightleftarrows A_3 \rightleftarrows A_2$$

is reversible, is of deficiency one, and each linkage class is tree-like and of deficiency zero. Moreover, s=2 and N=3. The inequality systems generated by the algorithm are:

$$\mu_1 > 2\mu_1 > \mu_2 > \mu_3 > \mu_1 + \mu_2 \tag{7.15a}$$

and

$$2\mu_1 > \mu_1 > \mu_1 + \mu_2 > \mu_3 > \mu_2 \,. \tag{7.15b}$$

The reaction vectors for the Edelstein network are

$$\{\underset{\sim}{e}_1, -\underset{\sim}{e}_1, \underset{\sim}{e}_3 - (\underset{\sim}{e}_1 + \underset{\sim}{e}_2),\ \underset{\sim}{e}_1 + \underset{\sim}{e}_2 - \underset{\sim}{e}_3,\ \underset{\sim}{e}_2 - \underset{\sim}{e}_3,\ \underset{\sim}{e}_3 - \underset{\sim}{e}_2\} \,.$$

It is not difficult to confirm that $\underset{\sim}{m} = \underset{\sim}{e}_2 + \underset{\sim}{e}_3 = [0,1,1]$ is orthogonal to every reaction vector and that $\underset{\sim}{\mu} = [5,-1,3]$ is a solution to (7.15b). With $\underset{\sim}{m}$ and $\underset{\sim}{\mu}$ chosen this way we have $m_1\mu_1 = (0)(5) = 0$, $m_2\mu_2 = (1)(-1) = -1$, and $m_3\mu_3 = (1)(3) = 3$. Since the set $\{m_1\mu_1, m_2\mu_2, m_3\mu_3\}$ contains a positive and a negative element, we can assert that there exists for the Edelstein network a set of rate constants such that the induced differential equations admit multiple positive equilibria within a stoichiometric compatibility class. Indeed, such a set of rate constants was displayed in §4.2.

<u>Example</u>. Let us consider the <u>reversible</u> Pikios-Luss network (Table I, network 15):

$$A_3 \rightleftharpoons A_1 \rightleftharpoons A_2$$

$$A_3 + A_2 \rightleftharpoons 2A_1$$

Here again the network is reversible, is of deficiency one, and each linkage class is tree-like and of deficiency zero. Moreover, s=2 and N=3. The inequality systems generated by the algorithm are[*]:

$$\mu_3 > \mu_1 < \mu_2 \qquad \mu_1 > \mu_2 + \mu_3 > 2\mu_1 \qquad\qquad (7.16a)$$

and

$$\mu_3 < \mu_1 > \mu_2 \qquad 2\mu_1 > \mu_2 + \mu_3 \qquad \mu_3 > 2\mu_1 \qquad \mu_2 > 2\mu_1 \quad (7.16b)$$

The reaction vectors are

$$\{\underset{\sim}{e}_3 - \underset{\sim}{e}_1, \ \underset{\sim}{e}_1 - \underset{\sim}{e}_3, \ \underset{\sim}{e}_2 - \underset{\sim}{e}_1, \ \underset{\sim}{e}_1 - \underset{\sim}{e}_2, \ \underset{\sim}{e}_3 + \underset{\sim}{e}_2 - 2\underset{\sim}{e}_1, \ 2\underset{\sim}{e}_1 - \underset{\sim}{e}_2 - \underset{\sim}{e}_3\}.$$

It is not difficult to confirm that the positive vector $\underset{\sim}{m} = \underset{\sim}{e}_1 + \underset{\sim}{e}_2 + \underset{\sim}{e}_3 = [1,1,1]$ is orthogonal to each of these, nor is it difficult to confirm that the only solutions to (7.16a) and (7.16b) are of the kind for which μ_1, μ_2 and μ_3 are negative. Consequently, for every solution to (7.16a) or (7.16b) it must be the case that, with $\underset{\sim}{m} = [1,1,1]$, the set $\{m_1\mu_1, m_2\mu_2, m_3\mu_3\}$ consists entirely of negative numbers. We can assert therefore that the reversible Pikio-Luss network, when taken with mass action kinetics, cannot generate multiple positive equilibria within a stoichiometric compatibility class.

---

[*] In (7.16a) the inequality $\mu_2 + \mu_3 > 2\mu_1$ is redundant, as is $2\mu_1 > \mu_2 + \mu_3$ in (7.16b). These have been displayed explicitly so that the reader might not be perplexed by the absence of the linear form $\mu_2 + \mu_3$.

It is certainly true that the examples provided in the section were simple ones, ones for which the capacity of the network to generate multiple positive equilibria could, with a small amount of discomfort, be decided from direct study of the induced differential equations. However, for networks which are only slightly larger ad hoc study of those differential equations becomes a far less attractive option. It is for such networks that we expect the methods described here to hold real value.

It is anticipated that the theoretical basis for results reported in this section will be made available in [4]. Although we have restricted our discussion to reversible deficiency one networks composed of two tree-like linkage classes (each of deficiency zero), there is no serious obstacle to generalization aimed at similar networks composed of several linkage classes. Our restriction to two linkage classes was intended to avoid a slightly more cumbersome exposition. There is, however, some difficulty in relaxing the requirement that the linkage classes be tree-like. The extent of this difficulty is as yet difficult to assess.

8. OUTLOOK

Reaction networks have both algebraic and graphical character. In rough terms, the algebraic character of a network derives from the nature of its complexes, while the manner in which those complexes are linked by reaction arrows lends a network its character as a directed graph. To say that properties of the induced differential equations should perhaps depend upon an interplay of both aspects of network structure is to say the obvious.

Yet, a remarkable feature of Theorems 5.1 and 6.1 is the extent to which their hypotheses are insensitive to the graphical character of the networks described. This is not to say that the reaction arrows play no role, for they certainly decide the weak reversibility of the network under study and the manner in which complexes are partitioned into linkage classes. But, as far as Theorems 5.1 and 6.1 are concerned, these are the only roles the reaction arrows play; the detailed manner in which complexes are linked by reaction arrows is of little consequence. In a sense, then, the conditions set by the hypotheses of Theorems 5.1 and 6.1 are, for the most part, algebraic.

We have hardly begun to construct a complete picture of the relationship between reaction network structure and properties of the induced differential equations. In the end that picture will be complicated and will almost certainly be dominated at its center by more substantial interplay between algebraic and graphical aspects of network structure. Should this be true, then Theorems 5.1 and 6.1 represent important details painted in far to one side of the canvas, a side heavily algebraic in tone.

If we cannot see the center of the picture clearly, we might at least fill in some detail at its other end. The fact is that there exist networks which, solely by virtue of their graphical structure, induce differential equations which enjoy properties virtually identical to those described in part (c) of the Zero Deficiency Theorem.

We shall call a reversible network star-like if there exists a single fixed complex that resides either at the head or at the tail of every reaction arrow. That is, a reversible network is star-like if all its reaction arrows point away from or toward some "central" complex as illustrated in (8.1):

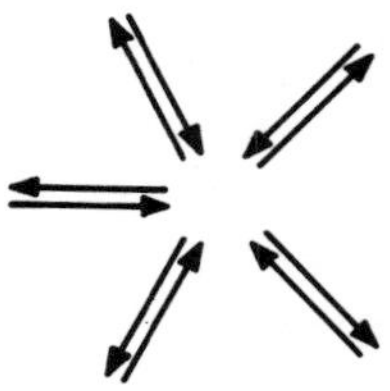

(8.1)

Properties of differential equations induced by star-like networks are described in our final theorem [9].

Theorem 8.1. Consider a reversible star-like network endowed with mass action kinetics. Regardless of what the complexes are and regardless of what values the rate constants take, the induced differential equations admit precisely one positive equilibrium in each stoichiometric compatibility class; every positive equilibrium is asymptotically stable; and there can exist no (non-trivial) cyclic composition trajectory which passes through a state wherein all species concentrations are positive.*

Where the hypothesis of Theorems 5.1 and 6.1 were heavily algebraic in tone, the hypothesis of Theorem 8.1 is purely graphical. Although it seems unlikely that Theorem 8.1 will see much direct application, it nevertheless reveals in a dramatic way the extent to which graphical aspects of network structure can influence properties of the induced differential equations. Indeed, it is possible to construct reversible star-like networks of arbitrarily high deficiency, but these will induce differential equations with properties like those induced by arbitrarily shaped weakly reversible networks of deficiency zero.

If Theorems 5.1 and 6.1 lie at one side of the picture we seek and Theorem 8.1 lies at the other, how should these join at its center? It is difficult to say. Although the algorithm presented in Section 7 is sensitive to both the algebraic and graphical aspects of network structure, it is too limited in scope to suggest how theory might look for arbitrarily shaped networks of high deficiency. A forthcoming article by Clarke [1] deals extensively with network structure and the stability of equilibria. The results he presents will no doubt help bring into focus some important details we have yet to see clearly.

*Theorem 8.1 admits generalization to networks composed of several reversible star-like linkage classes provided the deficiencies of the separate linkage classes sum to the deficiency of the network. In fact, when this last condition holds the conclusion Theorem 8.1 obtains so long as the separate linkage classes are either reversible star-like or weakly reversible and of deficiency zero.

REFERENCES

1. Clarke, B., Stability of complex reaction networks, Adv. Chem. Phys. (to appear - 1980).
2. Edelstein, B., A biochemical model with multiple steady states and hysteresis, J. Theor. Biol. 29 (1970), 57.
3. Feinberg, M., Mathematical aspects of mass action kinetics, Ch. 1 of Chemical Reactor Theory: A Review (L. Lapidus and N. Amundson, eds.) Prentice-Hall, Englewood Cliffs, 1977, 1-74.
4. ____________, Lectures on Chemical Reaction Networks, to be issued as a technical report of the Mathematics Research Center, University of Wisconsin - Madison.
5. ____________, Complex balancing in general kinetic systems, Arch. Rational Mech. Anal. 49 (1972), 187.
6. ____________ and F. J. M. Horn, Dynamics of open chemical systems and the algebraic structure of the underlying reaction network, Chem. Eng. Sci. 29 (1974), 775.
7. ____________ and F. Horn, Chemical mechanism structure and the coincidence of the stoichiometric and kinetic subspaces, Arch. Rational Mech. Anal. 66 (1977), 83.
8. Feinberg, M., Existence and Uniqueness of Equilibria for Chemical Reaction Networks of Positive Deficiency, in preparation for Arch. Rational Mech. Anal.
9. ____________, Dynamical properties of star-like reaction networks, in preparation.
10. Field, R. J. and R. M. Noyes, Oscillations in chemical systems. IV. Limit cycle behavior in a model of a real chemical reaction, J. Chem. Phys. 60 (1974), 1877.
11. Glansdorff, P. and I. Prigogine, Thermodynamic Theory of Structure, Stability, and Fluctuations, pp. 232-241, Wiley-Interscience, New York (1971).
12. Hastings, S. P. and J. D. Murray, The existence of oscillatory solutions in the Field-Noyes model for the Belousov-Zhabotinskii reaction, SIAM J. Appl. Math. 28 (1975) 678.
13. Higgins, J., The theory of oscillating reactions, Ind. Eng. Chem. 59 (1967), 18.
14. Horn, F. J. M. and R. Jackson, General mass action kinetics, Arch. Rational Mech. Anal. 47 (1972), 81.

15. Horn, F. J. M., Necessary and sufficient conditions for complex balancing in chemical kinetics, Arch. Rational Mech. Anal. 49 (1972), 172.

16. Hsu, I.-D. and N. D. Kazarinoff, An applicable Hopf bifurcation formula and instability of small periodic solutions of the Field-Noyes model, J. Math. Anal. Appl. 55, (1976), 61.

17. Lotka, A., Undamped oscillations derived from the law of mass action, J. Am. Chem. Soc. 42 (1920), 1595.

18. Pikios, C. A. and Dan Luss, Isothermal concentration oscillations on catalytic surfaces, Chem. Eng. Sci. 32 (1977), 191.

19. Shapiro, A. and F. Horn, On the possibility of sustained oscillations, multiple steady states, and asymmetric steady states in multicell reaction systems, Math. Biosciences 44 (1979), 19.

The research described here was supported by a Camille and Henry Dreyfus Teacher-Scholar Grant and by NSF Grant ENG 78-09242. Early work by Roy Jackson and Fritz Horn was not only crucial to results described in Section 5 but also influenced the more recent results described in Sections 6-8. Finally, I wish to thank Paul Berner, Charles Conley, Neil Fenichel and Chris Jones for stimulating discussions.

Department of Chemical Engineering
University of Rochester
Rochester, NY 14627

Steady-State Multiplicity and Uniqueness Criteria for Chemically Reacting Systems

Dan Luss

INTRODUCTION.

It is well known that the coupling between chemical and physical rate processes may lead to steady-state multiplicity in chemical reactors. The ability to predict in terms of observable quantities the reactions and conditions for which this multiplicity may occur is most useful for the development of rational control and start-up procedures.

In the last decade various criteria have been developed for predicting the conditions under which steady-state multiplicity may be encountered in several lumped and distributed parameter systems. Comprehensive reviews of the criteria can be found in the monographs by Aris [3], Denn [5], Perlmutter [16] and the chapters by Varma and Aris [24] and Luss [9]. I shall not attempt to summarize the work presented in these reviews or to present a comprehensive literature survey, but shall discuss some recent studies and developments.

Most previous studies of steady-state multiplicity were of systems in which a single chemical reaction took place. However, most practical control and start-up problems associated with steady-state multiplicity are encountered in systems in which several chemical reactions occur simultaneously. These problems are due to transient conditions which cause an undesired, exothermic reaction, whose rate is negligible under normal operating conditions, to take over and proceed at a high rate. This event is usually accompanied by a high rate of heat release and is referred to as a runaway.

ISBN 0-12-669550-4

An infinite number of multi-reaction networks exist and it is impossible to analyze and classify the behavior of each. I shall attempt to outline the insight that might be gained about the qualitative multiplicity features of lumped parameter systems in which two chemical reactions occur from information about the behavior of the single reaction case, and to point out some special multiplicity features of the two reaction networks. I shall later discuss how the exact criteria for a lumped parameter system may be utilized to attain sufficient uniqueness and multiplicity criteria for a distributed parameter system in which the same chemical reaction occurs.

We shall start with a review of the steady-state multiplicity of a lumped parameter system in which a single chemical reaction occurs.

A SINGLE REACTION IN A LUMPED PARAMETER SYSTEM.

Consider a single, n-th order ($n \geq 0$) irreversible chemical reaction occurring in a lumped parameter system. A classical example of such a system is a continuously stirred tank reactor (CSTR), described by the following steady-state species and energy balances [24] [see the notation for definition of the parameters and variables];

$$q(C_o - C) - V\hat{k}(T)C^n = 0 \tag{1}$$

$$q\rho C_p(T_o - T) + (-\Delta H)V\hat{k}(T)C^n - Ua(T - T_c) = 0. \tag{2}$$

Introducing the reference temperature [11]

$$T_m = \frac{T_o + HT_c}{1+H} \qquad H = \frac{Ua}{q\rho C_p}, \tag{3}$$

and the dimensionless variables

$$u = C/C_o \qquad y = T/T_m$$

$$\beta = \frac{(-\Delta H)C_o}{\rho C_p T_m(1+H)} \qquad Da = \frac{V\hat{k}(T_m)C_o^{n-1}}{q} \tag{4}$$

$$\gamma = \frac{E}{RT_m} \qquad X = \frac{\hat{k}(T)}{\hat{k}(T_m)} = \exp[\gamma(1 - \frac{1}{y})],$$

equations (1-2) become

$$1-u - Dau^{n}X = 0 \tag{5}$$

$$1-y + \beta Dau^{n}X = 0. \tag{6}$$

Multiplying (5) by β and adding it to (6) gives

$$u = (1+\beta-y)/\beta. \tag{7}$$

Substitution of (7) into (6) enables one to describe the steady state by the single equation

$$y - 1 = \beta^{1-n}Da\ f(y), \tag{8}$$

where

$$f(y) \triangleq (1+\beta-y)^{n} \exp[\gamma(1-1/y)]. \tag{9}$$

In adiabatic operation the reactor is not cooled (H=0) and $T_m = T_o$ so that

$$y = \frac{T}{T_o}, \qquad \gamma = \frac{E}{RT_o}, \qquad \beta = \frac{(-\Delta H)C_o}{\rho C_p T_o}. \tag{10}$$

Equation (8) may also be used to describe other systems. For example, the species and energy conservation balances describing catalytic pellets in which a chemical reaction occurs with negligible intraparticle concentration and temperature gradients are:

$$k_c S_x (C_o - C) - V_p \hat{k}(T) C^n = 0 \tag{11}$$

$$h\ S_x (T_o - T) + (-\Delta H) V_p \hat{k}(T) C^n = 0. \tag{12}$$

The corresponding steady states are again the solutions of (8) if we define

$$\beta = \frac{k_c C_o (-\Delta H)}{hT_o} \qquad Da = \frac{V_p \hat{k}(T_o) C_o^{n-1}}{S_x k_c}. \tag{13}$$

For an endothermic reaction $\beta < 0$ and the ℓ.h.s. of (8) is a monotonic increasing function of y, while the r.h.s. is a monotonic decreasing function of y. These two functions can intersect only once, and we conclude that for an endothermic n-th order reaction a unique steady state exists for all Da. We consider from now on only the exothermic case ($\beta > 0$).

The steady-state equation (8) may be rewritten as

$$F(y) \triangleq \frac{f(y)}{y-1} = \frac{\beta^{n-1}}{Da} . \tag{14}$$

The Damköhler number is always positive and all the solutions of (14) must be in $(1,1+\beta)$. A unique solution exists for all Da if and only if $F(y)$ is a monotonic decreasing function in this region. Before analyzing the general case we examine the special cases of first and zeroth order reaction. Here, $F(y)$ is a monotonic decreasing function of y if and only if for all y in $(1,1+\beta)$

$$y^2(\beta+\gamma) - y\gamma(2+\beta) + \gamma(1+\beta) \geq 0 \qquad \text{for } n = 1, \tag{15}$$

$$y^2/(y-1) \geq \gamma \qquad \text{for } n = 0. \tag{16}$$

Thus, uniqueness is guaranteed for all Da and n = 1 if and only if

$$\beta\gamma \leq 4(1+\beta). \tag{17}$$

The corresponding condition for n = 0 is

$$\gamma \leq \left\{ \begin{matrix} (1+\beta)^2/\beta & \text{for } \beta < 1 \\ 4 & \text{for } \beta \geq 1 \end{matrix} \right\}. \tag{18}$$

For an n-th order reaction ($n > 0$) monotonicity of $F(y)$ is assured if and only if for all y in $(1,1+\beta)$

$$\Gamma(y) \triangleq (n-1)y^3 + y^2(\beta+\gamma+1-n) - y\gamma(2+\beta) + \gamma(1+\beta) \geq 0. \tag{19}$$

Both $\Gamma(1) = \beta$ and $\Gamma(1+\beta) = n\beta(1+\beta)^2$ are positive for $n > 0$ and a schematic of $\Gamma(y)$ is shown in Figure 1. Tsotsis and Schmitz [21] have pointed out that condition (19) is violated, i.e., steady-state multiplicity exists for some Da, if and only if;

a. The cubic equation $\Gamma(y) = 0$ has three real roots, and

b. $\Gamma(y)$ has a local minimum in $(1,1+\beta)$.

Condition a is satisfied if and only if

$$\phi(\beta,\gamma,n) \triangleq -\gamma^3\beta^2+2\gamma^2 g_1(\beta,n)-\gamma g_2(\beta,n) + 4(1+\beta-n)^3(1+\beta) < 0, \tag{20}$$

where

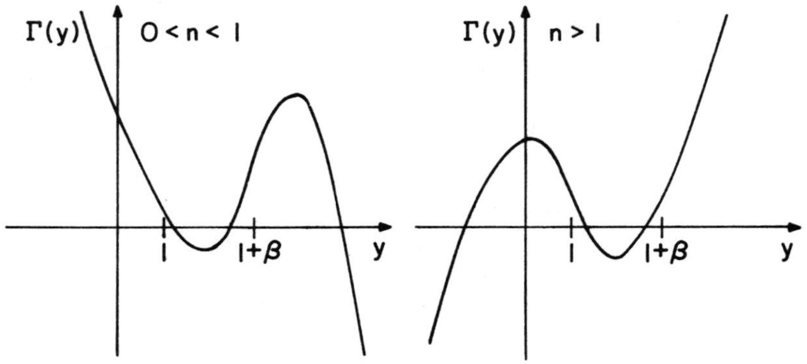

Figure 1. Schematic of $\Gamma(y)$ [21].

$$g_1(\beta,n) = (1-2n)\beta^2 + 2(2-n)\beta + 1 + n$$

$$g_2(\beta,n) = \beta^4 + 4(3-5n)\beta^3 - 2(4n^2+11n-11)\beta^2 + 4\beta(1-n)(3+2n) + (1-n)^3. \quad (21)$$

For $0 < n < 1$ condition (b) is

$$1 < \frac{(\gamma+\beta+1-n) - \sqrt{(\gamma+\beta+1-n)^2-3(1-n)(\beta+2)\gamma}}{3(1-n)} < 1 + \beta, \quad (22a)$$

while for $n > 1$ this condition becomes

$$\gamma\beta > 2\beta + n - 1. \quad (22b)$$

A simultaneous solution of (20) and (22) determines the region in the parameter space in which multiplicity occurs for some Da. Recently, Leib and Luss [8] have attained a simpler condition, proving that for any β and $n \geq 0$ multiplicity exists for some Da if and only if

$$\gamma \geq \gamma_m \quad (23)$$

where γ_m is the largest real root of $\phi(\beta,\gamma,n) = 0$.

Van den Bosch and Luss [22] attained very strong upper and lower bounds on the value of γ for which the bifurcation occurs. The advantage of the scheme is that these bounds on γ are expressed by simple, explicit functions of β and n.

Figure 2 describes the regions in the parameter space in which either uniqueness for all Da or multiplicity for some

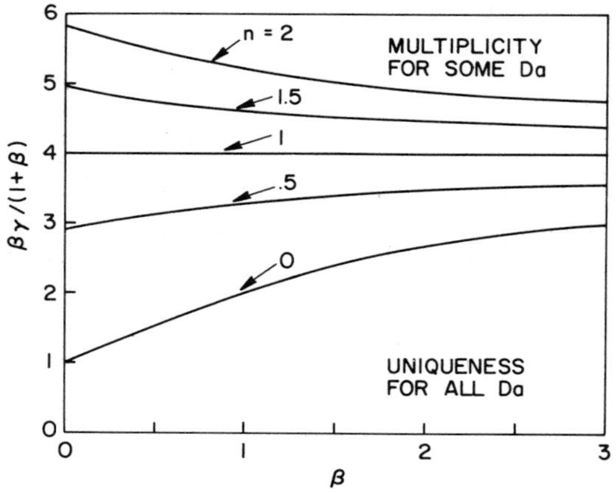

Figure 2. Regions of uniqueness and multiplicity for n-th order reactions [22].

Da exists. It is seen that the region in the parameter space for which multiplicity exists for some Da shrinks upon an increase in the reaction order. For sufficiently high values of β all the graphs approach the asymptote of $\beta\gamma/(1+\beta) = 4$.

Table 1. Values of γ and β for industrial catalytic reactions [22].

Reaction	β	γ	$\beta\gamma/(1+\beta)$
Oxidation of napthalene	.041	22	.88
Hydrogenation of benzene	.21	14	2.4
Vinylacetate synthesis	.42	19	5.5
Partial oxidation of ethylene	.58	18	6.5
CO oxidation in mufflers	.52	26	8.8
Ammonia synthesis	.59	28	10
Methanol synthesis	1.1	24	12
Higher alcohol from $CO+H_2$	1.2	28	15

A recent survey of experimental studies of steady-state multiplicity was presented by Schmitz [19]. Values of γ and β for several industrial catalytic reactions are reported in Table 1 [22]. For several of these reactions the values of $\beta\gamma/(1+\beta)$ are sufficiently high to cause multiplicity for some Da.

It should be noted that a reduction in the coolant temperature (keeping the other parameters fixed) increases the values of γ and β. This in turn increases the likelihood of steady-state multiplicity.

The above criteria are valid only for an n-th order reaction. Thus, the application of these criteria to reactions whose rates are described by different kinetic rate expressions may lead to erroneous predictions. For example, it is well known that steady-state multiplicity may occur in an isothermal system in which the reaction rate expression is of the Langmuir-Hinshelwood form [11,18,20], even though a unique steady-state exists for any isothermal n-th order reaction.

Chang and Calo [4] have recently used the catastrophe theory to determine the exact parameters for which steady-state multiplicity may occur for some Da for an n-th order chemical reaction in a non-adiabatic CSTR. An unfortunate choice of T_o instead of T_m as the reference temperature introduced two additional parameters into the steady-state equation and prevented a compact presentation of the results.

When the criterion for uniqueness for all Da is not satisfied for an n-th order reaction ($n > 0$), then F(y) has local minimal and maximal values in $(1,1+\beta)$ as illustrated by the schematic Figure 3. In this case three steady-state solutions exist for any Damköhler number in the finite region

$$\frac{F(y_{min})}{\beta^{n-1}} < \frac{1}{Da} < \frac{F(y_{max})}{\beta^{n-1}} . \qquad (24)$$

The two extremal values of y are the two zeros of $\Gamma(y)$ in $(1,1+\beta)$. [It can be easily shown that the third zero is always outside $(1,1+\beta)$]. For example, for a first order reaction

$$y_{max,min} = \frac{\gamma(2+\beta) \pm \sqrt{\gamma\beta[\gamma\beta-4(1+\beta)]}}{2(1+\beta)} . \qquad (25)$$

For a zeroth order reaction the multiplicity pattern may be more intricate, as the maximal value of F(y) may occur for $y > 1 + \beta$. An analysis of the behavior for the adiabatic case was presented by Van den Bosch and Luss [22,23] and the cooled CSTR case was examined by Chang and Calo [4].

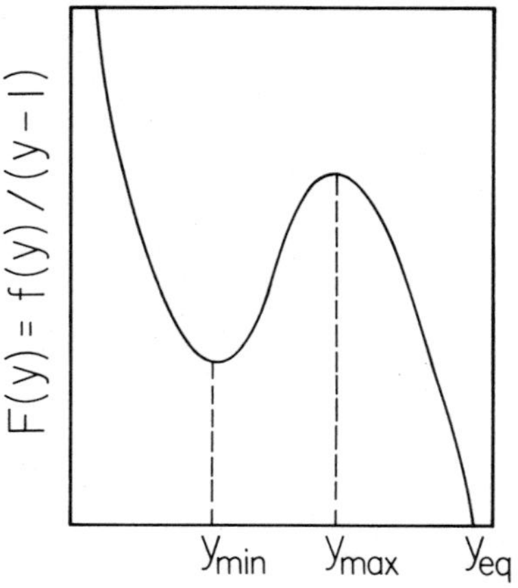

Figure 3. Schematic of F(y) for the case that multiple solutions exist for some Da.

TWO SIMULTANEOUS FIRST-ORDER REACTIONS.

Consider now an adiabatic CSTR in which two simultaneous reactions consume the same reactant

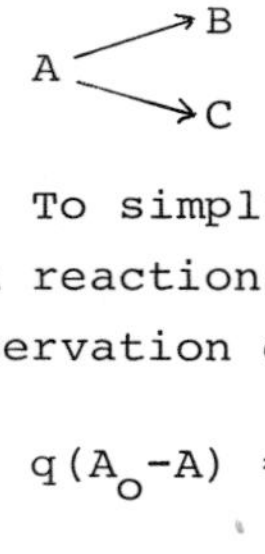

To simplify the algebraic manipulations we assume that both reactions are of first order. The species and energy conservation equations are:

$$q(A_o - A) = V[\hat{k}_1(T) + \hat{k}_2(T)]A \tag{26}$$

$$q\rho C_p(T - T_o) = (-\Delta H_1)V\hat{k}_1(T)A + (-\Delta H_2)V\hat{k}_2(T)A. \tag{27}$$

Without any loss of generality we define the reactions so that the first corresponds to the one with the lower activation energy. Introducing the dimensionless variables:

$$\begin{aligned} Da_1 &= Vk_1(T_o)/q & Da_2 &= V\hat{k}_2(T_o) \\ \sigma &= Da_2/Da_1 & \gamma_2 &= E_2RT_o \\ \beta_1 &= (-\Delta H_1)A_o/(\rho C_pT_o) & \beta_2 &= (-\Delta H_2)A_o/(\rho C_pT_o) \\ \mu &= E_2/E_1 & X &= \exp[\gamma_1(1-1/y)], \end{aligned} \tag{28}$$

equations (26) and (27) become

$$\frac{A}{A_o} = \frac{1}{1+Da_1(X+\sigma X^{\mu})} \tag{29}$$

$$\frac{y-1}{Da_1} = (1+\beta_1-y)X + \sigma(1+\beta_2-y)X^{\mu} \triangleq f(y). \tag{30}$$

It is convenient to rewrite (30) as

$$F(y) \triangleq \frac{f(y)}{y-1} = \frac{(1+\beta_1-y)X}{y-1} + \sigma\frac{(1+\beta_2-y)X^{\mu}}{y-1}$$

$$\triangleq F_1(y) + \sigma F_2(y) = \frac{1}{Da_1}. \tag{31}$$

Inspection indicates that both $F_1(y)$ and $F_2(y)$ describe a single first order reaction occurring in a CSTR. When both reactions are exothermic a unique steady state exists for all σ and Da_1 provided that both $F_1(y)$ and $F_2(y)$ are monotonic decreasing functions of y. Multiple steady states exist for some σ and Da_1 if and only if $F_1(y)$ and/or $F_2(y)$ are non-monotonic functions of y, i.e., steady state multiplicity cannot be caused by the interaction of two exothermic reactions unless at least one of them may cause multiplicity by itself.

It follows from the definition of f(y) by (30) that when both reactions are exothermic ($\beta_1, \beta_2 > 0$)

$$f(1+\beta_1)\ f(1+\beta_2) = -\sigma(\beta_2-\beta_1)^2 X(1+\beta_2)[X(1+\beta_1)]^{\mu} \leq 0. \tag{32}$$

Hence, if $\beta_1 \neq \beta_2$) F(y) as defined by (31) vanishes at least once in $(1+\beta_1, 1+\beta_2)$. If $\beta_1 = \beta_2$ F(y) vanishes just at $1+\beta_1$.

Figure 4 illustrates four different patterns the graphs of F(y) may take when F(y) vanishes just once in $(1+\beta_1, 1+\beta_2)$. Case a describes a situation in which F(y) is a monotonic decreasing function of y so that a unique solution exists for all Da_1. Cases (b-d) in Figure 4 illustrate multiplicity patterns of the form 1-3-1, 1-3-1-3-1 and 1-3-5-3-1, respectively. The novel multiplicity features shown in cases c and d are caused by the second reaction, whose rate is negligible at the ambient conditons ($\sigma = 10^{-7}$).

For all the cases shown in Figure 4 steady-state multiplicity occurs only in a finite range of Damköhler numbers.

A novel multiplicity feature, which has no analog in the single reaction case, is attained when F(y) vanishes more than once in $(1+\beta_1, 1+\beta_2)$. Figure 5 illustrates such a situation and it is seen that while there always exists a limiting Da_1 below which uniqueness is assured, there is no upper bound on Da_1 above which a unique steady state exists.

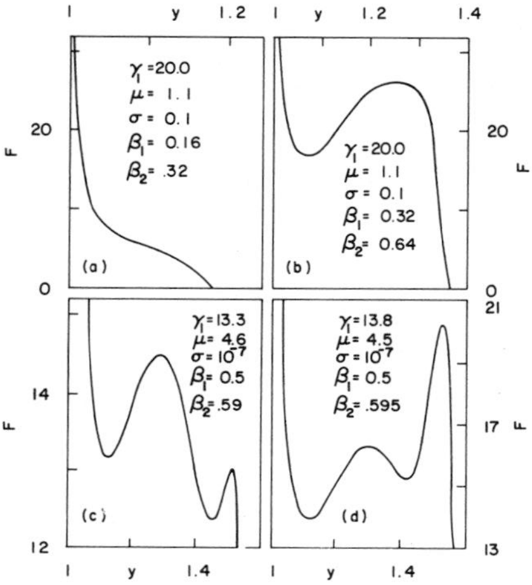

Figure 4. Typical F(y) graphs for two exothermic simultaneous reactions [10].

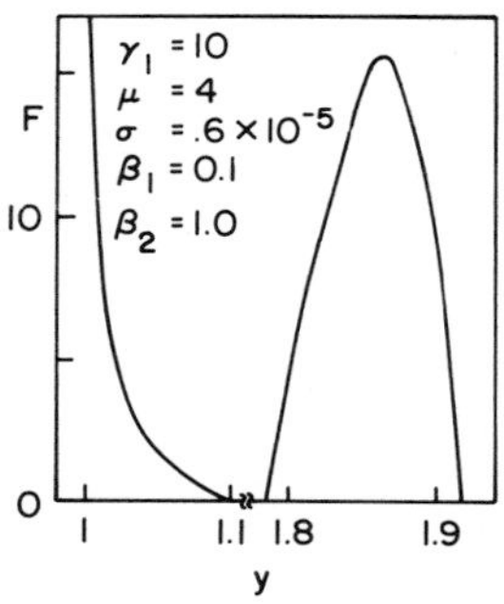

Figure 5. F(y) graph for two exothermic, simultaneous reactions when F(y) has three zeros [10].

A unique steady-state solution exists whenever a single, endothermic, n-th order reaction is carried out in a lumped-parameter system. We examine now whether multiplicity may occur when two simultaneous, first-order reactions, one or both of which are endothermic, are carried out in a lumped-parameter system. The treatment follows the elegant analysis by Michelsen [13].

A necessary and sufficient condition for multiplicity is that F(y) defined by (31) has a positive extremum. At this extremum point

$$\sigma = \frac{1-z_2}{1-z_1} \frac{P_1(g)}{P_2(g)} \exp[(\gamma_1-\gamma_2)g], \tag{33}$$

where we define

$$z_1 = \beta_1/(1+\beta_1) \qquad z_2 = \beta_2/(1+\beta_2)$$
$$g = 1-1/y, \tag{34}$$

and

$$P_1(g) = \gamma_1 g^2 - \gamma_1 z_1 g + z_1 \tag{35}$$

$$P_2(g) = -(\gamma_2 g^2 - \gamma_2 z_2 g + z_2) \;. \tag{36}$$

To simplify the analysis we consider only the practical cases for which β_1 and β_2 are larger than -1.

Substitution of σ from (33) into (31) yields

$$(Da_1)_{ext} = \frac{P_2(g)}{P_3(g)} (1-z_1) \exp(-\gamma_1 g) \tag{37}$$

$$(Da_2)_{ext} = \sigma(Da_1)_{ext} = \frac{P_1(g)}{P_3(g)} (1-z_2) \exp(-\gamma_2 g) \tag{38}$$

where

$$P_3(g) = (\gamma_2-\gamma_1)[g^2-(z_1+z_2)g+z_1 z_2] + z_2-z_1 . \tag{39}$$

A unique solution exists if no positive extremal Damköhler numbers exist. Thus, if there exists no g for which the three polynomials P_1, P_2 and P_3 have the same sign, it follows from (37) and (38) that a unique solution exists for all σ and Da_1.

The three polynomials have different signs at $g = 0$. Sign changes of the polynomials occur only at their zeros. Hence, the three polynomials may have the same sign for some $g \neq 0$ only if at least one of them has two real roots.

The quadratic polynomials P_1, P_2 and P_3 have real roots if and only if the following conditions are satisfied, respectively:

$$\beta_1\gamma_1 > 4(1+\beta_1) \tag{40}$$

$$\beta_2\gamma_2 > 4(1+\beta_2) \tag{41}$$

$$(\gamma_2-\gamma_1)(\beta_2-\beta_1) > 4(1+\beta_1)(1+\beta_2) . \tag{42}$$

The three polynomials satisfy the relation

$$(g-z_2)P_1 + (g-z_1)P_2 + gP_3 = 0. \tag{43}$$

Hence, at any zero of P_1, say $g = g_1$

$$\frac{P_2(g_1)}{P_3(g_1)} = -\frac{g_1}{g_1-z_1} = -\frac{g_1(g_1-z_1)}{(g_1-z_1)^2} = \frac{z_1}{(g_1-z_1)^2\gamma_1} \quad . \tag{44}$$

Similarly, at the zeros of P_2 and P_3

$$P_2(g_2) = 0 \Rightarrow \frac{P_1(g_2)}{P_3(g_2)} = \frac{z_2}{(g_2-z_2)^2\gamma_2} \quad . \tag{45}$$

$$P_3(g_3) = 0 \Rightarrow \frac{P_1(g_3)}{P_2(g_3)} = \frac{z_2-z_1}{(g_3-z_3)^2(\gamma_2-\gamma_1)} \quad . \tag{46}$$

Let us assume that at a zero of one of the polynomials the signs of the two remaining polynomials are identical. Then the three polynomials must have the same sign for all g values in the region between this zero and an adjacent zero either to the left or to the right. On the other hand, if at all the zeros the remaining two polynomials are of opposite sign, the three polynomials will never have the same sign and a unique solution exists for all σ and Da_1.

It follows from (44) that P_2 and P_3 have identical signs at the zeros of P_1 provided that

$$z_1/\gamma_1 > 0. \tag{47}$$

$P_1(g)$ has real roots if and only if condition (40) is satisfied. Clearly, (47) is satisfied whenever (40) is. Thus, condition (40) guarantees that multiplicity exists for some σ and Da_1. Similar considerations concerning the zeros of P_2 and P_3 indicate that multiplicity occurs provided either (41) and/or (42) are satisfied.

We conclude that a unique steady-state solution exists for any σ and Da_1 provided that neither of (40-42) is satisfied. Multiple steady-state solutions exist for some σ and Da_1 provided that one or more of (40-42) are satisfied.

The above analysis leads to the following conclusions:

a) If both reactions are exothermic, steady-state multiplicity occurs for some σ and Da_1 if and only if either (40) and/or (41) are satisfied, i.e., multiplicity may be encountered only if at least one of the two reactions can cause multiplicity for some Da by itself.

b) If one reaction is exothermic and the other is endothermic steady-state multiplicity for some σ and Da_1 may be attained in either one of two situations. The first is when the multiplicity is due to the exothermic reaction. In such a case either (40) or (41) is satisfied. The second is when the multiplicity is due to the interaction between the two reactions. Here, neither (40) nor (41) is satisfied but (42) is. This surprising behavior may occur only if the activation energy of the exothermic reaction exceeds that of the endothermic reaction.

c) If both reactions are endothermic, steady-state multiplicity may exist for some σ and Da_1 if and only if (42) is satisfied. This condition may be satisfied only if the reaction with the higher activation energy is less endothermic than the other reaction, i.e., $\beta_2 > \beta_1$.

The above mentioned multiplicities which occur only if (42) is satisfied are most surprising and cannot be deduced from the single reaction case.

The region in the parameter space for which multiplicity exists depends strongly on which of inequalities (40-42) are satisfied. The entire mutliplicity region may be mapped by plotting $(Da_2)_{ext}$ (from (38)) or σ (from (33)) vs $(Da_1)_{ext}$ (from (37)). The extremal values of σ may be determined by a logarithmic differentiation of (33)

$$\frac{1}{\sigma}\frac{d\sigma}{dg} = (\gamma_1-\gamma_2) + \frac{P_1'(g)}{P_1(g)} - \frac{P_2'(g)}{P_2(g)} . \tag{48}$$

Algebraic manipulations of (48) yield the result that

$$\frac{d\sigma}{dg} = 0 \implies \quad g\,P_4(g) = 0, \tag{49}$$

where

$$P_4(g) = g^3 - (z_1 + z_2)g^2 + (z_1z_2 + \frac{z_2 - z_1}{\gamma_2 - \gamma_1} + \frac{z_1}{\gamma_1} + \frac{z_2}{\gamma_2})g$$
$$-[z_1z_2(\frac{1}{\gamma_1} + \frac{1}{\gamma_2}) + \frac{2}{\gamma_2 + \gamma_1}(\frac{z_2}{\gamma_2} - \frac{z_1}{\gamma_1})]. \quad (50)$$

Similarly,

$$\frac{d}{dg}(Da_1)_{ext} = 0 \Longrightarrow (g - z_2)P_4(g) = 0. \quad (51)$$

$$\frac{d}{dg}(Da_2)_{ext} = 0 \Longrightarrow (g - z_1)P_4(g) = 0. \quad (52)$$

It follows that the extremal values of σ, Da_1 and Da_2 must be located either on the boundaries of the region(s) for which the three polynomials P_1, P_2 and P_3 have the same sign or at the roots of (49), (51) and (52), which are within these regions.

Possible patterns of F(y) when both reactions are endothermic are shown in Figure (6). Case a illustrates a situation in which $\beta_1 > \beta_2$ and a unique solution exists for all Da_1 in agreement with the analysis. Condition (42) is satisfied for cases b-d in Figure (6) and steady-state multiplicity occurs for a range of Damköhler numbers. In cases c and d F(y) (defined by (31)) has three zeros and there exists no upper bound on the Damköhler number above which a unique steady-state exists. Additional examples and criteria concerning this reaction network can be found in [10].

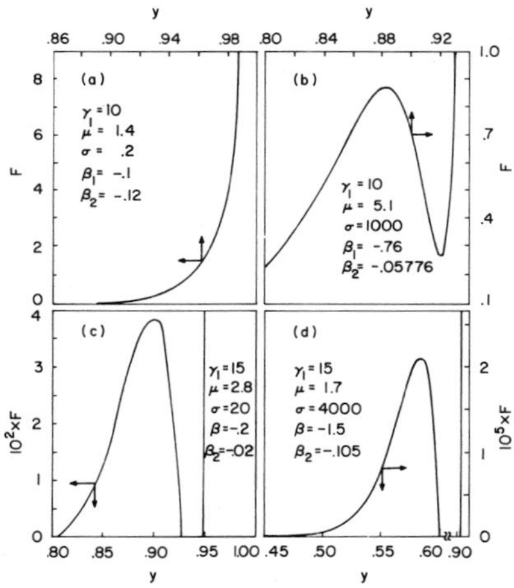

Figure 6. Different patterns of F(y) for two endothermic, simultaneous reactions [10].

TWO CONSECUTIVE OR PARALLEL FIRST-ORDER REACTIONS.

We examine now the multiplicity features of a lumped-parameter system in which two consecutive first-order reactions occur;

$$A \xrightarrow{\hat{k}_1} B \xrightarrow{\hat{k}_2} C.$$

Typical industrial examples of this reaction network are partial hydrogenation or oxidation reactions. Consider a mixture containing these reactants in an inert diluent with reaction on the surface of a non-porous catalytic wire or pellet. The steady-state species and energy balances are:

$$k_{CA}(A_o - A_s) = \hat{k}_1(T_s)A_s \tag{53}$$

$$k_{CB}(B_s - B_o) = \hat{k}_1(T_s)A_s - \hat{k}_2(T_s)B_s \tag{54}$$

$$h(T_s - T_o) = (-\Delta H_1)\hat{k}_1(T_s)A_s + (-\Delta H_2)\hat{k}_2(T_s)B_s \tag{55}$$

Introducing the dimensionless variables:

$$\begin{aligned}
Da_1 &= \frac{\hat{k}_1(T_o)}{k_{CA}} & Da_2 &= \frac{\hat{k}_2(T_o)}{k_{CB}} \\
\gamma_1 &= \frac{E_1}{RT_o} & \mu &= \frac{E_2}{E_1} \\
\beta_1 &= \frac{(-\Delta H_1)k_{CA}A_o}{hT_o} & \beta_2 &= \frac{(-\Delta H_2)k_{CB}A_o}{hT_o} \\
y_s &= T_s/T_o & \alpha &= B_o/A_o \\
\nu &= k_{CA}/k_{CB} & X &= \exp[\gamma_1(1-1/y)],
\end{aligned} \tag{56}$$

allows rewriting of (53) and (54) as

$$\frac{A_s}{A_o} = \frac{1}{1+Da_1X} \tag{57}$$

$$\frac{B_s}{A_o} = \frac{1}{1+Da_2X^{\mu}}\left[\alpha + \frac{\nu Da_1X}{1+Da_1X}\right]. \tag{58}$$

Substitution of (57) and (58) into (55) gives

$$y-1 = \frac{\beta_1 Da_1X}{1+Da_1X} + \frac{\beta_2 Da_2X^{\mu}}{1+Da_2X^{\mu}}\left(\alpha + \frac{\nu Da_1X}{1+Da_1X}\right) \triangleq f(y). \tag{59}$$

Equation (53) may be used also as a model of two consecutive reactions in a CSTR. In this case ν is unity and the definition of the dimensionless parameters is somewhat different from that of (56).

In many applications in the petrochemical industry several independent reactions occur in parallel. Consider a lumped-parameter system in which two first-order, parallel reactions occur

$$A \xrightarrow{\hat{k}_1} P_1$$

$$B \xrightarrow{\hat{k}_2} P_2 \; .$$

The corresponding dimensionless steady-state equation is identical with (59) but does not contain the term multiplied by ν. Thus, (59) permits a compact representation of both reaction networks. When $\nu \neq 0$ (59) describes the case of two consecutive reactions, while if $\nu = 0$ and $\alpha \neq 0$ (59) describes two parallel independent reactions.

If both reactions are non-exothermic ($\beta_1, \beta_2 \leq 0$) f(y) is a monotonic decreasing function of y, while the ℓ.h.s. of (59) is a linear increasing function of y. In this case a unique solution exists for any values of the remaining parameters, similar to the single reaction case. Thus, we examine only cases for which at least one of the reactions is exothermic.

The steady-state equation contains eight parameters. It may be rewritten in various forms, so that one of the parameters is equal to a function of the dimensionless temperature y and the remaining seven parameters. By separating the terms multiplying Da_2 the following expression is attained

$$\frac{1}{Da_2} = X^{\mu} \frac{1+\alpha\beta_2-y+Da_1X(1+\delta-y)}{y-1-Da_1X(1+\beta_1-y)} \triangleq X^{\mu} \frac{N(y)}{D(y)} \triangleq F(y), \tag{60}$$

where

$$\delta = \beta_1 + (\alpha+\nu)\beta_2. \tag{61}$$

Inspection of (60) indicates that F(y) vanishes at the zeros of N(y) and becomes unbounded at the zeros of D(y). These zeros are the steady-state solutions of (60) in the limit of very small and very large Da_2. Thus, important information about the multiplicity pattern may be extracted from the knowledge of the number of these roots. The zeros of D(y) and N(y) describe a lumped system in which a first-order reaction occurs. Thus, the method described previously for the single reaction case can be used to determine the exact number of the solutions of either N(y) or D(y).

We shall consider first the case that both reactions are exothermic. The schematic Figure (7) illustrates five different shapes the graph of F(y) may take when D(y) has a unique zero, i.e. a unique steady state exists for sufficiently small Da_2. The number of solutions for very large Da_2 is equal to the number of zeros of N(y). Thus, when N(y) has three solutions (cases c and e in Figure 7) multiple solutions exist for all Da_2 larger than some critical value. In certain cases (d and e) up to five solutions exist for some intermediate values of Da_2.

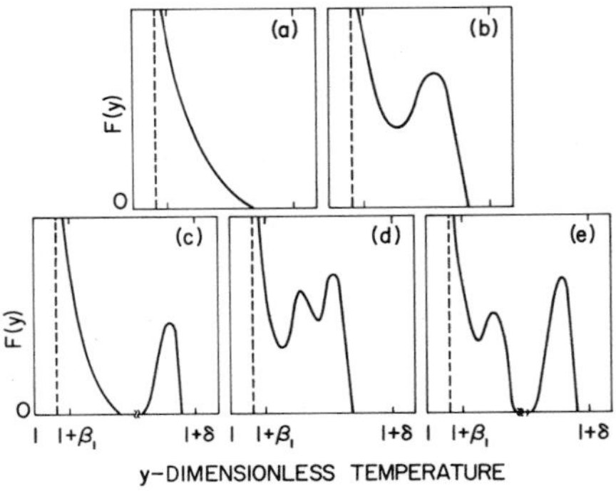

Figure 7. Multiplicity patterns for two exothermic reactions when F(y) has a unique asymptote [17].

Obviously, the shape of F(y) for intermediate values of Da_2 cannot be deduced from the asymptotic behavior of very large and very small Da_2.

Figure 8 describes eight patterns the graph of F(y) may take when D(y) has three zeros, i.e., when three solutions exist for very small Da_2. In this case multiple solutions exist for all Da_2 if N(y) has multiple solutions (cases b, d, f and h in Figure 8). This is a novel feature, which has no analog in either the single reaction or the two simultaneous reactions case. The graphs illustrate that up to five solutions may exist for intermediate values of Da_2.

When the second reaction is isothermal ($\beta_2 = 0$) no more than three solutions may exist for any set of parameters. Figure 9 describes the possible multiplicity patterns when the first reaction is isothermal ($\beta_1 = 0$), while the second

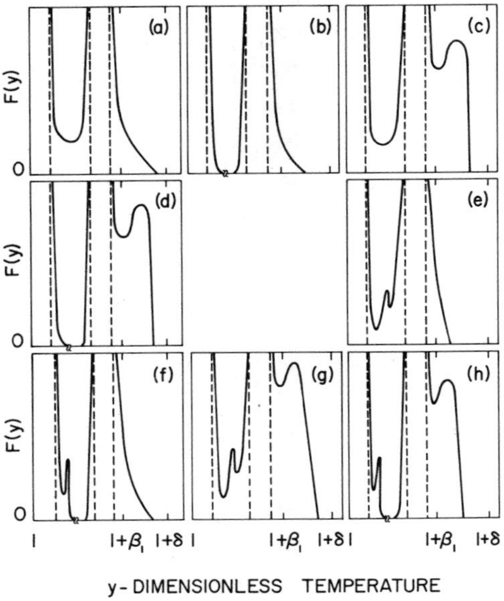

Figure 8. Multiplicity patterns for two exothermic reactions when F(y) has three asymptotes [17].

is exothermic. It is interesting to note that for certain parameters up to five solutions exist in this case.

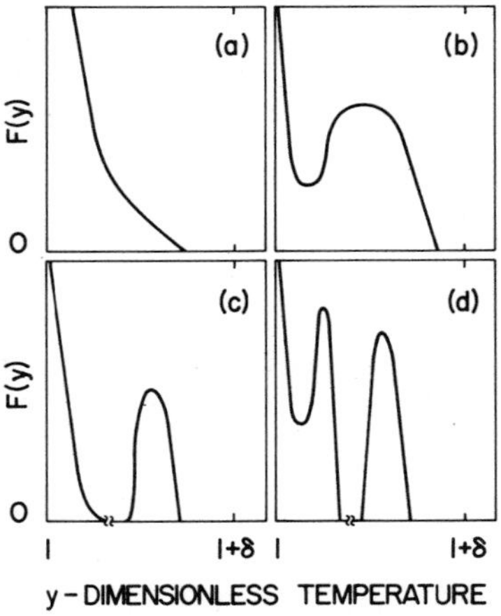

Figure 9. Multiplicity patterns for one isothermal and one endothermic reaction [17].

Interesting behavioral features may be encountered when one reaction is exothermic while the second is endothermic. When the desired normal reaction is endothermic, as for example catalytic reforming, heat is supplied to the reactor and the control scheme is usually not capable of cooling the reactor rapidly. Here a runaway may be caused by a transient which shifts the system to an undesired steady state at which the rate of the exothermic reaction is appreciable. Thus, to avoid unexpected pitfalls in operation, it is essential to be aware of all the possible steady states.

In order to gain insight into the structure of F(y) for the case that one reaction is exothermic and one endothermic, we determine first the region in which it is positive, so that (60) has a physically meaningful solution ($Da_2 \geq 0$).

According to the definition of N(y) and D(y)

$$N(y) + D(y) = \beta_2[\alpha+(\nu+\alpha)Da_1X]. \tag{62}$$

It follows that when $\beta_1 > 0$ and $\beta_2 < 0$

$$N(y) + D(y) < 0. \tag{63}$$

Thus, when D(y) is positive, both N(y) and F(y) are negative. We conclude that F(y) may be positive only in the region in which D(y) is negative. Similarly, it can be shown that when $\beta_1 < 0$ and $\beta_2 > 0$, F(y) may be positive only in the region in which D(y) is positive.

When β_1 is negative, the equation D(y) = 0 describes a lumped parameter system in which a single, endothermic reaction occurs. Thus, F(y) has a single asymptote at the unique zero of D(y), and is positive only to the right of this asymptote. Cases a-c in Figure 10 illustrate the various features of F(y) in this case.

When β_1 is positive, D(y) = 0 describes a lumped system in which a single, exothermic reaction occurs. The maximum number of zeros of D(y) is three and F(y) has an asymptote at each of these zeros. Cases d-f in Figure 10 describe the multiplicity features when D(y) has a unique zero, while cases g-i illustrate the behavior when D(y) has three zeros. When $\beta_1 + \nu\beta_2 > 0$, N(y) describes a lumped system in which a single exothermic reaction occurs and it may have up to three solutions (cases g-i).

A more detailed discussion of the multiplicity features of this reaction network and corresponding uniqueness criteria may be found in [17].

A POROUS CATALYST WITH UNIFORM INTRAPARTICLE TEMPERATURE.

In practice distributed-parameter systems are often described by simplified lumped models. It is of interest to determine the influence of the lumping on the occurrence of

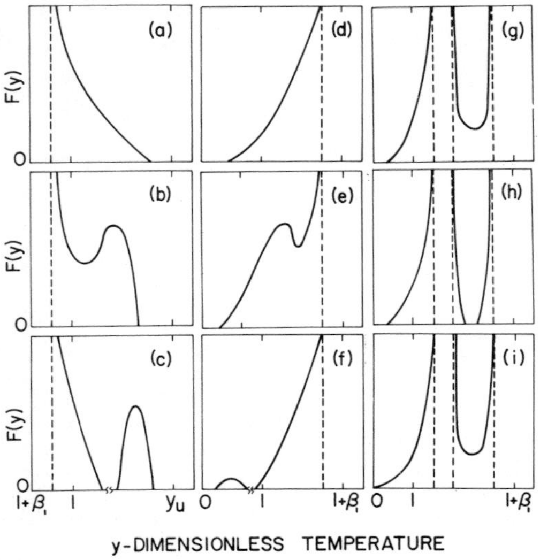

Figure 10. Multiplicity patterns for one exothermic and one endothermic reaction [17].

steady-state multiplicity. To obtain a partial answer we consider a single, n-th order reaction occurring in a porous, catalytic pellet. When a lumped-parameter model is used to describe this situation exact uniqueness criteria are attained from the requirement that F(y) be a monotonic function of y. In many applications the concentration profile within the catalyst is not uniform and needs to be described by a distributed-parameter model. Experience indicates that a very suitable model is one which assumes that the temperature within the pellet is uniform but different from that of the ambient fluid, while the concentration profile is determined by the interaction between the diffusion and reaction in the pellet [6]. We describe here a method of utilizing the exact bounds derived for the lumped system to attain sufficient uniqueness and multiplicity criteria for the distributed model.

The species and energy balances in this case are:

$$S_x k_c (C_o - C_s) = \int_{V_p} \hat{k}(T_s) C^n dv \tag{64}$$

$$S_x h (T_s - T_o) = (-\Delta H) \int_{V_p} \hat{k}(T_s) C^n dv. \tag{65}$$

Combining (64) and (65) gives

$$(-\Delta H) k_c (C_o - C_s) = h (T_s - T_o), \tag{66}$$

which may be expressed by the following dimensionless relation

$$u_s = (1+\beta - y_s)/\beta. \tag{67}$$

The dimensionless concentration within the isothermal pellet satisfies the equation [2]

$$\nabla^2 u = \phi^2 u^n \qquad \text{in } V_p \tag{68}$$

subject to the boundary condition

$$u = u_s \tag{69}$$

where

$$\phi = \frac{V_p}{S_x} \sqrt{\frac{\hat{k}(y_s) C_s^{n-1}}{D_e}} . \tag{70}$$

The Thiele modulus is the ratio between the characteristic time for diffusion and for reaction in the pellet. When the Thiele modulus is small the concentration throughout the pellet is essentially uniform. However, when ϕ is larger than unity, sharp concentration gradients exist within the pellet.

It is customary to express the ratio between the actual rate and that attained when no diffusional resistance exists within the pellet by an effectiveness factor defined as

$$\eta = \frac{1}{V_p} \int_{V_p} \left(\frac{u}{u_s} \right)^n dv. \tag{71}$$

For an n-th order reaction the effectiveness factor depends only on the Thiele modulus, and may be expressed for a slab catalyst by a hypergeometric function [2, p. 144]. While explicit analytical solutions are not available for the general case, it is well known that η decreases monotonically from an asymptotic value of unity for small ϕ to another asymptote with a slope of -1 on logarithmic scales for large ϕ.

Using (67) and (71) the energy balance (65) may be rewritten as

$$Sh = \frac{(1+\beta-y_s)\eta\phi^2(y_s)}{y_s-1} \triangleq \frac{f(y_s)}{y_s-1} \triangleq F(y_s), \tag{72}$$

where

$$Sh = k_c V_p/(D_e S_x)$$

$$\phi^2(y_s) = \left(\frac{V_p}{S_x} \right)^2 \frac{\hat{k}(y_s) C_o^{n-1} (1+\beta-y_s)^{n-1}}{D_e \, \beta^{n-1}} . \tag{73}$$

When $\beta < 0$ (endothermic reaction)

$$\frac{df}{dy_s} = (1+\beta-y_s) \frac{d(\eta\phi^2)}{d\phi^2} \frac{d\phi^2}{dy_s} - \eta\phi^2 < 0, \tag{74}$$

and a unique solution exists for all Sh. When the reaction is exothermic ($\beta > 0$) uniqueness is assured for all Sh if and only if $F(y_s)$ is a monotonic decreasing function of y_s in $(1,1+\beta)$, i.e.,

$$(y_s-1)\left[\frac{d\ \ell n[\phi^2(1+\beta-y_s)]}{dy_s} + \frac{d\ \ell n\ \eta}{dy_s} \right] < 1 \qquad 1 \le y_s \le 1+\beta. \tag{75}$$

For an n-th order, isothermal reaction η is a unique continuous function of ϕ and

$$\frac{d\ \ell n\ \eta}{dy_s} = \frac{1}{2} \frac{d\ \ell n\ \eta}{d\ \ell n\ \phi} \frac{d\ \ell n\ \phi^2}{dy_s}$$

$$= \frac{1}{2} \frac{d\ \ell n\ \eta}{d\ \ell n\ \phi} \left[\frac{d\ \ell n[\phi^2(1+\beta-y_s)]}{dy_s} + \frac{1}{1+\beta-y_s} \right]. \tag{76}$$

Using (76) condition (75) becomes

$$(y_s-1)\left[\frac{d\ell n\phi^2(1+\beta-y_s)}{dy_s} \left(1 + \frac{1}{2} \frac{d\ell n\eta}{d\ell n\phi}\right) + \frac{0.5}{1+\beta-y_s} \frac{d\ell n\eta}{d\ell n\phi} \right] < 1. \tag{77}$$

Assuming that $d\ell n\eta/d\ell n\phi = 0$, condition (77) becomes identical to the uniqueness criterion for the corresponding lumped system, i.e., the one in which $u = u_s$ everywhere within the pellet. When $d\ell n\eta/d\ell n\phi$ is negative, uniqueness criterion (77) is satisfied for some values of β, γ and n for which multiplicity exists in the lumped parameter system, i.e. when (77) is violated if $d\ell n\eta/d\ell n\phi = 0$. We conclude that accounting for the intraparticle concentration gradients reduces the region in the parameter space for which steady-state multiplicity is possible for some Sh.

It will now be shown how a simple redefinition of the reaction order and activation energy can transform the exact bounds for the lumped system into sufficient uniqueness and multiplicity criteria for the porous pellet.

Suppose that the following a priori bounds can be established

$$b_1 \geq \frac{d\ell n\eta}{d\ell n\phi} \geq b_2. \tag{78}$$

Using the maximum principle it can be proven that $b_1 \leq 0$, while numerical calculations indicate that $b_2 \geq -1$ (this bound is derived in [22] for slab, cylindrical and spherical pellets). Replacing $d\ell n\eta/d\ell n\phi$ in (77) by its upper bound b_1 yields the conservative result that uniqueness is assured for all Sh if for all y_s in $(1,1+\beta)$

$$(n_d-1)y_s^3 + y_s^2(\beta+\gamma_d-\eta_d+1] - y_s\gamma_d(2+\beta) + \gamma_d(1+\beta) \geq 0 \tag{79}$$

where

$$n_d-1 = (n-1)(1+0.5b_1) \tag{80}$$

$$\gamma_d = \gamma(1+0.5b_1). \tag{81}$$

The sufficient uniqueness criterion (79) is identical to the uniqueness criterion (19) for the lumped-parameter system with n_d and γ_d replacing n and γ. Application of the conservative bound $b_1 = 0$ gives $n_d = n$ and $\gamma_d = \gamma$. This implies that the conditions which guarantee uniqueness for all Sh for the lumped model guarantee uniqueness also for the distributed model.

Similar arguments may be used to prove that if we define

$$n^d-1 = (n-1)(1+0.5b_2) \tag{82}$$

$$\gamma^d-1 = \gamma(1+0.5b_2) \tag{83}$$

then multiplicity occurs for some Sh whenever the exact multiplicity bound for the lumped system is satisfied for n^d and γ^d. Use of the conservative lower bound $b_2 = -1$ gives

$$n^d = 0.5\ (n+1) \tag{84}$$

$$\gamma^d = 0.5\gamma \tag{85}$$

According to the above analysis uniqueness is assured for a first-order reaction and all Sh provided

$$\beta\gamma < 4(1+\beta), \tag{86}$$

while multiplicity occurs for some Sh if

$$\beta\gamma > 8(1+\beta). \tag{87}$$

The conservative bounds of $b_1 = 0$ and $b_2 = -1$ eliminate the dependence of the criteria on pellet geometry and $\phi(1)$ at the expense of increased conservatism. To attain a quantitative gauge of the region in the parameter space for which these criteria yield no conclusive information we define

$$r = 1 - \frac{\gamma_+(\beta,n)}{\gamma_-(\beta,n)} \tag{88}$$

where $\gamma_+(\beta,n)$ and $\gamma_-(\beta,n)$ refer to the values of γ which satisfy, for some β and n, the uniqueness and multiplicity criteria using $b_1 = 0$ and $b_2 = -1$. Clearly, the smaller r is, the better are the criteria. Figure (11) indicates that r is a monotonic decreasing function of n and that it approaches asymptotically 0.5 for all n and large β. Sharper criteria may be attained by use of stricter bounds on b_1 and b_2 [22].

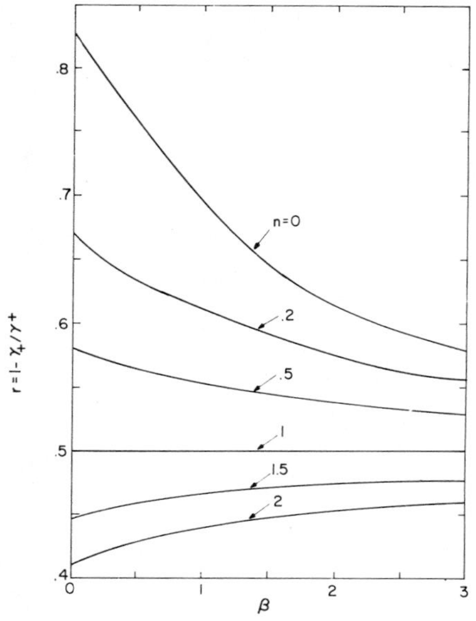

Figure 11. The dependence of r, defined by (88), on β and the reaction order [22].

The exact transition from uniqueness to multiplicity occurs when the maximum value of dF/dy in $(1,1+\beta)$ is zero. Thus, for an n-th order reaction $(n > 0)$ the bifurcation occurs for parameters which satisfy both

$$\frac{dF(y_s^*,\gamma,\beta,n,\phi(1))}{dy_s} = 0 \tag{89}$$

and

$$\frac{d^2F(y_s^*,\gamma,\beta,n,\phi(1))}{dy_s^2} = 0 \tag{90}$$

where y_s^* is in $(1,1+\beta)$. In the special case of a first-order reaction η and ϕ are independent of β and depend only on y_s, γ and $\phi(1)$, and (89) and (90) are linear in β. Thus, one can combine (89) and (90) to obtain a single equation independent of β. This equation can be solved for y_s for any specified value of γ and $\phi(1)$. Then y_s can be substituted back into (89) to give the corresponding β. Chang and Calo [4] used this procedure to determine the exact uniqueness bounds for a spherical pellet [Figure (12)].

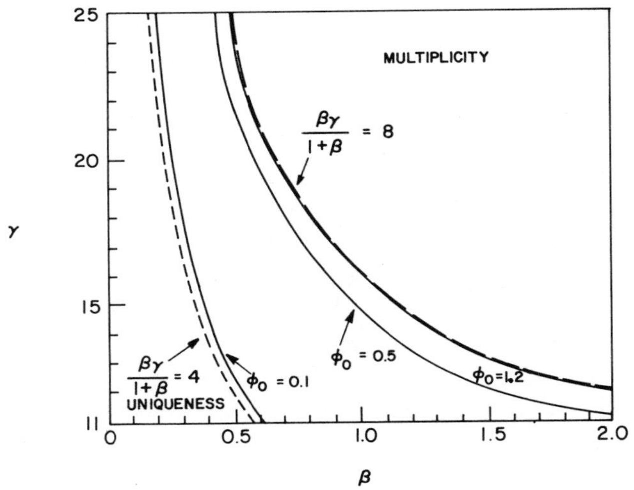

Figure 12. The exact bound between region of uniqueness and multiplicity for a first-order reaction in a spherical porous catalytic pellet as a function of ϕ_0 [4].

The procedure of solving (89) and (90) simultaneously can be used in principle to determine the exact bifurcation for reactions of any order. However, due to the highly transcendental nature of the equation for $n \neq 1$ a cumbersome iterative scheme needs to be used.

When steady-state multiplicity exists for some Sh, the bifurcation occurs for y_s values satisfying (89). The corresponding Sh can be determined from (72). Periera et al. [14], used a similar scheme to determine the multiplicity region for a first-order reaction in a slab catalyst.

The results presented here are valid only for an n-th order reaction. More exotic multiplicity features are found for Langmuir-Hinshelwood kinetics for which the reaction rate may decrease with an increase in the concentration of the limiting reactant [18,20]. Practical examples of this behavior are the oxidation of carbon monoxide in automotive catalysts and the methanation of carbon monoxide over nickel catalysts. Techniques of predicting the conditions under which multiplicity may be caused by intraparticle diffusional resistances are reviewed in [3] and [9]. Ho [7] and Periera and Varma [15] have presented recently some new results. Andersen and Michelsen [1] presented an interesting study of steady-state multiplicity due to intraparticle gradients for a porous pellet in which two simultaneous reactions occur.

CONCLUDING REMARKS

The results presented here indicate that considerable insight into the multiplicity patterns of multi-reaction networks may be gained from knowledge of the behavioral features of the single reaction case. However, certain features of these multi-reaction systems have no analog in the single reaction case and can be determined only by a careful and lengthy analysis. This unexpected behavior could lead to pitfalls in the development of control and start-up policies. Thus, it is of practical importance to develop a better knowledge and understanding of the special features associated with these reaction networks and the conditions under which they occur.

In practice distributed systems are often described by simplified lumped models. The analysis of the single, n-th order reaction in a porous pellet, whose temperature is uniform but different from that of the surroundings, indicates that the region in the parameter space for which multiplicity may occur for the distributed model is smaller than that attained for the corresponding lumped-parameter model. It would be of interest to determine whether this result is true in general for single and multi-reaction networks.

NOTATION

a	Area through which heat exchange occurs
b_i	Bounds on $d\ell n\eta/d\ell n\phi$
C_p	Heat capacity
C	Concentration
Da	Damköhler number
D_e	Effective diffusivity

$D(y)$	Denominator of equation (60)
$f(y)$	Heat generation function
$F(y)$	Function, $f(y)/(y-1)$
g	Function of dimensionless temperature, $1-1/y$
h	Heat transfer coefficient
H	Dimensionless heat exchange parameter, defined by (3)
$-\Delta H$	Heat of reaction
$\hat{k}$	Reaction rate constant
k_c	Mass transfer coefficient
n	Reaction order
n_d	Apparent order defined by (80)
n^d	Apparent order defined by (82)
$N(y)$	Numerator of equation (60)
P	Polynomial of g
q	Volumetric flow rate
r	Ratio defined by (88)
R	Universal gas constant
S_x	External surface area
T	Temperature
u	Dimensionless concentration
U	Overall heat transfer coefficiet
v	Volume element
V	Volume of reactor
V_p	Volume of pellet
X	Exponential temperature dependence of rate constant
y	Dimensionless temperature
z	Function of β defined by (34)
β	Dimensionless heat of reaction
γ	Dimensionless activation energy
γ_d	Apparent dimensionless activation energy, defined by (81)
γ^d	Apparent dimensionless activation energy, defined by (83)
δ	Quantity defined by (61)
η	Effectiveness factor, defined by (71)
μ	Ratio of activation energies, E_2/E_1
ν	Ratio of mass transfer coefficients, k_{CA}/k_{CB}

ρ Density

σ Ratio of two Damköhler numbers

ϕ Thiele modulus, defined by (70)

REFERENCES

1. Andersen, A. S. and M. L. Michelsen, in Chemical Engineering with Per Søltoft, Teknisk Forlag a-s, Copenhagen, 1977, 39-47.
2. Aris, R., The Mathematical Theory of Diffusion and Reaction in Permeable Catalysts, Vol. 1, Clarendon Press, Oxford, 1975.
3. ____________, The Mathematical Theory of Diffusion and Reaction in Permeable Catalysts, Vol. 2, Clarendon Press, Oxford, 1975.
4. Chang, H. C. and J. M. Calo, Chem. Eng. Sci., 34 (1978), 285-299.
5. Denn, M. M., Stability of Reaction and Transport Processes, Prentice-Hall, Englewood Cliffs, N.J., 1975.
6. Hlavacek, V. and M. Kubicek, Chem. Eng. Sci., 25 (1970), 1761-1771.
7. Ho, T. C., Chem. Eng. Sci., 31 (1976), 235-240.
8. Leib, T. M. and D. Luss, Chem. Eng. Sci., 35 (1980), submitted for publication.
9. Luss, D., Steady State and Dynamic Behavior of a Single Catalytic Pellet, in Chemical Reactor Theory (L. Lapidus and N. R. Amundson, eds.), Prentice-Hall, N.J., 1977, 181-268.
10. ____________ and G. T. Chen, Chem. Eng. Sci., 30 (1975), 1483-1495.
11. Kauschus, W., J. Demont and K. Hartmann, Chem. Eng. Sci., 33 (1978), 1283-1285.
12. Matsuura, T. and M. Kato, Chem. Eng. Sci., 22 (1967), 171-184.
13. Michelsen, M. L., Chem. Eng. Sci., 32 (1977), 454-456.
14. Periera, C. J., J. J. Carberry and A. Varma, Chem. Eng. Sci., 34 (1978), 249-255.
15. Periera, C. J. and A. Varma, Chem. Eng. Sci., 33 (1978), 1645-1657.
16. Perlmutter, D. D., Stability of Chemical Reactors, Prentice-Hall, Englewood Cliffs, N.J., 1972.

17. Pikios, C. A. and D. Luss, Chem. Eng. Sci., 34 (1979), 919-927.

18. Roberts, G. W. and C. N. Satterfield, Ind. Eng. Chem. Fundls., 5 (1966), 317-325.

19. Schmitz, R. A., Adv. Chem. Sci., 148 (1975), 154-211.

20. Schneider, P. and P. Mitschka, Coll. Czech. Chem. Comm., 31 (1966), 3677-3701.

21. Tsotsis, T. T., and R. A. Schmitz, Chem. Eng. Sci., 34 (1979), 135-137.

22. Van den Bosch, B. and D. Luss, Chem. Eng. Sci., 32 (1977), 203-212.

23. Van den Bosch, B. and D. Luss, Chem. Eng. Sci., 32 (1977), 560-562.

24. Varma, A. and R. Aris, Stirred Pots and Empty Tubes, in Chemical Reactor Theory (L. Lapidus and N. R. Amundson, eds.), Prentice-Hall, N.J., 1977, 78-155.

The author wishes to thank S. Wedel and T. Leib for helpful comments and discussions. The author was partially supported by the National Science Foundation Grant ENG 5336.

Department of Chemical Engineering
University of Houston
Houston, Texas 77004

Density-Dependent Interaction–Diffusion Systems

Donald G. Aronson

0. INTRODUCTION.

In the past several years, reaction-diffusion systems have attracted a great deal of attention from mathematicians and other scientists. Typical problems which arise in chemical engineering are discussed throughout this volume. In many of these problems, one is concerned with the interaction between some reaction mechanism and diffusive transport. The problems are generally nonlinear, with the nonlinearity occurring in the reaction mechanism. In this lecture I want to explore some consequences of nonlinearity in the diffusion process. I shall do this by discussing several examples.

The first example is gas flow in a porous medium. The equation governing such a flow is a nonlinear diffusion equation without any reaction terms. This example serves to introduce the principal effect of the nonlinearity in the diffusion: the finite speed of propagation of disturbances from rest. The second example is a Gurtin-MacCamy population model [15] which combines the porous medium flow mechanism with various nonlinear growth terms. For this model, I shall discuss the existence of traveling waves and compare the results with what is known for the corresponding linear diffusion model (Fisher's equation). The last example is a predator-prey interaction system in which the tendency of

ISBN 0-12-669550-4

each species to migrate can depend on the local densities of both species. Since this system is currently being actively studied, I shall only discuss its derivation.

Unfortunately, none of the examples which I consider involves a chemical reaction. There is, however, some material on density dependent diffusion in the chemical literature. For example, the work of Zabusky et. al. [9], [23] on proton diffusion through an immobilized protein membrane, and the work of Garg and Ruthven [12] on zeolites (see also [22]). I hope this lecture will stimulate further interest.

I am indebted to Hans Weinberger for many helpful discussions in the course of this work and for his permission to present some of our unpublished research in this lecture.

1. POROUS MEDIUM FLOW.

Consider an ideal gas flowing in a homogeneous porous medium. If the flow is *isentropic* then the equation of state is

$$\gamma = \gamma_0 p^{\alpha} . \tag{1.1}$$

Here γ is the density and p is the pressure of the gas, while $\gamma_0 > 0$ and $\alpha \in (0,1]$ are constants. The dynamics are described by the conservation of mass

$$\operatorname{div}(\gamma \underset{\sim}{v}) = -f \frac{\partial \gamma}{\partial t} \tag{1.2}$$

and Darcy's law

$$\underset{\sim}{v} = - \frac{\kappa}{\mu} \operatorname{grad} p, \tag{1.3}$$

where $\underset{\sim}{v}$ is the velocity vector, f is the porosity of the medium (that is, the volume fraction accessible to the gas), κ is the permeability of the medium, and μ is the viscosity of the gas [19]. Darcy's law is an empirically derived relationship which characterizes porous medium flow.

Set

$$m = 1 + \frac{1}{\alpha}$$

and use equations (1.1) and (1.3) to eliminate p and $\underset{\sim}{v}$ in (1.2) to obtain

$$\frac{\partial \gamma}{\partial t} = \frac{\kappa}{\mu \gamma^{1/\alpha}(1+\alpha) f} \Delta(\gamma^m) .$$

By changing the time or distance scale, one can reduce the constant to unity. The result is the porous medium equation

$$\frac{\partial u}{\partial t} = \Delta(u^m) \qquad (1.4)$$

where u is the scaled density. In the porous medium flow problem $m \in [2, +\infty)$. However, equation (1.4) arises in other applications, notably in plasma physics, with various values of $m > 0$ [6]. Throughout this work I shall assume that

$$m > 1 .$$

Note that since u represents a density, it is natural to require that

$$u \geq 0 .$$

If one computes the Laplacian

$$\Delta(u^m) = \operatorname{div}(mu^{m-1}\operatorname{grad} u) = mu^{m-1}\,\Delta u + m(m-1)u^{m-2}|\operatorname{grad} u|^2$$

one sees that equation (1.4) is uniformly parabolic in any region where u is bounded away from zero, but that it is degenerate in the neighborhood of any point where $u = 0$. In terms of Fickian diffusion, the diffusivity mu^{m-1} vanishes with u. The most striking manifestation of this degeneracy is that in porous medium flow there is a finite speed of propagation of disturbances from rest. This should be contrasted with the heat conduction case, $m = 1$, where there is an infinite speed of propagation. (Indeed, if you take the equation of heat conduction literally, it tells you that when you strike a match you feel the heat before you see the light!)

Consider, for example, the initial value problem for flow in a one-dimensional porous medium

$$\begin{cases} u_t = (u^m)_{xx} & \text{in } \mathbb{R} \times \mathbb{R}^+ \\ u(x,0) = u_o(x) & \text{in } \mathbb{R}, \end{cases} \qquad (1.5)$$

where

$$u_o(x) \begin{cases} > 0 & \text{for } x \in J \equiv (a_1, a_2) \\ = 0 & \text{for } x \in \mathbb{R}\setminus J \end{cases}$$

with $u_0^m \in \mathrm{Lip}(\mathbb{R})$ and $-\infty < a_1 < a_2 < +\infty$. Strictly speaking, this problem has no solution in the classical sense (that is, with a suitable number of continuous derivatives). However, there is a perfectly well defined generalized solution which possesses all the properties which I shall need [21]. Henceforth, the phrase "solution of problem (1.5)" will refer to this generalized solution.

Let

$$P[u] = \{(x,t) \in \mathbb{R} \times \mathbb{R}^+ : u(x,t) > 0\}$$

where u is the solution of problem (1.5). Then $P[u]$ is bounded by the segment J of the x-axis and the curves $x = \zeta_i(t)$ for $i = 1$ and 2, where $\zeta_i(0) = a_i$, $\zeta_1(t)$ is a Lipschitz continuous nonincreasing function of t, and $\zeta_2(t)$ is a Lipschitz continuous nondecreasing function of

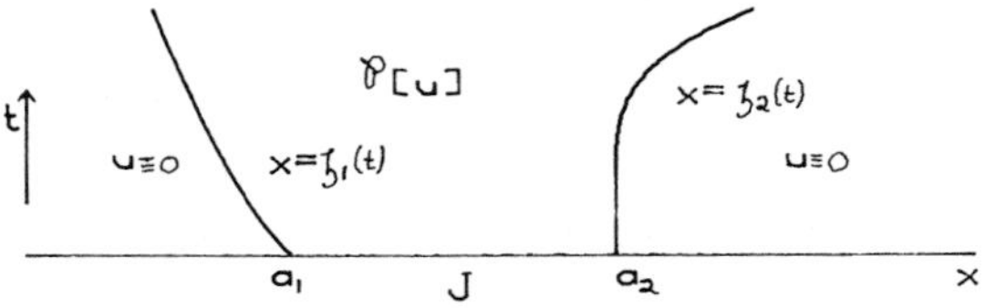

Figure 1.

t. There exist numbers $t_i^* \in [0,+\infty)$ for $i = 1$ and 2 such that

$$\zeta_i(t) \text{ is strictly monotone and } C^1 \text{ for } t \in (t_i^*,+\infty)$$

and

$$\zeta_i(t) \equiv a_i \text{ for } t \in [0,t_i^*] .$$

Moreover, if $v = u^{m-1}$ then

$$\dot{\zeta}_i(t) = -\frac{m}{m-1} \lim_{x \to \zeta_i(t)\pm} v_x(x,t) \quad \text{for all } t \neq t_i^* \tag{1.6}$$

where $x \to \zeta_i(t)\pm$ means that $\zeta_i(t)$ is to be approached from inside $P[u]$ ([7], [17], [3]).

Note that v is essentially the pressure of the gas and (1.6) simply says that the motion of the interface $x = \zeta_i(t)$ is governed by Darcy's law. If $t_i^* > 0$ then the interface $x = \zeta_i(t)$ remains stationary for t_i^* units of time. Every interface must ultimately begin to move and, once in motion, it can never stop. The case $t_i^* > 0$ can actually occur. Indeed one has

$$t_i^* = 0 \quad \text{if} \quad u_o(x) \geq c|x - a_i|^\gamma \quad \text{for some} \quad c > 0 \quad \text{and} \quad \gamma \in (0,2)$$

and

$$t_i^* > 0 \quad \text{if} \quad u_o(x) \leq c(x - a_i)^2 \quad \text{for some} \quad c > 0 \ .$$

For example, if $m = 2$ and

$$u_o(x) = \begin{cases} \cos^2 x & \text{for} \quad |x| \leq \pi/2 \\ 0 & \text{for} \quad |x| > \pi/2 \end{cases}$$

Then $t_i^* = 1/12$ for $i = 1$ and 2 ([17], [2]).

It is known that (1.6) holds at $t = t_i^*$ if $\cdot$ is interpreted as the right-hand derivative [17]. However, the existence of $\dot{\zeta}_i(t_i^*)$ in the ordinary sense is an open question. A sufficient condition has been obtained in reference [4]. For example, consider the case $m = 2$ with

$$u_o(x) = \begin{cases} (1 - \theta)\cos^2 x + \theta\cos^4 x & \text{for} \quad |x| \leq \pi/2 \\ 0 & \text{for} \quad |x| > \pi/2 \ . \end{cases}$$

Then the result proved in [4] shows that $\zeta_i \in C^1(\mathbb{R}^+)$ for $\theta \in [0,1/4]$, but gives no information for $\theta > 1/4$.

For flows in higher dimensional porous media, the qualitative results about finite speed of propagation still hold. However, very little is known about the regularity properties of the interface. The best result to date is the Caffarelli-Friedman proof of the Hölder continuity of the interface [8].

2. SINGLE SPECIES POPULATION MODELS.

In reference [15], Gurtin and MacCamy consider the growth and spread of a spatially distributed biological population whose tendency to migrate is governed by the local population density. In particular, the individuals in their populations do not like crowds and so tend to move away from high density regions. The Gurtin-MacCamy model is a continuum model and is described by the equation

$$\frac{\partial u}{\partial t} = \Delta\varphi(u) + f(u) \ , \qquad (2.1)$$

where $\varphi(0) = 0$, $\varphi'(0) = 0$, and $\varphi'(u) > 0$ for $u > 0$. The growth function $f(u)$ describes the underlying (nondiffusive) population dynamics. Gurney and Nisbet [13], [14] derive a similar equation from probabilistic considerations.

For the most part reference [15] is concerned with a model problem with $\varphi(u) = u^m$ for $m \geq 2$ and Malthusian growth $f(u) = \mu u$, where $\mu \in \mathbb{R}$ is constant. In this case, equation (2.1) can be transformed into the porous medium equation (1.4) and Gurtin and MacCamy use the known properties of porous medium flow to deduce properties of solutions to equation (2.1).

Recently H. F. Weinberger and I looked at the question of existence of plane wave solutions to equation (2.1) under various assumptions on φ and f. To describe our results in a specific case, consider the density dependent analog of Fisher's equation

$$\frac{\partial u}{\partial t} = \Delta(u^m) + u(1 - u) \tag{2.2}$$

with $m \geq 1$.

To obtain plane wave solutions to equation (2.2) it suffices to find traveling wave solutions to the one-dimensional equation

$$u_t = (u^m)_{xx} + u(1 - u) . \tag{2.3}$$

A travelling wave solution to (2.3) is a solution of the form

$$u(x,t) = q(x - ct)$$

for some constant $c \in \mathbb{R}^+$, where $q = q(\xi)$ satisfies

$$(q^m)'' + cq' + q(1 - q) = 0 \quad \text{in } (-\infty,\omega) \tag{2.4}$$

and

$$q(-\infty) = 1, \; q > 0 \quad \text{and} \quad q' < 0 \quad \text{in} \quad (-\infty,\omega), \; q(\omega) = 0$$

for some $\omega \in (-\infty,+\infty]$.

Before discussing the results for $m > 1$, I shall briefly recall what is known in the case $m = 1$ for Fisher's equation. Fisher [10] and Kolmogorov, Petrovsky and Piscunov [18] have shown the existence of traveling wave solutions for all speeds $c \geq c^* = 2$ and that there are no traveling waves for $0 \leq c < c^*$. These waves are all of <u>change of phase</u> type (Figure 2b), that is, $\omega = +\infty$. The minimal wave speed c^* plays a special role. It is the <u>asymptotic speed of propagation of disturbances from rest.</u> To explain this, let u denote the solution of the initial value problem

$$\begin{cases} u_t = u_{xx} + u(1-u) & \text{in } \mathbb{R} \times \mathbb{R}^+ \\ u(x,0) = u_o(x) & \text{in } \mathbb{R} \end{cases} \tag{2.5}$$

where $u_o \geq 0$, $u_o \not\equiv 0$, and $u_o(x) = 0$ for all $x \geq x_o$ for some $x_o \in \mathbb{R}$. Then for every $x \in \mathbb{R}$,

$$\lim_{t\uparrow+\infty} u(x+ct,t) = \begin{cases} 0 & \text{if } c > c^* \\ 1 & \text{if } 0 \leq c \leq c^* \end{cases}$$

([5]). Roughly speaking, if you start from a point on the x-axis and run toward $+\infty$ with speed c, then you will outrun the solution of problem (2.5) if $c > c^*$, but it will outrun you if $c \leq c^*$. Note that the fact that c^* is the asymptotic speed of propagation of disturbances from rest does not imply that disturbances from rest are propagated with a finite speed. In fact, for Fisher's equation, as for the heat conduction equation, all disturbances are propagated with infinite speed.

For $m > 1$, Weinberger and I have shown the existence of a minimal wave speed $c^* = c^*(m) > 0$. The traveling waves with speeds $c > c^*$ are all of change of phase type (Figure 2b).

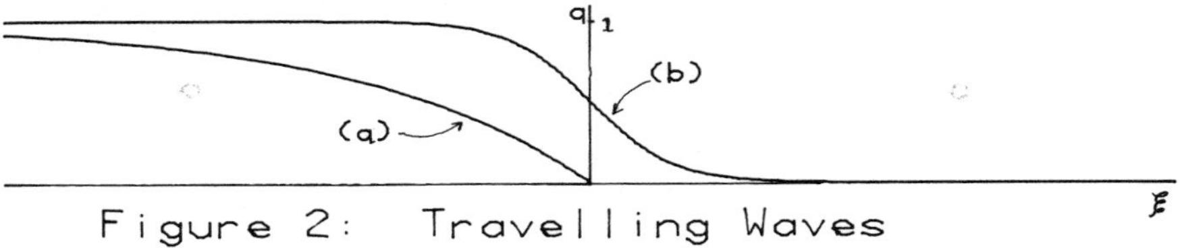

Figure 2: Travelling Waves

However, for $m > 1$, the wave with minimal speed c^* is <u>sharp</u> (Figure 2a), that is, $\omega < +\infty$. There are no traveling waves with speeds $c \in [0,c^*)$. For example, when $m = 2$ one can show that $c^*(2) = 1$ and find the explicit solution

$$q(\xi) = [1 - \tfrac{1}{2}e^{-\xi/2}]^+. \tag{2.6}$$

This solution was also discovered independently by W. Newman (see [20]). The minimal wave speed is again the asymptotic speed of propagation of disturbances from rest in the sense described in the preceding paragraph.

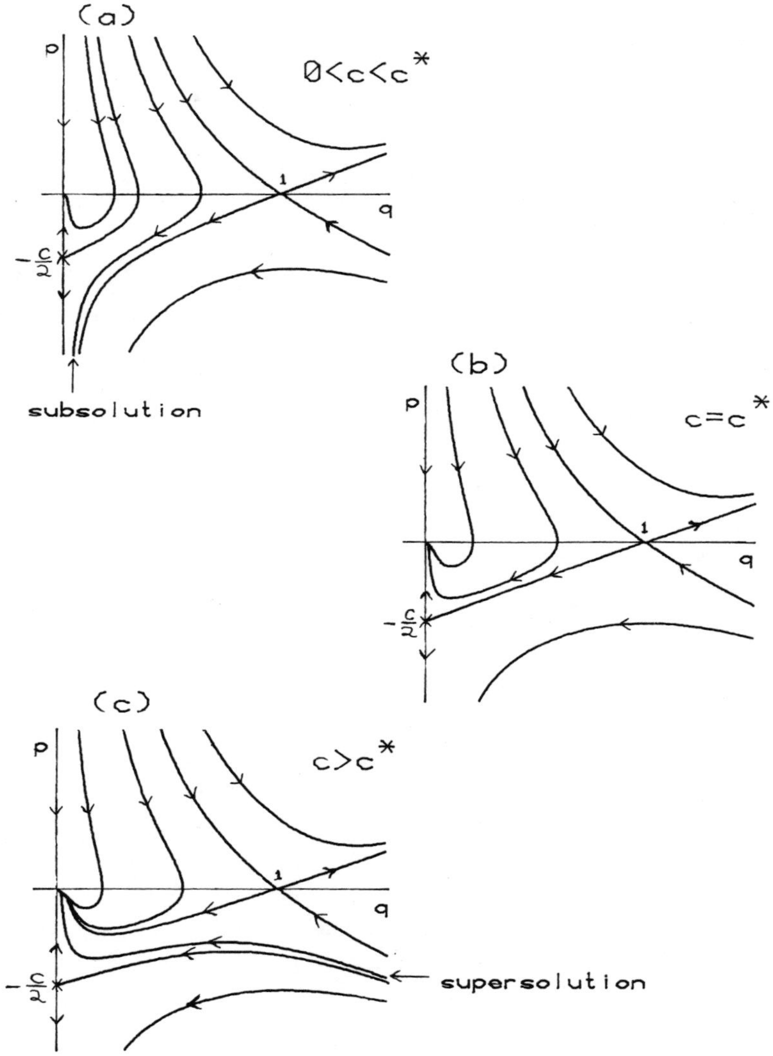

Figure 3: Phase Portraits for m=2

To prove the existence of traveling wave solutions to equation (2.3) it is natural to study the phase portrait for the wave equation (2.4). For this purpose, set $p = q'$ and rewrite equation (2.4) as a pair of first order equations. In the case $m = 2$ the resulting system is

$$\begin{cases} q' = p \\ p' = -\dfrac{p^2}{q} - \dfrac{cp}{2q} - \dfrac{f(q)}{2q} \end{cases} \tag{2.7}$$

where $f(q) = q(1 - q)$. To eliminate the singularity at $q = 0$, introduce the new independent variable τ defined by the symbolic equation

$$\frac{d}{d\xi} = \frac{1}{q}\frac{d}{d\tau} . \tag{2.8}$$

Then, if $\cdot$ denotes differentiation with respect to τ, the system (2.7) becomes

$$\begin{cases} \dot{q} = pq \\ \dot{p} = -p^2 - \frac{c}{2}p - \frac{1}{2}f(q) . \end{cases} \tag{2.9}$$

The system (2.9) is not singular and has critical points at

$$(q,p) = (0,0),\ (0,-c/2),\ (1,0) .$$

Thus the effect of the change of variables (2.8) is to resolve the singularity at $q = 0$ into the critical points (0,0) and (0,−c/2).

The phase portraits for the system (2.9) are drawn in Figure 3. When c is small the trajectory in the right half plane which approaches the saddle point (0,−c/2) and the trajectory in the lower half plane which leaves the saddle point (1,0) are as shown in Figure 3a. As c increases, these two trajectories move toward one another, and at the critical value $c = c^*$ they coincide (Figure 3b). The resulting "saddle-saddle" connection corresponds to the sharp traveling wave. For $c > c^*$ the configuration is as shown in Figure 3c. The trajectory from (1,0) now approaches (0,0) and the resulting "saddle-node" connection corresponds to a change of phase wave. Note that the saddle-saddle connection in Figure 3b is a straight line. From this observation, which is due to our computer graphics terminal, it is an easy matter to derive the explicit solution given in (2.6).

The proof of existence of traveling wave solutions to equation (2.3) when $m = 2$ carries over with only slight alterations to the equation

$$u_t = \{\varphi(u)\}_{xx} + u(1 - u), \tag{2.10}$$

where $\varphi(0) = 0$, $\varphi'(0) = 0$, and $\varphi'(u) > 0$ for $u > 0$. Indeed, one simply carries through the indicated arguments for the equation satisfied by

$$v \equiv \varphi'(u)/\varphi'(1).$$

For example, if $\varphi(u) = u^m$ then $v = u^{m-1}$ satisfies

$$v_t = m\, v_{xx} + \frac{m}{m-1} v_x^2 + h(v)$$

where

$$h(v) = (m - 1)v\,(1 - v^{1/(m-1)}),$$

and the corresponding wave equation is

$$mqq'' + \frac{m}{m-1} q'^2 + cq' + h(q) = 0 .$$

A somewhat different proof of the existence of traveling wave solutions to equation (2.10) and others for $c > c^*$ has been given by Atkinson, Reuter, and Ridler-Rowe [1].

The proof that c^* is the asymptotic speed of propagation of disturbances from rest is similar in outline to the proof of the corresponding result for $m = 1$ given in reference [5]. The key ingredients are the existence of subsolutions for $c \in (0,c^*)$ and of supersolutions for $c > c^*$ together with a suitable comparison principle. The sub- and super- solutions are provided by the solutions to the wave equation (2.4) corresponding to the trajectories indicated in Figures 3a and 3c respectively.

There are, of course, corresponding results for other choices of the growth function $f(u)$. For example, for the density dependent analog of Nagumo's equation

$$u_t = (u^m)_{xx} + u(1 - u)(u - a) \tag{2.11}$$

with $a \in (0,1/2)$ and $m > 1$ there is a number $c^* = c^*(m) > 0$ such that (2.11) possesses a sharp traveling wave solution with $c = c^*$, but has no traveling wave solutions for $c \neq c^*$.

3. INTERACTING POPULATIONS.

So far I have discussed problems which lead to a single nonlinear diffusion equation. Now I want to give an example of a problem which leads to a system of equations. Since I want to emphasize the modeling process rather than the model itself, I shall make various simplifying assumptions which may indeed be unrealistic. However, it will be clear that more realistic and elaborate models can be built in a similar manner.

It is reasonable to expect density dependent migration behavior in spatially distributed predator-prey interactions. For example, there is evidence that some insect predators will tend to stay in the vicinity of their last meal, at least for some time [11]. In this section I shall formulate a model of such an interaction. Since I am thinking about insect populations with discrete synchronized generations it is conceptually simpler to formulate a model which is discrete in both space and time. The model will lead to a rather cumbersome system of difference equations. In the diffusion approximation, this system yields an interaction-diffusion system with density dependent diffusion.

For simplicity I shall consider only a sedentary prey and restrict space to one dimension. Consider a doubly infinite array of loci indexed by $j = 0, \pm 1, \pm 2, \ldots$. Let

u_n^j = the number of prey at the j-th locus in the nth generation

and

v_n^j = the number of predators at the j-th locus in the n-th generation,

where $n = 0,1,2,\ldots$. For each n , the n-th generation is subdivided into three age classes indexed by n, $n + \frac{1}{3}$, and $n + \frac{2}{3}$. The transition from the n-th to the (n+1)-st generation is given by the flow chart in Figure 4.

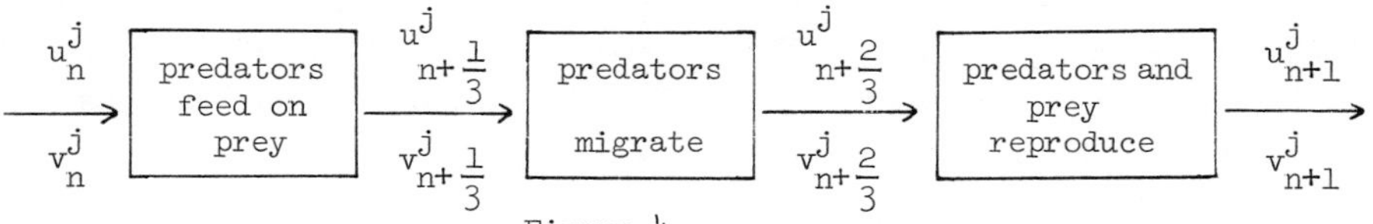

Figure 4

In the interest of simplicity, I shall assume that the predation mechanism is given by the Nicholson-Bailey model [11]. Thus

$$\begin{cases} u^j_{n+\frac{1}{3}} = u^j_n \exp\{-av^j_n\} \\ v^j_{n+\frac{1}{3}} = v^j_n \end{cases} \tag{3.1}$$

where $a \in \mathbb{R}^+$ is the attack rate. To describe the migration mechanism, I assume that there is given a doubly infinite array of numbers $\{p_{\ell j}(u)\}$, where

$p_{\ell j}(u)$ = the probability that a predator will jump from the ℓ-th locus to the j-th locus given that the number of prey at the ℓ-th locus is u.

Naturally, it is assumed that $p_{\ell j}(u) \geq 0$ and that

$$\Sigma_j \, p_{\ell j}(u) = 1 \quad \text{for all} \quad \ell = 0, \pm 1, \pm 2, \ldots \quad .$$

Note that, in general, $p_{\ell\ell}(u) \neq 0$. The migration process is described by the equations

$$\begin{cases} u^j_{n+\frac{2}{3}} = u^j_{n+\frac{1}{3}} \\ v^j_{n+\frac{2}{3}} = \sum_\ell p_{\ell j}(u^\ell_n) v^\ell_n \, . \end{cases} \tag{3.2}$$

Finally I assume that the prey are governed by an exponential logistic growth law in the absence of the predator, and that the predators decline exponentially in the absence of the prey. Thus the reproduction phase of the life cycle is given by

$$\begin{cases} u^j_{n+1} = u^j_{n+\frac{2}{3}} \exp[r(1 - \frac{1}{K} u^j_{n+\frac{2}{3}})] \\ v^j_{n+1} = \sum_\ell p_{\ell j}(u^\ell_n)[(1 - m)v^\ell_n + \gamma u^\ell_n(1 - \exp\{-a\, v^\ell_n\})], \end{cases} \tag{3.3}$$

where r = prey birth rate, K = prey carrying capacity, m = predator death rate, and γ = predator's conversion efficiency. The age structure can now be ignored, since by substituting from equations (3.1) and (3.2) one obtains

$$\begin{cases} u^j_{n+1} = u^j_n \exp[-a\ v^j_n + r(1 - \frac{1}{K}\ u^j_n \exp\{-a\ v^j_n\})] \\ v^j_{n+1} = \sum_\ell p_{\ell j}(u^\ell_n)[(1 - m)v^\ell_n + \gamma u^\ell_n(1 - \exp\{-a\ v^j_n\})]\ . \end{cases} \tag{3.4}$$

Although the system (3.4) is well suited to numerical studies it does not yield much in the way of analytic insights. In an attempt to gain some insight, I will replace the system (3.4) by its diffusion approximation. For this purpose, I assume that $u^j_n = u(jh, n\tau)$ and $v^j_n = v(jh, n\tau)$ for some smooth functions u and v, where h is the distance between adjacent loci and τ is the generation time. Then, setting

$$a = \alpha\tau\ ,\ m = \mu\tau\ ,\ \text{and}\ \ r = \rho\tau$$

I look for the limit as h and τ tend to zero with $jh \to x$ and $\tau n \to t$. It is easy to verify that under this limiting process the first equation in the system (3.4) yields the differential equation

$$\frac{\partial u}{\partial t} = \rho u\ (1 - \frac{1}{K}\ u) - \alpha\ u\ v. \tag{3.5}$$

Assuming that the $p_{\ell j}(u)$ are sufficiently smooth and applying Taylor's theorem twice, one obtains

$$p_{\ell j}(u^\ell_n) = p_{\ell j}(u + h(\ell-j)u_x + \frac{h^2}{2}(\ell-j)^2 u_{xx} + O(h^3)) = p_{\ell j}(u)$$

$$+ \{h(\ell-j)u_x + \frac{h^2}{2}(\ell-j)^2 u_{xx}\}p'_{\ell j}(u) + \frac{h^2}{2}(\ell-j)^2 u_x^2 p''_{\ell j}(u) + O(h^3)$$

where $u = u(jh, n\tau)$. Using Taylor's theorem again, the second equation in (3.4) can be written as

$$\begin{aligned} v^j_{n+1} - v^j_n = {} & \{\sum_\ell p_{\ell j}(u) - 1\}v + \tau\sum_\ell p_{\ell j}(u)\{-\mu v + \alpha\gamma uv + O(h)\} \\ & + h\sum_\ell(\ell-j)\{(1 - \mu\tau)p_{\ell j}(u)v_x + p'_{\ell j}(u)u_x\ v\} \\ & + \frac{h^2}{2}\ \sum_\ell(\ell-j)^2\{p_{\ell j}(u)v_{xx} + 2(1-\mu\tau)p'_{\ell j}(u)u_x v_x \\ & + p'_{\ell j}(u)u_{xx}v + p''_{\ell j}(u)u_x^2 v\} + O(h^3) \end{aligned} \tag{3.6}$$

where $v = v(jh, n\tau)$. Now assume that $\frac{h^2}{\tau} \to \lambda > 0$ so that $\tau h = O(h^3)$ and rewrite (3.6) in the form

$$\frac{v_{n+1}^j - v_n^j}{\tau} = \frac{\partial^2}{\partial x^2}\{\frac{h^2}{2\tau}\sum_\ell(\ell-j)^2 p_{\ell j}(u)v\} + \frac{\partial}{\partial x}\{\frac{h}{\tau}\sum_\ell(\ell-j)p_{\ell j}(u)v\}$$

$$+ \sum_\ell p_{\ell j}(u)\{-\mu v+\alpha\gamma uv\} + \frac{1}{\tau}\{\sum_\ell p_{\ell j}(u) - 1\}v + 0(h).$$

Following the lead of Fritz John's classic paper [16] on finite difference approximations to parabolic equations, I assume that

$$\frac{1}{\tau}\{\sum_\ell p_{\ell j}(u) - 1\} \to 0 , \tag{3.7}$$

$$\frac{h}{\tau}\sum_\ell (\ell-j)p_{\ell j}(u) \to 0 , \tag{3.8}$$

and that there exists a function $\Phi(u)$ such that

$$\frac{h^2}{2\tau}\sum_\ell (\ell-j)^2 p_{\ell j}(u) \to \Phi(u). \tag{3.9}$$

Note that (3.7) implies that

$$\sum_\ell p_{\ell j}(u) \to 1 .$$

These conditions can be interpreted in terms of the mean and variance of the distribution of the $\{p_{\ell j}\}$. Strictly speaking, I should assume that (3.7) holds in the sense of C^0, (3.8) holds in the sense of C^1, and (3.9) holds in the sense of C^2. With these assumptions the diffusion approximation to the second equation in (3.4) is

$$\frac{\partial v}{\partial t} = \frac{\partial^2}{\partial x^2}\{\Phi(u)v\} + v(\alpha\gamma u-\mu). \tag{3.10}$$

The system (3.5), (3.10), particularly in the special case

$$\Phi(u) = D\{(1 - \frac{1}{K}u)^p\}^+$$

with constants $D > 0$ and $p \geq 0$, is currently being studied. An account of the results will be published elsewhere.

REFERENCES

1. C. Atkinson, G. E. H. Reuter and C. J. Ridler-Rowe, Traveling wave solutions for some nonlinear diffusion equations, to appear.
2. D. G. Aronson, Regularity properties of flows through porous media: a counterexample, SIAM J. Appl. Math., 19 (1970), 299-307.
3. D. G. Aronson, Regularity properties of flows through porous media: the interface, Arch. Rat. Mech. Anal., 37 (1970), 1-10.
4. D. G. Aronson, L. A. Caffarelli and S. Kamin, How an initially vertical interface in porous medium flow begins to move, in preparation.
5. D. G. Aronson and H. F. Weinberger, Nonlinear diffusions in population genetics, combustion, and nerve propagation, Partial Differential Equations and Related Topics, Lecture Notes in Math. vol. 446, Springer, 1975.
6. J. G. Berryman and C. J. Holland, Stability of the separable solution for fast diffusion, to appear.
7. L. A. Caffarelli and A. Friedman, Regularity of the free boundary for the one-dimensional flow of a gas in a porous medium, Amer. J. Math., to appear.
8. L. A. Caffarelli and A. Friedman, Regularity of the free boundary of a gas flow in an n-dimensional porous medium, to appear.
9. G. S. Deem, N. J. Zabusky and H. Steinlicht, Association-dissociation time-scale factor for proton transport in immobilized protein membranes, J. Membrane Sci., 4 (1978), 61-80.
10. R. A. Fisher, The wave of advance of advantageous genes, Ann. of Eugenics, 7 (1937), 355-369.
11. M. P. Hassell, Arthropod Predator-Prey Systems, Princeton, 1978.
12. D. R. Garg and D. M. Ruthven, The effect of the concentration dependence of diffusivity on zeolitic sorption curves, Chem. Eng. Sci., 27 (1972), 417-423.

13. W. S. C. Gurney and R. M. Nisbet, The regulation of inhomogeneous populations, J. Theor. Biol. 52 (1975), 441-457.

14. W. S. C. Gurney and R. M. Nisbet, A note on non-linear population transport, J. Theor. Biol, 56 (1976), 249-251.

15 M. E. Gurtin and R. C. MacCamy, On the diffusion of biological populations, Math. Biosci., 33 (1977), 35-49.

16. F. John, On the integration of parabolic equations by difference methods, Comm. Pure Appl. Math., 5 (1952), 155-211.

17. B. F. Knerr, The porous medium equation in one dimension, Trans. Amer. Math. Soc., 234 (1977), 381-415.

18. A. Kolmogoroff, I. Petrovsky and N. Piscounoff, Étude de l'équation de la diffusion avec croissance de la quantité de matierè et son application à un probleme biologique, Bull. Univ. Moskou, Ser. Internat., Sec. A, 1 (1937) #6, 1-25.

19. M. Muskat, The Flow of Homogeneous Fluids Through Porous Media, McGraw-Hill, 1937.

20. W. I. Newman and C. Sagan, Galactic civilizations: population dynamics and interstellar diffusion, Icarus, to appear.

21. O. A. Oleinik, A.S. Kalashnikov and Chzou Yui-Lin, The Cauchy problem and boundary problems for equations of the type of unsteady filtration, Izv. Akad. Nauk SSSR Ser. Mat., 22 (1958), 667-704.

22. C.J. Pereira and A. Varma, Effectiveness factors for the case of mildly concentration-dependent diffusion coefficients, Chem. Eng. Sci., 33 (1978), 396-399.

23. N. J. Zabusky and G. S. Deems, Intrinsic electric fields and proton diffusion in immobilzed protein membrane, Biophys. J., 25 (1979), 1-16.

This work was supported in part by the National Science Foundation Grant MCS78-02158. Figures 1,2 and 3 were drawn by G. R. Hall using the computer graphics equipment in the University of Minnesota School of Mathematics Dynamical Systems Laboratory. Mr. Hall and the Laboratory are supported by National Science Foundation Grant MSC78-02173 and a Grant from the Graduate School of the University of Minnesota.

School of Mathematics
University of Minnesota
Minneapolis, Minnesota 55455

Strange Oscillations in Chemical Reactions—Observations and Models

Roger A. Schmitz, Gary T. Renola and Anthony P. Zioudas

1. INTRODUCTION.

Open chemically reacting systems are known to exhibit a variety of intriguing behavioral features including hysteresis phenomena and sustained, self-generated oscillations. Examples of such systems are most abundant in biological processes, but occurrences in nonbiological reacting systems are not rare by any means. In fact, an observation of simple hysteresis behavior or of a single-peak-per-cycle periodicity in a reacting system is often no longer noteworthy. Recent interest has shifted to more complex oscillations, which include multipeak periodic cycles and nonperiodic or "chaotic" behavior.

To the chemical engineer the occurrences of hysteretic and oscillatory behavior are much more than a curiousity. An elucidation of the causative rate processes, to the extent that mathematical models can be formulated, is important because the rational design of reactors and procedures for their operation and control rely on the predictive capabilities of such models.

There is no intention here to give an exhaustive review of this general subject area. Several recent reviews are available [1-5]. Instead, the focus will be directed shortly to recent findings in our studies of a particular solid-catalyzed reaction -- findings which have received very little exposure thus far in the literature. It seems worthwhile, however, first to give a brief summary, primarily from the

ISBN 0-12-669550-4

perspective of the chemical engineer, of the developments in the study of oscillatory reaction processes which have led to the present level of interest.

In the chemical engineering literature the earliest investigations into oscillatory reactions were devoted to the classic problem of a single exothermic reaction in a continuous-flow, stirred tank reactor (CSTR). The mathematical description of this problem is reducible to two nonlinear ordinary differential equations which result from species and energy balance considerations. Bifurcation theorems and methods of phase plane analysis have been applied to this problem. The early works of Van Heerden [6,7], Bilous and Amundson [8] and Aris and Amundson [9] have been nicely summarized and amplified in recent publications by Uppal, Ray and Poore [10,11]. Presently the theoretical behavior in parameter space for this classic problem is completely illumined. Furthermore, the theoretical predictions have been put to test in a number of experimental studies to which references are given in the aforementioned review papers.

The mathematical description of this classic CSTR problem extends to a number of other types of open reacting systems, including reactions on catalytic surfaces (even certain isothermal models) as shown in reference [3].

Where sustained oscillations are concerned, the model of the classic CSTR problem predicts simple oscillations; that is, simple closed curves as limit cycles in the phase plane.

The celebrated Belousov-Zhabotinskii (B-Z) reaction† has served as the principal vehicle leading to studies of more complex oscillatory behavior in nonbiological systems. Apparently the first studies of this reaction in a open system (that is, in a CSTR) were experimental studies reported in references [12] and [13]. Among the experimental observations were sustained oscillations of a complex nature, including multipeak periodic patterns and apparent chaos. Similar experimental observations have been reported very recently by Hudson et al. [14]. In none of these did the authors present a specific mathematical model, but pointed out that their

†The B-Z reaction is the oxidation of malonic acid by bromate, homogeneously catalyzed by a metal ion.

experimental observations suggested a model consisting of two nonlinear ordinary differential equations which would describe trajectories moving on a folded manifold in the state space. Models have since been proposed for the B-Z reaction which indeed exhibit complex oscillatory behavior of the type observed [15,16]. Rössler had earlier introduced the possibility of the existence of chaotic states in reacting systems using theoretical results from prototype reactions [17-19].

2. SOLID-CATALYZED REACTIONS

The subject of oscillatory chemical reactions is particularly interesting in cases of reactions which are catalyzed on solid surfaces. The reasons are: (1) solid-catalyzed reactions are of major importance in chemical process applications, and (2) the reaction kinetic models generally considered to be descriptive of catalytic rate processes fail to describe oscillatory behavior. It appears, therefore, that new and important information regarding catalytic reaction mechanisms and models might result from research which couples experimental observations with theoretical predictions.

The remainder of this paper reports on some of our recent work with the nickel-catalyzed oxidation of hydrogen. Particular attention is given to oscillatory behavior including multipeak periodic and chaotic oscillations. Other experimental work with the oscillatory behavior of this reaction is cited in the aforementioned review papers. A recent study by Zuniga and Luss [20] also reports experimental observations of complex oscillations in the catalytic oxidation of hydrogen on platinum. Aside from the modelling work reported herein, a mathematical model for catalytic reactions on polycrystalline surfaces presented by Jensen and Ray [21] predicts complex oscillatory behavior.

3. THE EXPERIMENTAL SYSTEM

A description of the experimental portion of this research has been given in an earlier publication [22]. A brief summary is warranted here because the experimental results provided the motivation for our theoretical study to which succeeding sections of this paper are devoted.

A mixture of hydrogen and oxygen was fed continuously to the bottom of a Pyrex tube, packed over most of its length with Pyrex beads and wrapped with electical heating tape.

The catalyst specimen, here a disk of polycrystalline nickel foil (diameter of 1 cm and thickness of 0.075 mm) oriented perpendicularly to the tube axis, was positioned in a void space in the tube about 2 cm above the top of the packed section. The signal from a thermocouple, welded to the downstream side of the foil, was the only response variable measured in the experiments. The principal experimental parameters were the composition and temperature of the gas mixture. Observations were satisfactorily reproducible for a given catalyst sample.

Results of numerous tests conducted with a gas temperature of 231°C and with a certain nickel sample over a period of several weeks are particularly illustrative of the behavior of interest in this paper. In these tests, self-generated sustained oscillatory catalyst temperatures were recorded for gas compositions in the range 1 to 5% O_2. Furthermore, at fixed experimental parameters, the oscillations became increasingly complex in time eventually taking on an apparent chaotic nature -- i.e. time-dependent but not periodic. In the range of about 1.75 to 3% O_2, the oscillations typically developed through a well-defined sequence of periodic patterns before becoming chaotic. An example of one such sequence is shown for an O_2 concentration of 2.5% in Fig. 1, which presents sections of temperature recordings over the time interval of 4 to 15 hours from the start of a certain test. Evident in the figure are bifurcations from single-peak to two-peak and from two-peak to four-peak "pseudo-steady" periodic patterns and finally to apparent chaos. One such test was run continuously for 122 hours with chaotic responses persisting throughout -- although gradual changes in the time scale of the oscillations occurred during that time. Slow adjustments in the physical and/or chemical nature of the catalytic material appear to be taking place in the early stages of a test which cause passages through bifurcation conditions.

This experimental facility was set up for exploratory studies of catalyst dynamics, and as such it was not well-suited for the systematic study of a number of factors which one suspects might influence the complex dynamics which were observed. Questions come to mind, for example, as to the importance of such factors as (1) nonuniform transfer

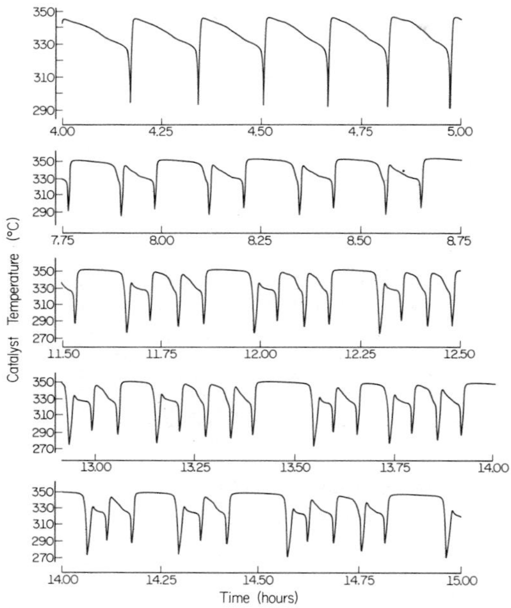

Fig. 1. Experimental Temperature Recordings for $[O_{2,g}] = 2.5\%$.

coefficients over the catalyst surface, (2) the presence of a thermocouple attached to the catalyst sample, (3) macroscopic variations in catalyst activity over the sample, and (4) impurities present in the feed streams or present on the catalyst initially. Presumably a more refined experimental system, presently being constructed, will enable us to study such factors in the laboratory or to eliminate them from consideration. We felt that considerable insight could be gained presently by studying the predicted behavior of a mechanistic model. In the theoretical analysis, outlined in the following section, we sought an answer to one main question: could a simple, yet plausible, model which would account for the obvious physical and chemical rate processes, and which would ignore the complicating factors mentioned above, predict intrinsic dynamic complexities reminiscent of those observed experimentally? Quantitative comparisons of theoretical with experimental results and model verification were not primarily of interest at this stage.

4. THE MATHEMATICAL MODEL

A schematic representation of the physical picture on which the derivation of equations were based is shown in Fig. 2. The physical rate processes are those of heat and mass

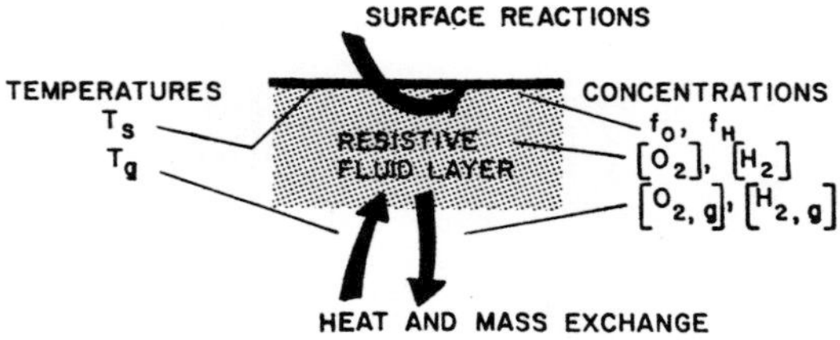

Fig. 2. Schematic Representation of Mechanistic Model.

transport through a "resistive" fluid layer covering the catalyst surface. For the sake of simplicity in the form of the model, the resistive layer was assumed to be of uniform composition. It plays a role in the system dynamics because it provides a capacity, along with the catalytic surface, for chemical components. The thermal capacity of the layer was assumed to be negligible in comparison to the capacity of the catalyst material. The supposition of the uniform fluid layer leads to a mathematical description analogous to that of a continuous flow gradientless catalytic reactor.

The following surface reactions are assumed to occur:

$$2M + O_2 \longrightarrow 2M^* \qquad (R1)$$

$$2H^* \longrightarrow M^* + M + H_2O \qquad (R2)$$

$$2M^* + O_2 \rightleftharpoons 2\,O^* \qquad (R3)$$

$$2M^* + H_2 \rightleftharpoons 2\,H^* \qquad (R4)$$

$$2H^* + O^* \longrightarrow H_2O + 3M^* \qquad (R5)$$

The forms of reactions (R1) and (R2), which account for the oxidation and reduction of surface sites, were inferred by us from various literature sources many of which were contradictory and not directly applicable to the situation at hand.

Reactions (R3)-(R5) comprise the familiar Langmuir-Hinshelwood mechanism for chemisorption and surface reaction. The reduced metal surface is assumed to play a passive role in the sense that the catalytic activity expressed in reactions (R3)-(R5) is assumed to be confined to oxidized sites. It seems that the evidence to support such an assumption is weak, but for our purposes it serves as a reasonable starting point. The rate expressions used for these reactions were the mass-action Arrhenius forms given in Eqs. (1)-(6) below.

$$R_1 = k_1 f_M{}^2 [O_2] \tag{1}$$

$$R_2 = k_2 f_H^2 \tag{2}$$

$$R_3 = k_3 f^2 [O_2] - k_{-3} f_o{}^2 \tag{3}$$

$$R_4 = k_4 f^2 [H_2] - k_{-4} f_H{}^2 \tag{4}$$

$$R_5 = k_5 f_H{}^2 f_o \tag{5}$$

where

$$k_j = k_{j,o} \exp(-E_j/R_g T_s) \tag{6}$$

$$j = (-3,-4,1,2,3,4,5)$$

Autocatalytic effects, commonly felt to be an important component in oscillating reactions, are provided in this reaction model by two mechanisms: (1) the inhibition of a strongly chemisorbed reactant, and (2) thermal feedback from the exothermic processes.

The basic equations, given below, consist of material balance expressions for oxidized sites, gas phase O_2, gas phase H_2, and chemisorbed H_2 and an enthalpy balance. The latter incorporates the assumption that reaction (R5) is the only significant contributor to reaction enthalpy effects.

$$N\, df/dt = A(2R_1 + R_2 - 2R_3 - 2R_4 + 3R_5) \tag{7}$$

$$v_e\, d[O_2]/dt = A[k_g([O_{2,g}] - [O_2]) - R_1 - R_3] \tag{8}$$

$$N\, df_o/dt = A(2R_3 - R_5) \tag{9}$$

$$v_e\, d[H_2]/dt = A[k_{gH}([H_{2,g}] - [H_2]) - R_4] \tag{10}$$

$$N\, df_H/dt = A\,(-2R_2 + 2R_4 - 2R_5) \tag{11}$$

$$m_s C_{ps}\, dT_s/dt = A[h(T_g - T_s) + (-\Delta H) R_5] \tag{12}$$

An additional equation expressing the assumption that the total number of surface sites is constant is given by

$$f_M + f + f_O + f_H = 1 \tag{13}$$

The simplified system of three differential equations given below in Eqs. (14)-(16) was obtained from those given above by invoking the addition assumptions that (1) reactions (R3) and (R4) are each in equilibrium; i.e. $f_O = f\sqrt{K_3[O_2]}$ and $f_H = f\sqrt{K_4[H_2]}$, (2) the equilibrium constants K_3 and K_4 are constant, and (3) $[H_2] = [H_{2,g}]$. As with all previously stated assumptions, we justify these on the bases that they are not unreasonable and that they lead to a more easily managed mathematical description. The derivation of Eqs. (14)-(16) further involves the substitution of the rate expressions from Eqs. (1)-(6) and the substitution for f_M from Eq. (13). Eq. (14) below is obtained by algebraically combining Eqs. (8) and (9) to eliminate R_3, and Eq. (15) is obtained by combining Eqs. (7), (9) and (11) to eliminate R_3 and R_4. In physical terms the following equations correspond to balances on total oxygen, Eq. (14), on reduced sites, Eq. (15) and an enthalpy balance, Eq. (16).

$$\begin{aligned}(N\sqrt{K_3[O_2]})\, df/dt + [2v_e + Nf\sqrt{K_3} / 2\sqrt{[O_2]}]\, d[O_2]dt \\ = 2AK_g([O_{2,g}]-[O_2])-2AK_1[O_2][1-f(1+\sqrt{K_4[H_{2,g}]}+\sqrt{K_3[O_2]})]^2 \\ - Ak_5f^3\, K_4[H_{2,g}]\sqrt{K_3[O_2]}\end{aligned} \tag{14}$$

$$\begin{aligned}N[1 + \sqrt{K_3[O_2]} + \sqrt{K_4[H_{2,g}]}]\, df/dt + (Nf\, \sqrt{K_3}/2\sqrt{[O_2]})\, d[O_2]/dt \\ = 2Ak_1[O_2][1 - f\,(1 + \sqrt{K_4[H_{2,g}]} + \sqrt{K_3[O_2]})]^2 \\ - Ak_2\, f^2\, K_4[H_{2,g}]\end{aligned} \tag{15}$$

$$m_sC_{ps}dT_s/dt = hA\,(T_g - T_s) + (-\Delta H)\, Ak_5\sqrt{K_3[O_2]}\, K_4[H_{2,g}]f^3 \tag{16}$$

For convenience, we used the following dimensionless forms of the preceding equations:

$$\begin{aligned}(\alpha_A\sqrt{\omega})df/d\tau + (\alpha_F + \alpha_A f/2\sqrt{\omega})d\omega/d\tau \\ = 1 - \omega - \gamma_1\Gamma_1(\theta)\omega[1 - f\,(1 + \beta_H + \beta_O\sqrt{\omega}]^2 - \gamma_5\Gamma_5(\theta)\sqrt{\omega}f^3\end{aligned} \tag{17}$$

$$\begin{aligned}\alpha_R(1 + \beta_H + \beta_O\sqrt{\omega})\, df/d\tau + (\alpha_R\beta_O f/2\sqrt{\omega})\, d\omega/d\tau \\ = \omega\gamma_{1,2}\Gamma_1(\theta)[1-f(1 + \beta_H + \beta_O\sqrt{\omega})]^2 - \Gamma_2(\theta)f^2\end{aligned} \tag{18}$$

$$d\theta/d\tau = -\theta + \gamma_5 \Gamma_5(\theta)\sqrt{\omega}\, f^3 \tag{19}$$

where

$$\Gamma_j(\theta) = \exp\,[-T_{A,j} / (1 + T_{ad}\theta)];\ j = 1, 2, 5 \tag{20}$$

The coefficients of the derivatives in Eqs. (17)-(19) comprise an asymmetric capacitance matrix which is known to have a significant effect on the nature of the eigenvalues of the linearized system [23].

5. RESULTS OF ANALYSIS AND SIMULATIONS

Our analysis of the preceding equations consisted of computer simulations and of examinations of steady-state solutions and of eigenvalues of the linearized transient equations. The most interesting discovery was that of the existence of unique and unstable steady states which, as computer simulations showed, gave rise to complex oscillations over a portion of the parameter space. Because the dimensionless equations contain twelve unspecified parameters, an investigation of the behavioral features over the complete parameter space was not feasible. From limited searching of that space, however, we found that sustained oscillatory solutions existed in the vicinity of the set of parameter values listed in Table I below.

Table 1. Parameter Values for Simulations

$T_{ad} = 0.25$	$T_{A,5} = 10$	$\alpha_R = 1.2 \times 10^{-4}$	$\gamma_1 = 30$
$T_{A,1} = 10$	$\alpha_A = 5$	$\beta_H = 0.1$	$\gamma_{1,2} = 0.08$
$T_{A,2} = 15$	$\alpha_F = 0.5$	$\beta_o = 2.0$	γ_5 as specified

The unspecified parameter γ_5 might be viewed as a catalyst activity parameter, the effect of which was examined with particular interest in our simulation work.

Simulations yielded sustained oscillatory responses for $1.37 \times 10^5 \leq \gamma_5 \leq 7 \times 10^5$. Both complex periodic and chaotic states were obtained as shown in Figs. 3 and 4. Curve (3) of Fig. 3 is an example of an apparent chaotic state. Such states were found in the range $5.96 \times 10^5 \leq \gamma_5 \leq 6.10 \times 10^5$. The range of chaotic states for the above set of parameters is relatively small; presumably it could be enlarged by using other sets. (Incidentally, a time span of 100 units of dimensionless time τ corresponds roughly to 5 minutes in our experimental system.)

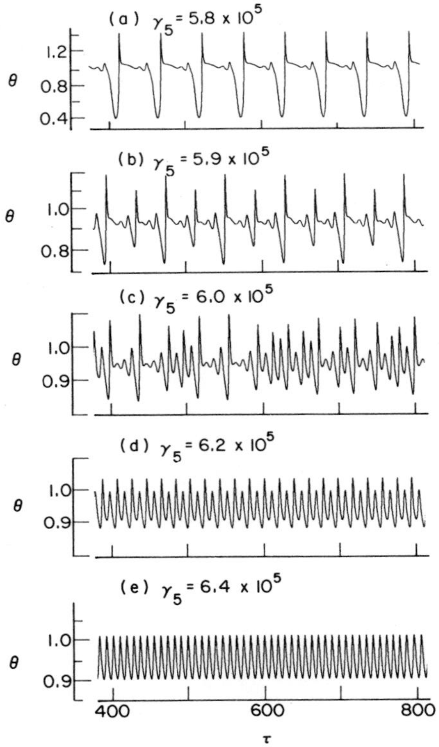

Fig. 3. Simulated Oscillations.

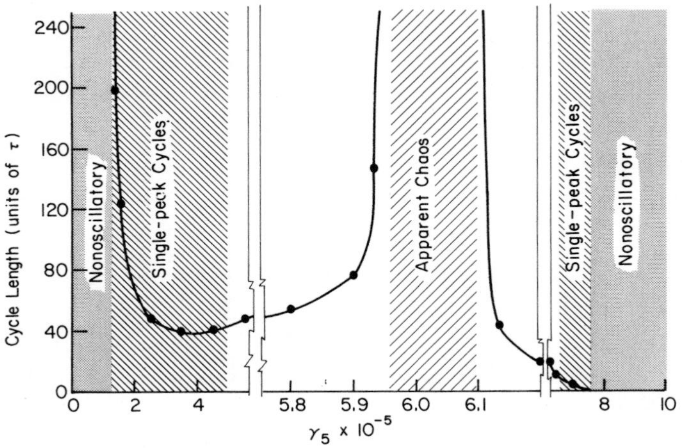

Fig. 4. Summary of Results of Simulations.

As shown in Fig. 4 simulated oscillations were of the single-peak type in the vicinity of both boundaries of the oscillatory range. The bifurcation at the left boundary was "hard" in the sense that those oscillatory states immediately to the right of the boundary were of large amplitudes and of the relaxation type. Oscillations near the bifurcation point at the right end of the oscillatory regime, on the other hand, were of very small amplitudes and of harmonic nature. The stable states just to the left of the oscillatory regime were at relatively low values of θ (virtually at the trough of the oscillations of curve (a) in Fig. 3) while those just to the right of the regime were at relatively high values θ (virtually at the average of the oscillations of curve (e) in Fig. 3.)

Figs. 5 and 6 show projections of the three-dimensional phase trajectories onto the coordinate planes for the cases (a) and (c) of Fig. 3. Fig. 5 shows the projections of a

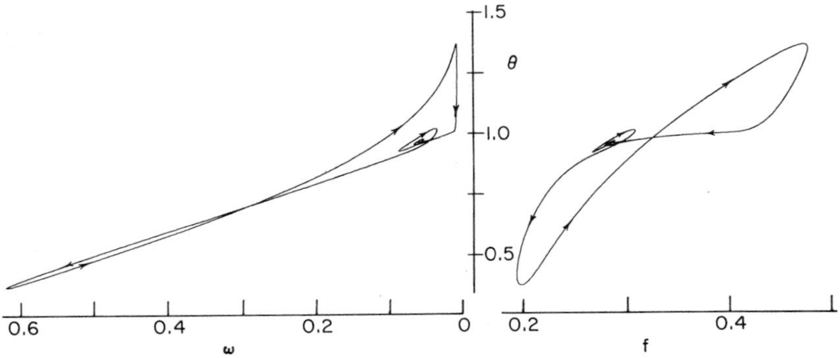

Fig. 5. Profections of Limit Cycle for $\gamma_5 = 5.8 \times 10^5$.

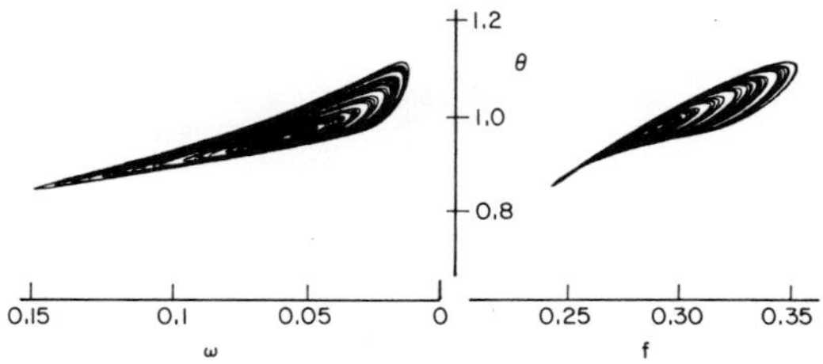

Fig. 6. Projections of Phase Trajectories for Chaotic State with $\gamma_5 = 6.0 \times 10^5$.

globally attracting limit cycle, while Fig. 6 shows the projections of a globally attracting volume for an apparent chaotic state. All trajectories are eventually captured by that volume. We were not able to determine from the simulations whether or not that volume becomes completely filled in the course of time.

We gave special attention to the accuracy of the numerical solutions because a recent publication by Showalter et al. [24] demonstrated that true periodic states could appear to be chaotic ones due to numerical inaccuracies. We carried out the numerical simulations using two different IMSL subroutines. One of these, called DVOGER, employed Gear's method, the other, called DREBS, used an extrapolation technique which is described as being better suited to high accuracy demands. We tested the accuracy of a number of the simulations by comparing the output of the two routines specifying error tolerances of 10^{-14} (a relative step error) in DVOGER and 10^{-13} (an absolute step error) in DREBS. To the practical limit of our capabilities for numerical testing, the results presented in this paper are qualitatively accurate. Obviously the existence of an intrinsic chaotic state can never be proved through numerical solutions. We did find, as did Showalter et al. [24] that some periodic states would appear to be chaotic if the specified error tolerance was not sufficiently small. Even with a very small error tolerance, we encountered some anomalies that remain unsettled at the moment. For example we found an apparent periodic state at $\gamma_5 = 5.95 \times 10^5$ with a dimensionless period of about 36 -- a large departure from the trend shown in Fig. 4. We also found apparent periodic states at γ_5 values of 5.97×10^5 (period of about 100) and 6.05×10^5 (period of about 75) even though the responses at γ_5 values of 5.96, 5.98, and 6.09 (all x 10^5) were all apparently chaotic. Some tests of effects of initial conditions suggest that stable periodic and chaotic states coexist. There is also the possibility that the chaotic regime indicated in Fig. 5 is actually composed of small sections of chaotic states separated by sections of periodic ones.

Finally, we note that the <u>isothermal</u> version of this system also exhibits oscillatory states, although all oscillations in that version are of the simple single-peak periodic type. Isothermal oscillations in this reaction system have been observed experimentally [3,4].

6. CONCLUDING REMARKS

The results of our work with a solid-catalyzed reaction are both interesting and promising. The interesting feature is the strange nature of the observed oscillations. The promising aspect is that a simple, yet plausible, mathematical model predicts dynamic behavior which is somewhat similar to that observed. There is reason to believe that further work beyond this preliminary stage will be even more enlightening.

Once the behavior of a single catalyst particle is understood, questions will naturally be raised as to the behavior of assemblages of interacting particles, some of which may be oscillatory by themselves and others nonoscillatory. In many practical situations of interest to the chemical engineer, large assemblies of interacting particles are encountered -- as in a fixed bed reactor.

7. SYMBOLS

A	Total surface area
C_{ps}	Heat capacity of catalytic material
E_j	Activation energy for the jth reaction
f	Fraction of surface sites which are oxidized
f_H, f_O	Fraction of surface sites occupied by chemisorbed hydrogen atoms and by chemisorbed oxygen atoms, respectively
f_M	Fraction of surface sites which are reduced
h	Coefficient for the exchange of heat between external gas and resistive gas
H*	Chemisorbed hydrogen atom
$[H_2], [H_{2,g}]$	Concentration of hydrogen in resistive gas volume and in external gas, respectively
ΔH	Enthalpy of reaction (R5) per mole of H_2O
k_g, k_{gH}	Coefficients for exchange of oxygen and hydrogen, respectively, between external gas and resistive gas
k_j, k_{-j}	Rate coefficients for the jth reaction, and the reverse of the jth reaction, respectively

$k_{j,o}$	Pre-exponential factor for rate coefficient of jth reaction
K_3, K_4	Chemisorption equilibrium constants. k_3/k_{-3} for reaction (R3), and k_4/k_{-4} for reaction (R4), respectively
m_s	Mass of catalyst material
M	Reduced metal site
M*	Oxidized metal site
N	Total number of surface sites
$[O_2]$, $[O_{2,g}]$	Concentration of oxygen in resistive gas volume and in external gas, respectively
O*	Chemisorbed oxygen atom
R_g	Universal gas constant
R_j	Intrinsic rate per unit surface area for the jth reaction
t	Time
T_{ad}	Maximum dimensionless temperature rise of catalytic material at steady state, $k_g[O_{2,g}](-\Delta H)/hT_g$
T_g	Temperature of the external gas
T_s	Temperature of the catalytic material
$T_{A,j}$	Dimensionless Arrhenius temperature for jth reaction, E_j/R_gT_g
v_e	Effective volume of the resistive gas
α_A	Dimensionless time constant for chemisorption, $N\beta_O/2Ak_g[O_{2,g}]\alpha_T$
α_F	Dimensionless time constant for oxygen exchange, $v_e/Ak_g\alpha_T$
α_R	Dimensionless time constant for metal reduction, $N/Ak_{2,O}\beta_H^2\alpha_T$
α_T	Thermal time constant, m_sC_{ps}/hA
β_H, β_O	Dimensionless groups $\sqrt{K_4[H_{2,g}]}$ and $\sqrt{K_3[O_{2,g}]}$, respectively
γ_1, $\gamma_{1,2}$, γ_5	Dimensionless groups $k_{1,0}/k_g$, $2k_{1,0}[0_{2,g}]/k_{2,0}K_4[H_{2,g}]$ and $k_{5,0}\beta_H^2\beta_0/k_g\ O_{2,g}$, respectively
$\Gamma_j(\theta)$	Arrhenius exponential factor, $\exp[-T_{A,j}/(1+T_{ad}\theta)]$, for jth reaction
θ	Dimensionless catalyst temperature, $h(T_s-T_g)/k_g[O_{2,g}](-\Delta H)$

ω Reduced oxygen concentration in the resistive gas, $[O_2]/[O_{2,g}]$

τ Dimensionless time, t/α_T

REFERENCES

1. Nicolis, G., and J. Portnow, Chemical oscillations, Chem. Rev. 73 (1973), 365-384.
2. Schmitz, R. A., Multiplicity, stability, and sensitivity of states in chemically reacting systems, Adv. Chem. Ser. 148 (1975), 154.
3. Sheintuch, M., and R. A. Schmitz, Oscillations in catalytic reactions, Catal. Rev.-Sci. Eng. 15 (1977), 107.
4. Slin'ko, M. G., and M. M. Slin'ko, Self-oscillations of heterogeneous catalytical reaction rates, Catal. Rev.-Sci. Eng. 17 (1978), 119.
5. Schmitz, R. A., Stability and control of chemically reacting systems, Proceedings of the 1978 Joint Atomatic Control Conference Volume II, pp. 21-30, Instrument Society of America, Pittsburgh, 1978.
6. Van Heerden, C., Autothermic processes - properties and reactor design, Ind. Eng. Chem. 45 (1953), 1242.
7. ______________, The character of the stationary state of exothermic processes, Chem. Eng. Sci. 8 (1958), 133.
8. Bilous, O., and N. R. Amundson, Chemical reactor stability and sensitivity, AIChE J. 1 (1955), 513.
9. Aris, R., and N. R. Amundson, An analysis of chemical reactor stability and control, Chem. Eng. Sci. 7 (1978) 121.
10. Uppal, A., W. H. Ray, and A. B. Poore, On the dynamic behavior of continuous stirred tank reactors, Chem. Eng. Sci. 29 (1974), 967.
11. Uppal, A., W. H. Ray, and A. B. Poore, The classification of the dynamic behavior of continuous stirred tank reactors - influence of reactor residence time, Chem. Eng. Sci. 31 (1976), 205.
12. Graziani, K. R., J. L. Hudson, and R. A. Schmitz, The Belousov-Zhabotinskii reaction in a continuous flow reactor, Chem. Eng. J. 12 (1976), 9.

13. Schmitz, R. A., K. R. Graziani and J. L. Hudson, Experimental evidence of chaotic states in the Belousov-Zhabotinskii reaction, J. Chem. Phys. 67 (1977), 3040.

14. Hudson, J. L., M. Hart, and D. Marinko, An experimental study of multipeak periodic and nonperiodic oscillations in the Belousov-Zhabotinskii reaction, J. Chem. Phys. 71 (1979), 1601.

15. Tyson, John J., On the appearance of chaos in a model of the Belousov reaction, J. Math. Biology 5 (1978), 351.

16. Tomita, K., and I. Tsuda, Chaos in the Belousov-Zhabotinsky reaction in a flow system, Phys. Lett. A71 (1979), 489.

17. Rössler, O. E., Chaotic behavior in simple reaction systems, Z. Naturforsch. 31 (1976), 259.

18. Rössler, O. E., Chemical turbulence: Chaos in a simple reaction-diffusion system, Z. Naturforsch. 31 (1976), 1168.

19. Rössler, O. E., Chaos in abstract kinetics: Two prototypes, Bull. Math. Biol. 39 (1977), 275.

20. Zuniga, J. E., and D. Luss, Kinetic oscillations during the isothermal oxidation of hydrogen on platinum wires, J. Catal. 53 (1978), 312.

21. Jensen, K. F., and W. H. Ray, A new approach to modelling the dynamics of catalytic surfaces, Proc. Symp. on Physicochemical Oscillations, Aachen, Germany, Deutsche Bunsengesellschaft für Physikalische Chemie (1979).

22. Schmitz, R. A., G. T. Renola, and P. C. Garrigan, Observations of complex dynamic behavior in the H_2-O_2 reaction on nickel, Bifurcation Theory and Applications in Scientific Disciplines (O. Gurel and O. E. Rössler, eds.) Annals of the New York Academy of Sciences 316 (1979), 638.

23. Luss, D., The influence of capacitance terms on the stability of lumped and distributed parameter systems, Chem. Eng. Sci. 29 (1974), 1832.

24. Showalter, K., R. M. Noyes, and K. Bar-Eli, A modified oregonator model exhibiting complicated limit cycle behavior in a flow system, J. Chem. Phys. 69 (1978), 2514.

This research was partially supported by National Science Foundation Grant ENG 78-12298 and a grant from the Mobil Foundation.

Department of Chemical Engineering
University of Notre Dame
Notre Dame, IN 46556

Time-Periodic and Spatially Irregular Patterns

Louis N. Howard

1. In the last several years Kopell and I have been studying some mathematical models which we hope will illuminate certain aspects of the Zaiken-Zhabotinskii patterns of chemical waves occurring sometimes in the Belousov reaction. These models are all based on reaction-diffusion equations of the form

$$c_t = F(c) + K\nabla^2 c \;, \tag{1}$$

in which c stands for a vector of chemical concentrations and $F(c)$ gives the rate of production of these substances as a result of the various reactions involved, after certain reasonable approximations (associated with neglect of long-term changes for example) have been made. Diffusion is modelled by the last term in (1), K being a positive definite matrix of diffusivities. Our approach has been to somewhat idealize certain of the more striking features of the observed wave patterns, and try to see if mathematical models of these idealizations with at least the right qualitative character can be constructed on the basis of reaction-diffusion equations. We have generally considered cases in which the chemical reaction itself is oscillatory, assuming that the system of ordinary differential equations (the "kinetic equations") $c_t = F(c)$ has a stable limit-cycle solution. This seems to be a reasonable approximation in certain ranges of concentration of the major ingredients for the Belousov reaction,

ISBN 0-12-669550-4

though other possibilities (e.g. "excitable but not self-excited" or aperiodic oscillations) no doubt occur with this reaction and may well sometimes be associated with wave patterns.

In some cases it has been possible to carry this through in sufficient generality that in principle a quantitative comparison with experimental observations could be attempted. Such comparisons however do not seem to have been made yet; they are made difficult by (a) some uncertainty about the detailed chemistry - and perhaps some of the diffusivities - (b) the need for some non-trivial numerical calculations, and (c) the lack of sufficiently well-controlled experiments directed specifically to this end.

In other cases it has been necessary to be satisfied for the present with "models of models" - special forms of reaction-diffusion equations which are not chemically realistic but possess properties making them mathematically more tractable. Such models can be qualitatively interesting, but are not useful for quantitative comparisons. As will be seen, the main results to be described today are of this character.

2. The first idealization we considered was directed to understanding better why waves are even possible in parabolic systems like (1). It should perhaps be emphasized that by "waves" we do not here refer to an isolated propagating "pulse" or "front", but to a more or less extended train of concentration variations which retain a certain identity for some time, while propagating through the medium. (The word "wave" has an immense number of different meanings both in common and scientific usage; it is unlikely that anything can be done about this, of course.) The idealization of this is a periodic infinite plane wave train: do equations like (1) have solutions of the form

$$c = y(\sigma t - \alpha x), \tag{2}$$

where y is a 2π-periodic vector function of its argument? Under the assumption that the kinetic equations have a stable limit cycle solution, the answer to this question is "yes": for each value of the wave number in a certain range $0 \leq |\alpha| < \alpha_0$ there is a value of the angular frequency $\sigma = H(\alpha^2)$ and a locally unique (up to a phase shift) y such that (2) satisfies (1). It is true in some cases, and probably in all, that if K is sufficiently close to a scalar matrix then a subset of this range, $0 \leq |\alpha| < \alpha_s < \alpha_0$, corresponds to wave trains which are stable as solutions of (1). One method of demonstrating this (singular perturbation from the $\alpha = 0$ case, which is the limit cycle of the kinetics) is given in [1]. The waves given by this method have a wave-form y which for small α is close to that of the limit cycle; thus they may be of large amplitude and highly nonlinear. Probably the simplest examples of such occur for the models with just 2 concentrations and

$$K = k\,I$$

$$F(c) = \begin{bmatrix} \lambda & -\omega \\ \omega & \lambda \end{bmatrix} \begin{bmatrix} c_1 \\ c_2 \end{bmatrix} \qquad (3)$$

where λ and ω are functions of $r^2 = c_1^2 + c_2^2$, with $\lambda > 0$ for $0 \leq r < r_0$ and $\lambda < 0$ for $r > r_0$. [We here allow the "concentrations" to be negative. To be thought of in chemical terms they should be regarded as deviations from some base concentrations; but such models cannot really be very good approximations to real chemical kinetics.] Setting $c_1 + ic_2 = r\,e^{i\theta}$ one finds easily that the equations are satisfied when

$$r = r_* < r_0$$
$$\theta = \sigma t - \alpha x \qquad (4)$$

where $\sigma = \omega(r_*)$ and $k\alpha^2 = \lambda(r_*)$.

The idealization of an infinite periodic plane wave train is of greater usefulness than is immediately apparent because it can be used as a center of perturbation. In [2] we have investigated solutions of (1) which can be described as "slowly-varying waves"; at a given time the function c can be locally approximated by the wave-form of <u>some</u> plane wave, but this approximation is not uniform over large scales; the local wave number and frequency vary slowly. A fairly complete description of the evolution of such slowly-varying waves, so long as they remain slowly varying, can be given in terms of the properties of the idealized plane wave trains.

However, this description itself indicates that in general slowly-varying waves will not always remain slowly-varying. Eventually regions of increasingly rapid variation in local wave number will usually develop; in these the former description is no longer applicable. Something like this is often seen in the Zaiken-Zhabotinskii patterns. There are often large patches of the domain throughout most of which the behaviour seems locally much like a plane wave, but these patches meet one another along thin transition zones in which there is a rapid change from the one slowly-varying wave to the other. In the experiments, these transition zones are places toward which the wave crests appear to converge, when observed in a system moving with the transition zone. According to the theory, it is the (relative) group velocities, not the phase velocities, which should be directed toward the transition zone, but in the experimental situation it appears (also for other reasons) that the phase and group velocities are in the same direction.

3. The simplest idealization of a transition zone is a solution of (1) on $-\infty < x < \infty$ which for each x is periodic in t, with everywhere the same period $2\pi/\sigma$, and which for $x \to \pm\infty$ tends to the plane wave with angular frequency σ and wave number $\pm\alpha$ ($\sigma = H(\alpha^2)$, and assuming $H'(\alpha^2) > 0$, so that the phase and group velocities are similarly directed). Under suitable hypotheses, such solutions were shown to exist (rigorously in some cases, and formally in others) in [2], where also other idealizations, corresponding to (moving) transitions from one wave number to another, not necessarily its opposite, are considered.

These solutions have a strong mathematical analogy with shocks in viscous heat conducting gas dynamics, which are thin regions of rapid transition from one constant state to another. The plane waves are the analogs of the constant states - they are not constant, but are among the simplest special solutions of (1). For this reason we refer to these transition solutions as "shocks".

The ideal shock solutions can also be given another interpretation. From symmetry considerations it is easily seen that $\frac{\partial c}{\partial x} = 0$ at the center of a shock solution connecting two waves of opposite wave number. Thus on the semi-infinite domain the shock solution may be regarded as corresponding to a wave train propagating toward an impermeable boundary, where it appears to be "absorbed". [The absence of any kind of reflection from walls, or interpenetration of impinging wave trains, is a reminder that these chemical waves are really very different from linear or weakly nonlinear waves like electromagnetic or water waves.]

4. In the Zaiken-Zhabotinskii patterns, the patches of slowly varying waves are frequently actually filled with a pattern of concentric circular wave crests propagating outward. The centers of these patterns resemble the boundaries where they meet other patches in being regions which cannot be described as slowly varying waves. However they differ in an important respect: the waves are propagating away from them rather than toward them (what really counts is the group velocity); this difference we believe to be more important than the obvious geometrical difference. To investigate whether or not reaction-diffusion equations have solutions resembling these centers, the simplest idealization seems to be to consider just one space dimension, and seek solutions of (1) on $-\infty < x < \infty$ which are periodic in time and as $x \to \pm\infty$ tend to plane waves with wave numbers $\pm\alpha$. This is just like the ideal shock problem, except that the waves (really the group velocities) are going <u>outward</u>. It can be seen from the theory of the shock transitions that such solutions do not in general exist. However, it is possible that they may exist for certain exceptional values of the wave number α (and corresponding frequency σ). Such solutions, if they exist, have no analog in gas dynamics, where they would correspond to "expansion shocks". Because they do not exist for arbitrarily small α, it does not seem to be possible to find them by the same kind of method (a bifurcation method) we used in [2] to study the shocks. In a forthcoming paper [3] we study this problem in the case of

the special "λ-ω model" (3), using a singular perturbation method.

We set $K = 1$ and take $r_0 = 1$ (merely scaling) and let $\omega(r) = \varepsilon\Omega(r^2)$; thus we consider a family of problems parameterized by ε, and suppose ε small. Strictly speaking, ε should not be thought of as a measure of the size of ω compared with λ, but rather of $\omega'(1)$ compared with $\lambda'(1)$; small ε may be thought of as corresponding to rapid radial approach to the limit cycle. If we set $r = r(x)$ and $\theta = \omega(r_*)t + \theta_1(x)$ in the polar form of the reaction diffusion equations for this model, the result may be written

$$\left.\begin{aligned} r_{xx} - r\theta_{1x}^2 &= -\,r\lambda(r) \\ \frac{1}{r^2}(r^2\theta_{1x})_x &= \varepsilon\left(\Omega(r_*^2) - \Omega(r^2)\right) \end{aligned}\right\} \qquad (5)$$

(Thus the particular form of r and θ is consistent with the reaction-diffusion equations, and reduces the partial differential equations to ordinary differential equations; this is the great advantage of the $\lambda - \omega$ models.) The parameter r_*, which is taken to be < 1, corresponds to the frequency of the ideal "center-structure" we are seeking. Evidently the equations (5), if regarded as a third order system for r, r_x and $a = -\theta_{1x}$ (the "local wave number") have two critical points at $r = r_*$, $r_x = 0$, and $a = \pm\alpha = \pm\sqrt{\lambda(r_*)}$. These correspond to the two oppositely directed plane waves of frequency $\omega(r_*)$. The center solution we are seeking thus corresponds to a trajectory of this system going from the critical point with $a = -\alpha$ to the one with $a = +\alpha$; a trajectory joining these critical points in the other direction would correspond to a shock transition between the two plane waves. We assume λ' and ω' are negative, the first corresponding to a stable limit cycle and the second, as it turns out, to the phase and group velocities of plane waves being similarly directed. Investigation of the linearization at these critical points shows that the one at $-\alpha$ has a two-dimensional stable manifold and a one dimensional unstable manifold, while that at $+\alpha$ has a two-dimensional unstable and one-dimensional stable manifold. Thus one would expect the existence of a connecting trajectory of the shock type (and this is true at least for small enough α) but not in general one of the center type.

Now if $\varepsilon = 0$, equations (5) are the radial and azimuthal equations of motion of a particle in a central force field given by $-r\lambda(r)$. As such they have the angular momentum integral

$$j = -r^2\theta_{1x} = \text{const.} \tag{6}$$

and the energy integral

$$e = \frac{1}{2}(r_x^2 + r^2\theta_{1x}^2) + \int_1^r r'\lambda(r')dr' = \text{const.} \tag{7}$$

These suggest that in studying the equations for small ε it may be helpful to use these quantities themselves as dependent variables. It is also convenient to use $R = r^2$ instead of r itself, and since for the solutions of interest it turns out that $j = 0(\varepsilon)$ and $e = 0(\varepsilon^2)$ we introduce

$$J = j/\varepsilon \tag{8}$$

and

$$E = e/\varepsilon^2 = \frac{1}{4\varepsilon^2}\left[\left(\frac{1}{2}R_x^2 + 2\varepsilon^2J^2\right)/R + 2F(R)\right] \tag{9}$$

where $F(R) = 2\int_1^r r'\lambda(r')dr'\lambda$. ($F$ has a maximum of zero at $r = 1$.) Then $(5)^1$ becomes

$$R_{xx} + 2(RF(R))' = 4\varepsilon^2E \tag{10}$$

$$J_x = R[\Omega(R) - \Omega(R_*)] \tag{11}$$

(Here the prime means differentiation with respect to R.)

For some purposes it is convenient to differentiate (9) and use (10) and (11) to obtain

$$E_x = J[\Omega(R) - \Omega(R_*)] \tag{12}$$

(9) is then the equation of an invariant manifold of (10), (11) and (12), and we are interested only in solutions on this manifold. Note however that the singularity in (9) at $R = 0$ is not visible in (10), (11) and (12), which is why (12) is convenient.

In this form the two critical points are at $R = R_*$, $R_x = 0$, $J = \pm J_*$ (and $E = E_*$) where

$$\varepsilon^2 J_*^2 = R_*^2 \, R'(R_*) \tag{13}$$

and $$\varepsilon^2 E_* = \frac{1}{2}\,(RF(R))'_* \tag{14}$$

Since $F(1) = F'(1) = 0$ these indicate that if $R_* = (1-\varepsilon^2\rho_*)$, then J_* and E_* tend to finite non-zero limits as $\varepsilon \to 0$. We replace R_* by ρ_*, seeking for small ε a value of ρ_* such that (10) and (11) have a trajectory of the center type joining the two critical points. In this form the limiting problem for $\varepsilon \to 0$ (the "singular center") has a solution which is easy to find. At $\varepsilon = 0$ (10) becomes uncoupled from (11) and has the first integral

$$\frac{1}{2}R_x^2 + 2RF(R) = I \tag{15}$$

For the relevant kind of $F(R)$ - of which a representative example is $F = -\frac{1}{2}(1-R)^2$ (corresponding to $\lambda = 1-r^2$), the integral curves I = constant resemble those shown in

Fig. 1:

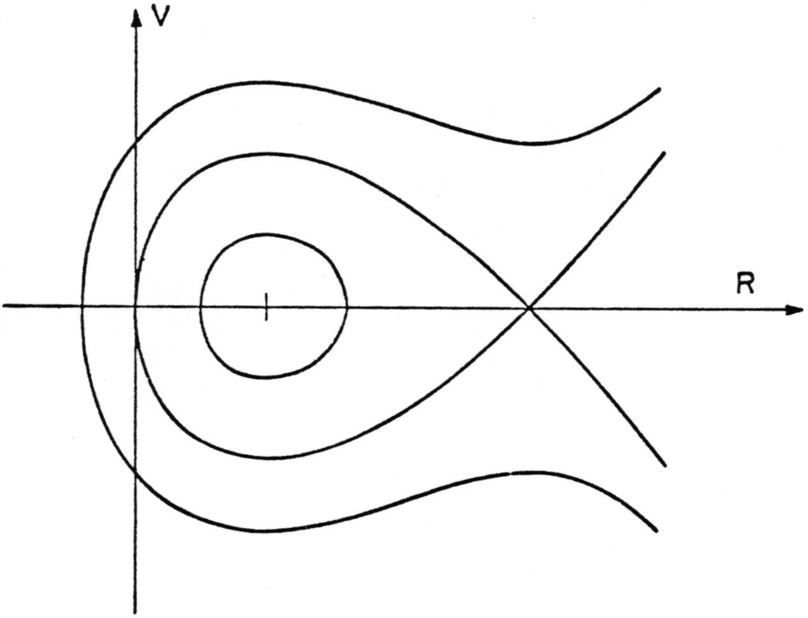

Fig. 1.

in particular $I = 0$ is a homoclinic trajectory joining the saddle point in the (R, R_x) phase plane to itself. If the corresponding solution is inserted into the $\varepsilon = 0$ form of (11), a corresponding $J(x)$ tending to $-J_*$ as $x \to -\infty$ can be obtained just by integration; the form of the corresponding trajectory in (R, R_x, J) space is shown in Fig. 2.

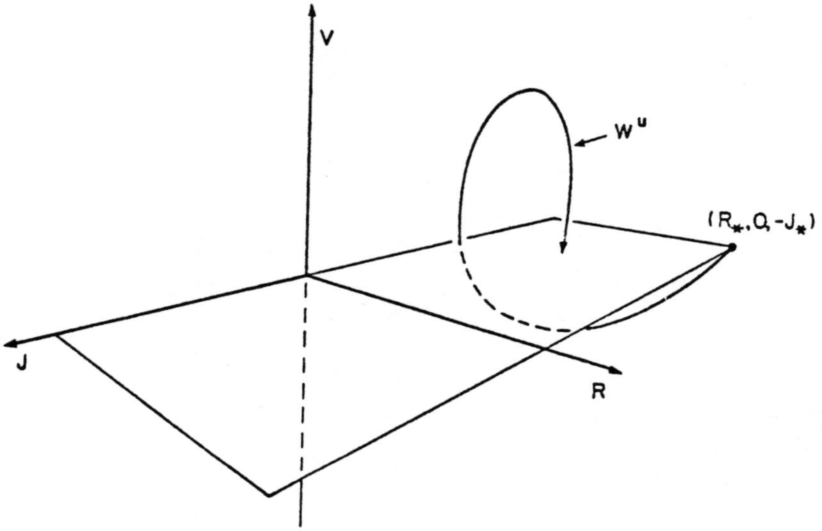

Fig. 2.

When just the lower half of the homoclinic trajectory in (R, R_x) has been traversed, the change in J is given by

$$\Delta J = \int J_x dx = \int R[\Omega(R) - \Omega(1)]\, dx$$

$$= \int_0^1 \frac{R[\Omega(R) - \Omega(1)]dR}{2\sqrt{-RF(R)}} = k, \quad \text{say.}$$

Thus halfway around we have $J = -J_* + k$. Now the value of J_* as $\varepsilon \to 0$ is seen from (13) to be given by $\varepsilon^2 J_*^2 = -F''(1)\varepsilon^2 \rho_*$ or $J_* = \sqrt{-F''(1)\rho_*}$. Thus if we choose ρ_* so that $J_* = k$, then by symmetry when one goes all the way around the homoclinic trajectory in (R, R_*), J will move from $-J_*$ to $+J_*$, giving us a "singular center" which is a plausible candidate to be the limit for $\varepsilon \to 0$ of a true center. With some effort, this can be shown to be really the case, i.e. for all small ε there is a value of ρ_*, near the singular value $k^2/(-F''(1))$, such that a trajectory joining the two critical points (in the "center structure" order) exists.

5. It is also possible to choose ρ_* so that when R goes halfway around the homoclinic trajectory in the (R,R_x) plane J changes by only $\frac{1}{2}J_*$; thus continuing on to a complete loop just brings J from $-J_*$ to zero. But then a second circuit of the homoclinic loop in the (R,R_x) plane will bring J up to $+J_*$, giving us another candidate for the limit of a center type connecting trajectory - the "singular second mode". It can also be shown that for small enough ε there really is such a second mode, with a value of ρ_* close to the singular value and a nearby trajectory. Fig. 3 shows a sketch in the R, J plane of a connecting trajectory of the "second mode" kind.

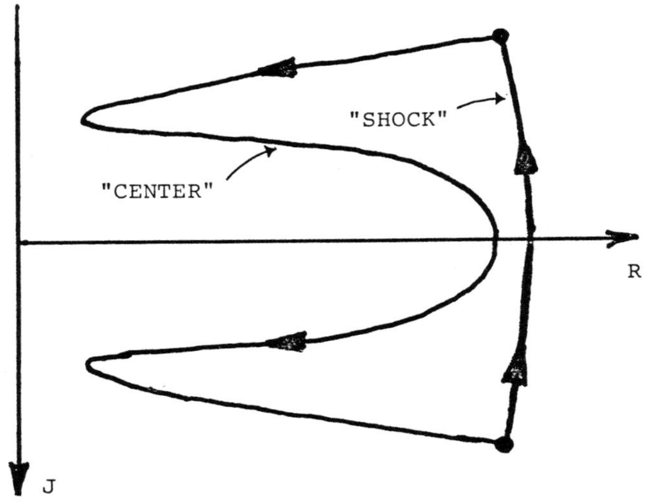

Fig. 3.

Similar higher modes also exist, if ε is small enough, but to show the existence of more and more modes it may be necessary to make ε smaller and smaller. Fig. 3 also shows the shock-type connecting trajectory projected on the (R,J) plane; the latter exists for more or less any J_* - in particular for the one corresponding to the second mode center-type connection. The two together form a kind of singular or limiting "periodic" orbit. This can itself be perturbed, and it can be shown that for somewhat smaller values of R_* (for which the trajectory leaving the critical point at $-J_*$ does not reach the one at $+J_*$) there exist periodic orbits of finite period in x which are rather close to the limiting one indicated in Fig. 3. Such orbits can be interpreted as corresponding to a center structure located at the center of a finite domain bounded by impermeable walls. Similar periodic orbits exist near the first mode center solution too, as well as for higher modes.

In order to show this, we essentially made use of the small ε singular perturbation to compute a Poincaré map from a portion of the $J = 0$ plane near $R = 1$, $R_x = 0$ back to this plane. For values of R_* near one corresponding to an infinite domain center structure solution, a trajectory starting in an appropriate part of the $J = 0$ plane travels from that plane toward the critical point at $-J_*$, near half of the shock-structure trajectory; passing near the critical point, it then follows roughly the center structure orbit to the neighborhood of the other critical point at $J = J_*$, and then returns to the plane $J = 0$ near the other half of the shock trajectory. (Not all orbits starting in $J = 0$ do this by any means - some may fly off and never return to $J = 0$ - but some do. By studying this Poincaré map and using the symmetry one can show the existence of a fixed point of the map, on the R axis, corresponding to the periodic orbit which gives the finite-domain center structure, for whichever mode is being considered.

Because these orbits pass close to the hyperbolic critical points, it turns out that this Poincaré map has a very large expansion in one direction, and contraction in another (transverse) one. Furthermore, it can be shown that for small enough ε there are certain strips in the $J = 0$ plane, one for each of the modes referred to above, whose images under the Poincaré map are other strips transversely cutting across the original strips. Fig. 4 shows a numerically computed example of this for the first two modes.

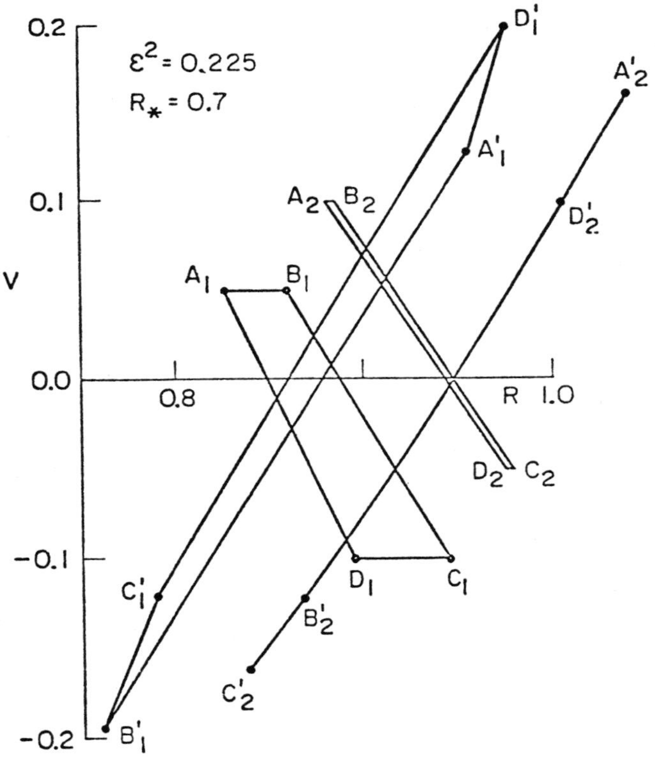

Fig. 4.

This shows that the Poincaré map contains a "horseshoe map" - cf. [4]. This implies the existence in the plane $J = 0$ of a Cantor set C of points invariant (as a whole) under the map, and on which the Poincaré map is equivalent to a "Bernoulli shift". For example, if we are considering a case of the first two modes such as that illustrated in Fig. 4, the points of C may be labelled by doubly infinite sequences of two symbols, with one distinguished one (e.g....121$\underline{2}$2112...). The action of the Poincaré map on such a point is to shift the mark one place to the left, say; the significance of the symbols is that the orbit joining the initial to the final point resembles a first or second mode depending on whether the marked symbol of the initial point is a "1" or a "2". Evidently a periodic sequence corresponds to a periodic orbit and so to a periodic pattern of waves, and such exist (lots of them in fact) for arbitrarily long periods. Furthermore, there is an uncountable number of aperiodic sequences, each of which corresponds to initial conditions for a bounded aperiodic orbit in (R,R_x,J) space and so to a spatially aperiodic pattern of chemical waves on $-\infty < x < \infty$. Each such pattern however goes with a temporally periodic solution of the reaction-diffusion equations. These solutions are evidently among the candidates for the long-time asymptotic behaviour of solutions, so there is certainly a considerable variety of such. As mentioned above, it is possible to show the existence of horseshoes of any finite order - i.e. shifts on any number of symbols - by making ε small enough.

References

[1] Kopell, N. and L. N. Howard, "Plane Wave Solutions to Reaction-Diffusion Equations". Studies in Applied Math. 52, 291-328 (1973).

[2] Howard, L. N. and N. Kopell, "Slowly Varying Waves and Shock Structures in Reaction-Diffusion Equations" Studies in Applied Math. 56, 95-145 (1977).

[3] Kopell, N. and L. N. Howard, "Target patterns and horseshoes from a perturbed central force problem: some temporally periodic solutions to reaction-diffusion equations." To appear in Studies in Applied Math. (1979 or 1980)

[4] Moser, J. "Stable and Random Motions in Dynamical Systems" Annals of Math. Studies, 77, Princeton Univ. Press (1973).

Massachusetts Institute of Technology
Department of Mathematics
Cambridge, Massachusetts 02139

Liesegang Rings and a Theory of Fast Reaction and Slow Diffusion

Joseph B. Keller

1. INTRODUCTION.

A chemically reacting system usually proceeds monotonically in time toward a state of completed reaction, a state of equilibrium, or a steady state. The ultimate state is generally homogeneous in space, as a consequence of diffusion, even though the initial state may be inhomogeneous. Therefore it is surprising that certain chemical systems do not behave in this way. Instead some of them exhibit temporal oscillations which do not decay in time, but which become temporally periodic. Other systems, or the same systems with different parameter values, perform oscillations which persist but remain chaotic rather than becoming periodic. Still others reach steady states which are not homogeneous in space, but which are spatially periodic or quasi-periodic. There are also systems which continue to oscillate in time and to be quasi-periodic in space.

Examples of such unusual behavior of chemically reacting systems have been known for a long time. One of them was discovered by R. E. Liesegang [1] in 1896 and studied by him for many years thereafter. The spatial pattern which occurs in this system consists of a family of concentric circular rings which are called Liesegang rings in his honor. It is this system and similar ones which we shall examine. In order to analyze them, first we shall have to describe in detail the physical and chemical processes which occur. This we shall do following the ideas of W. Ostwald [2] who, in 1897, outlined

ISBN 0-12-669550-4

an explanation involving precipitation and supersaturation. We shall extend that outline to a complete description of the various processes. Then we shall convert this verbal description into a mathematical theory, and formulate an appropriate mathematical problem corresponding to it. All of these considerations, and the subsequent analysis of the problem, are the joint work of S. I. Rubinow and the author, and are presented in greater detail in their paper [3].

In order to treat the mathematical problem we shall make certain appropriate simplifications. The most important of these is a consequence of the fact that the reaction process is fast compared to the diffusion process. This has led us to formulate and develop a general theory of reaction and diffusion processes involving fast reaction and slow diffusion. We shall describe this theory, which should have many other applications, and use it to simplify the present problem. It was worked out by P. S. Hagan and the author, and is contained in their paper [4] together with the related theory of some fast and some slow reactions.

The subsequent analysis of the simplified problem is carried out in [3] employing approximations like those in the related work of C. Wagner [5], S. Prager [6] and Zeldovich, Barenblatt and Salganik [7]. Some predictions of the theory are compared in [3] with certain relevant experimental results. More detailed comparison between theory and experiment will be possible when the equations of the theory are solved numerically. J.-M. Vanden-Broeck and the author [8] are solving them by using the method of finite differences to replace the partial differential equations by a finite set of algebraic equations. These equations, together with the initial, boundary, and jump conditions, are being solved by an appropriate iterative method.

Before turning to the study of Liesegang rings, we shall comment on the periodic or non-monotonic behavior of chemical systems in general. Let us first consider why chemical reactions are expected to proceed monotonically in time. The reason is that the rate of reaction is positive as long as the concentrations of the reactants are positive and the concentration of the reaction products is not too large. Furthermore there is no mechanism, such as inertia, to cause the reaction to overshoot the equilibrium state or the state of completed reaction. Therefore it is to be expected that a reaction will proceed in one direction at a decreasing rate. However when several reactions occur, there is no longer any justification for this belief. The presence of other processes, such as flow and diffusion, can also lead to destabilization of uniform steady states and the occurrence of temporal or spatial periodicity.

The interest in periodic and non-uniform states of reacting systems has been generated by scientific curiosity, by engineering necessity, and by potential applications in biology. The engineering reason is the need to control and eliminate oscillations which arise in stirred tank and other continuous flow reactors. The biological reasons include the possible explanation of biological clocks and a proposed explanation of morphogenesis. The latter of these, made by A. M. Turing in 1952, is that chemical reaction and diffusion are responsible for the creation of form in biological systems through the development of spatially nonuniform steady states. All of these reasons have led to extensive experimental and theoretical investigations of the non-monotonic and non-uniform behavior of chemical systems.

2. THE MECHANISM OF LIESEGANG RING FORMATION.

Liesegang, while experimenting with photographic materials, prepared a gel containing potassium dichromate. This gel was in the form of a thin layer on a glass plate. He placed a drop of silver nitrate solution on the gel and found, after some time, that concentric circular rings of silver chromate formed in the gel. The process whereby the silver chromate formed into visible solid particles is precipitation, so this phenomenon has sometimes been called "periodic" precipitation. However it is not periodic in either space or time, so that is a misnomer. We prefer to call it recurrent precipitation, since the rings form one after another with increasing spatial and temporal intervals between successive rings.

To explain this phenomenon we suppose that silver ions from the drop go into solution in the gel and diffuse radially outward. Within the gel they encounter chromate ions and react with them to form silver dichromate. As the chromate ions near the drop are consumed by the reaction, more chromate ions diffuse radially inward. Thus the two reactants diffuse toward one another and, upon coming together, continually produce silver dichromate. In this way an expanding circular region of silver dichromate is produced. This is the monotonic behavior to be expected in chemically reacting systems. The question then arises of why only separated rings of silver dichromate are visible at the end of the experiment?

To answer this question we assume that the silver dichromate is not visible until it precipitates out of solution and forms solid particles. Now it is well known that a dissolved substance cannot precipitate until its concentration c exceeds the saturation concentration c^s. Then the rate of precipitation is proportional to $c - c^s$, provided that some of the solid is present upon which it can precipitate. However if none of the solid is present the concentration c may have to reach a higher value $c^* > c^s$ before precipitation can start. Therefore, following Ostwald, we assume that precipitation starts only when $c \geq c^*$. After it has started, we assume that it continues as long as $c \geq c^s$.

In view of these assumptions, precipitation will begin in the gel at the location of the drop of silver nitrate solution, provided that the reactants are present in sufficiently high concentrations to produce silver dichromate with a concentration $c \geq c^*$. Precipitation will continue within a circle of increasing radius surrounding the drop until c falls below c^* at the outer boundary of this circle. Then the precipitation zone will stop growing.

The reason why c may fall below c^* is that the chromate ions in the gel are depleted by reaction and by diffusion toward the silver nitrate drop, while the concentration of silver ions is low far from the drop. As time goes on this latter concentration increases everywhere. A time may be reached when this silver ion concentration and that of the chromate ions is high enough somewhere in the gel to produce silver dichromate with a concentration $c \geq c^*$. Then a second zone of precipitation will start to form at that place. After a while this process will be repeated, forming a third zone, and so on. The resulting zones are the Liesegang rings.

Since Liesegang's first experiment, the same phenomenon has been found to occur in many other reactions. Furthermore, it has also been shown to occur in test tubes and capillary tubes, with the formation of planar bands at various positions along the tube instead of rings. It is this one dimensional case which we shall treat.

3. A MATHEMATICAL FORMULATION OF THE THEORY.

To formulate the preceding theory mathematically, we denote by a, b, c and d the molar concentrations of the four substances A, B, C and D. Here A is analogous to the silver ions, B to the chromate ions, C to the dissolved silver dichromate and D to the solid or precipitated silver dichromate. We assume that ν_A moles of A can combine with ν_B moles of B to form ν_C moles of C, and that this reaction is reversible. We also assume that C can precipitate to form D, but that D cannot dissolve to form C. In addition we suppose that A, B and C can diffuse with diffusion coefficients D_A, D_B and D_C, but that D cannot diffuse. Then the four concentrations, which depend upon the time t and the coordinate x along the tube, satisfy the following reaction-diffusion-precipitation equations within the tube $x > 0$:

$$a_t = D_A a_{xx} - \nu_A r \, , \tag{3.1}$$

$$b_t = D_B b_{xx} - \nu_B r \, , \tag{3.2}$$

$$c_t = D_C c_{xx} + \nu_C r - p(c,d) \, , \tag{3.3}$$

$$d_t = p(c,d) \, . \tag{3.4}$$

Here r is the reaction rate and p is the precipitation rate.

We shall assume that r is given by the law of mass action with rate constant k_+ for the forward reaction and k_- for the backward reaction. Thus

$$r(a,b,c) = k_+ a^{\nu_A} b^{\nu_B} - k_- c^{\nu_C} \; . \tag{3.5}$$

The precipitation rate was described in Section 2, and that description leads to the following expression for p:

$$\begin{aligned} p(c,d) &= 0 \; , && \text{if } c < c^* \text{ and } d = 0 \; , \\ &= q(c-c^s)_+ \; , && \text{if } c \geq c^* \text{ or } d > 0 \; . \end{aligned} \tag{3.6}$$

Here q is a rate constant and the subscript + on $c - c^s$ denotes the positive part of $c - c^s$.

Since we have assumed that the reactions occur in the tube occupying the region $x > 0$, we must specify the initial conditions within the tube and boundary conditions at the end $x = 0$. To correspond with Liesegang's experiment, we assume that only B is present initially. Thus we require that

$$a = c = d = 0 \; , \quad b = b_0 \text{ at } t = 0 \; , \quad x > 0 \; . \tag{3.7}$$

At the endpoint $x = 0$ we assume that A is kept at the concentration a_0 while B and C cannot leave the tube there. Thus

$$a = a_0 \; , \quad b_x = c_x = 0 \text{ at } x = 0 \; , \quad t > 0 \; . \tag{3.8}$$

In addition to these conditions, we require that a, b, c and their x derivatives be continuous throughout the tube for $t > 0$:

$$a, b, c, a_x, b_x, c_x \text{ continuous for } x > 0, \; t > 0 \; . \tag{3.9}$$

The preceding equations constitute our mathematical formulation of the theory of Liesegang band formation. The bands are the regions within which $d(x,t)$ is positive after a sufficiently long time. Thus the bands are defined by

$$d(x,\infty) > 0 \; . \tag{3.10}$$

This formulation is complete in the sense that the equations seem to have a unique solution. This solution determines exactly where the bands will be, the thickness of each band, the time when each band starts to form, the concentrations between the bands, etc. The solution will also show whether or not bands do form, and it will yield the sets of parameter values for which they form. The solution can be calculated numerically, but we have not calculated it. Instead we have first simplified the equations and then analyzed the resulting problem.

4. REDUCTION TO A SIMPLER PROBLEM.

In order to simplify the theory presented in Section 3, we shall proceed heuristically, making assumptions based upon intuitive chemical considerations. Later on we shall indicate how the same simplified theory results from a systematic application of the theory of fast reactions and slow diffusion. Our first assumption is that the reaction term $\nu_A r$ in (3.1) is negligible compared to the other terms in that equation. This assumption is valid when the concentration of B is sufficiently small. It yields

$$a_t = D_A a_{xx} \; . \tag{4.1}$$

Next we add ν_B times (3.3) to ν_C times (3.1) to eliminate r and obtain

$$\nu_C b_t + \nu_B c_t = \nu_C D_B b_{xx} + \nu_B D_C c_{xx} - \nu_B p(c,d) \; . \tag{4.2}$$

Then we assume that the reaction proceeds so rapidly that it is in equilibrium, which yields

$$r(a,b,c) = 0 \; . \tag{4.3}$$

We shall use (4.2) and (4.3) instead of (3.2) and (3.3). But then we must replace the two boundary conditions $b_x = c_x = 0$ by the single condition that B, both free and contained in C, cannot escape through the boundary:

$$\nu_C D_B b_x + \nu_B D_C c_x = 0 \quad \text{at } x = 0 \; , \quad t > 0 \; . \tag{4.4}$$

With these modifications the problem can be dealt with by first solving (4.1) for $a(x,t)$ with $a(x,0) = 0$ and $a(0,t)=a_o$. The solution is

$$a(x,t) = a_o \operatorname{erfc}[x/(4D_A t)^{1/2}] \; . \tag{4.5}$$

Then (4.3) can be solved for b in terms of a and c and the result can be used to eliminate b from (4.2) and (4.4). In this way (4.2) becomes an equation for c in which d seems to occur.

However both in the region where $p = 0$ and in the region where $p > 0$, d does not occur in the equation for c. Thus the entire problem can be reduced to one of determining c in each of these regions. The boundary between the two regions consists of the curve on which $c = c^*$ together with the upper halves of the tangents to this curve where it becomes vertical. Once c is found, d can be calculated from (3.4).

The reduction just described is carried out in detail in the work of Keller and Rubinow [3], where the resulting equations are analyzed and solved approximately. Therefore we shall not describe that analysis and its results. Instead we shall present the theory of fast reaction and slow diffusion, and show how it can be used to derive the reduced problem in a systematic way.

5. FAST REACTION AND SLOW DIFFUSION.

In many chemical systems, such as those considered in the preceding sections, both chemical reactions and diffusion occur, but at quite different rates. Often the reactions proceed much more rapidly than the diffusion. In such cases it is possible to simplify the analysis of the process by taking advantage of this difference in rates. One way of doing so is to assume that the reactions have reached equilibrium or completion at each point of space. Then the state at each point is constrained to satisfy the condition that the reaction rates vanish there. When this constraint is adjoined to the reaction-diffusion equations governing the system, an overdetermined set of equations results. Chemical and physical reasoning can then be used to delete some of the equations and thereby make the system determined.

Instead of using this intuitive procedure, it is possible to proceed in a systematic mathematical manner which we shall now describe. The basic idea is to introduce into the equations a small parameter ε which is the ratio of the time scale for reaction to the time scale for diffusion. Then the asymptotic expansion of the solution is sought for ε near zero. This expansion can be constructed either by the "two time" method or by the method of matched asymptotic expansions.

Both of these methods are utilized in [4]. Here we shall indicate how to obtain the desired results via the latter method.

Let us begin by considering the following system of reaction-diffusion equations:

$$u_t(x,t,\varepsilon) = f(u,x,t,\varepsilon t,\varepsilon) + \varepsilon D\Delta u \,, \tag{5.1}$$

$$u(x,0,\varepsilon) = g(x) \,. \tag{5.2}$$

Here $u(x,t,\varepsilon)$ is a vector of concentrations, temperature and possibly other dependent variables, $f(u,x,t,\varepsilon t,\varepsilon)$ is a vector of reaction rates, heat production rate, etc. and D is the diffusion coefficient matrix. The essential feature of (5.1) is that the diffusion term contains the factor ε, which is small. We have also permitted f to depend upon all the variables, and even to have a slow time dependence via the argument εt. In the initial condition (5.2) the function $g(x)$ is given.

To obtain the asymptotic expansion of u for ε small, we assume first that it is of the form

$$u(x,t,\varepsilon) \sim u^0(x,t) + \varepsilon u^{(1)}(x,t) + \cdots \,. \tag{5.3}$$

We shall call this the initial layer expansion of u. We now substitute (5.3) into (5.1) and (5.2) and set $\varepsilon = 0$ to obtain

$$u_t^0(x,t) = f(u^0,x,t,0,0) \,, \tag{5.4}$$

$$u^0(x,0) = g(x) \,. \tag{5.5}$$

These equations describe the evolution in time of u^0 at the point x. Further terms in (5.3) can be found by considering the coefficients of higher powers of ε in (5.1) and (5.2), but we shall not examine them. Instead we shall consider the behavior of $u^0(x,t)$ as $t \to \infty$.

To this end, we assume that $f(u^0,x,t,0,0)$ has a limit as $t \to \infty$, and then we may expect $u^0(x,t)$ to have a limit also. Let us denote it by $u^0(x,\infty)$. Thus we assume that

$$\lim_{t\to\infty} u^0(x,t) = u^0(x,\infty) \,. \tag{5.6}$$

Then $u^0(x,\infty)$ must satisfy the following equilibrium condition, which results from (5.4):

$$f[u^0(x,\infty),x,\infty,0,0] = 0 \ . \tag{5.7}$$

In general $u^0(x,\infty)$ is not uniquely determined by (5.7), but it is determined by the initial condition (5.5) and the reaction equation (5.4). It is this limit $u^0(x,\infty)$ of $u^0(x,t)$ which will be needed to determine the outer expansion of u, which we consider next.

The initial layer expansion (5.3) is presumably valid for fixed values of t as $\varepsilon \to 0$. In order to find the behavior of u for long times of order ε^{-1}, we must construct another expansion, called the outer expansion of u. In order to construct it we first introduce the new time variable $\tau = \varepsilon t$ and consider u to be a function of τ:

$$u(x,t,\varepsilon) = v(x,\tau,\varepsilon) \ , \qquad \tau = \varepsilon t \ . \tag{5.8}$$

Then (5.1) becomes the following equation for v:

$$\varepsilon v_\tau(x,\tau,\varepsilon) = f(v,x,\varepsilon^{-1}\tau,\tau,\varepsilon) + \varepsilon D\Delta v \ . \tag{5.9}$$

Next we assume that v has the expansion

$$v(x,\tau,\varepsilon) = v^0(x,\tau) + \varepsilon v^{(1)}(x,\tau) + \cdots \ . \tag{5.10}$$

This is the outer expansion of u.

Both the inner expansion (5.5) and the outer expansion (5.10) represent u. If there is a common region of validity of the two expansions, as we assume, then in it they must be asymptotic to one another. Thus we have in this region

$$u^0(x,t) + \varepsilon u^{(1)}(x,t) + \cdots \sim v^0(x,\varepsilon t) + \varepsilon v^{(1)}(x,\varepsilon t) + \cdots \ . \tag{5.11}$$

Suppose that (5.11) holds for $t = \varepsilon^{-1/2}$. Then by choosing this value for t and letting $\varepsilon \to 0$ in (5.11) we obtain

$$u^0(x,\infty) = v^0(x,0) \ . \tag{5.12}$$

This procedure for deriving (5.12) is called "matching" the initial layer expansion and the outer expansion. The result (5.12) yields the initial value of v^0 in terms of the "final" value of u^0.

We next substitute (5.1) into (5.9) and set $\varepsilon = 0$ to get

$$f[v^0(x,\tau),x,\infty,\tau,0] = 0 \ . \tag{5.13}$$

This equation shows that v^0 must lie on the equilibrium surface (or manifold) defined by $f = 0$, but it does not determine where v^0 is on this surface unless the surface is a single point. From (5.12) we know that at $\tau = 0$, v^0 starts at the point $u^0(x,\infty)$. To find how it evolves as τ increases, we consider the terms of order ε in (5.9), which yield the following linear equation for $v^{(1)}$:

$$f_v[v^0(x,\tau),x,\infty,\tau,0]v^{(1)} = v^0 - D\Delta v^0 - f_\varepsilon \ . \tag{5.14}$$

The gradient matrix f_v in (5.14), which is the coefficient of $v^{(1)}$, will generally be a singular matrix. Let us suppose that it has k left null-vectors $\ell_1,\ldots,\ell_k$. Then (5.14) is solvable for $v^{(1)}$ only if the right side satisfies k solvability conditions. They can be obtained by scalar multiplication of (5.14) on the left by each left null-vector. This yields the k conditions

$$\ell_j \cdot \{v_\tau^0(x,\tau) - D\Delta v^0(x,\tau) - f_\varepsilon[v^0(x,\tau),x,\infty,\tau,0]\} = 0 \ , \quad j = 1,\ldots,k \ . \tag{5.15}$$

This is a system of k equations that determine how $v^0(x,\tau)$ diffuses on the equilibrium surface upon which it is constrained to lie by (5.13). The initial value of v^0 is given by (5.12). Thus (5.12), (5.13) and (5.15) determine the slow diffusion of $v^0(x,\tau)$, the leading term in the outer expansion of $u(x,\varepsilon^{-1}\tau,\varepsilon)$. The fast reaction is governed by (5.4) and (5.5), which determine $u^0(x,t)$, the leading term in the initial layer expansion of $u(x,t,\varepsilon)$.

The system of equations (5.13) and (5.15), together with the initial condition (5.12), are the results of our analysis. We shall not examine them further here, nor shall we obtain more terms in the outer expansion. These and other matters are considered in [4].

6. APPLICATION TO THE THEORY OF LIESEGANG BANDS.

We shall now apply the theory of Section 5 to the equations of Section 3, which constitute our formulation of the theory of Liesegang bands. To do so we first rewrite these equations, introducing explicitly the small parameter ε to indicate the relative size of each term. In (3.1) for example, the diffusion term is small of order ε, say, but the reaction term is supposed to be smaller. Therefore we shall assume it to be of order ε^2. Thus we write (3.1) in the form

$$a_t = \varepsilon D_A a_{xx} - \varepsilon^2 \nu_A r \; . \tag{6.1}$$

Similarly we write (3.2) and (3.3) in the form

$$b_t = \varepsilon D_B b_{xx} - \nu_B r \; , \tag{6.2}$$

$$c_t = \varepsilon D_C c_{xx} + \nu_C r - \varepsilon p \; . \tag{6.3}$$

The system (6.1)-(6.3) is of the form (5.1) with the following identifications:

$$u = \begin{pmatrix} a \\ b \\ c \end{pmatrix} , \qquad f = \begin{pmatrix} -\varepsilon^2 \nu_A r \\ -\nu_B r \\ \nu_C r - \varepsilon p \end{pmatrix} ,$$

$$D = \begin{pmatrix} D_A & 0 & 0 \\ 0 & D_B & 0 \\ 0 & 0 & D_C \end{pmatrix} . \tag{6.4}$$

Thus the leading term of the initial layer expansion, $u^0 = (a^0, b^0, c^0)$, satisfies (5.4) which becomes

$$a_t^0 = 0 \; , \tag{6.5}$$

$$b_t^0 = -\nu_B r(a^0, b^0, c^0) \; , \tag{6.6}$$

$$c_t^0 = \nu_C r(a^0, b^0, c^0) \; . \tag{6.7}$$

From (3.7) the corresponding initial conditions are

$$a^0(x,0) = c^0(x,0) = 0 \; , \quad b^0(x,0) = b_0 \; , \quad x > 0 \; . \tag{6.8}$$

With these initial conditions and r given by (3.5), the solution of (6.5)-(6.7) is

$$a^0(x,t) = c^0(x,t) = 0 , \quad b^0(x,t) = b_0 , \quad x > 0 . \tag{6.9}$$

Thus nothing changes in the initial layer, to this order.

To treat the outer expansion we shall write

$$u(x,t,\varepsilon) = v(x,\tau,\varepsilon) = \begin{pmatrix} \alpha(x,\tau,\varepsilon) \\ \beta(x,\tau,\varepsilon) \\ \gamma(x,\tau,\varepsilon) \end{pmatrix} . \tag{6.10}$$

Then by using (6.4) for f in (5.13) we get the single equation

$$r[\alpha^0(x,\tau),\beta^0(x,\tau),\gamma^0(x,\tau)] = 0 . \tag{6.11}$$

Thus the leading term in the outer expansion satisfies (4.3), as we assumed in Section 4.

Next from (6.4) we compute f_v at $\varepsilon = 0$, and obtain

$$f_v = (f_a,f_b,f_c) = \begin{pmatrix} 0 & 0 & 0 \\ -\nu_B r_a & -\nu_B r_b & -\nu_B r_c \\ \nu_C r_a & \nu_C r_b & \nu_C r_c \end{pmatrix} . \tag{6.12}$$

This matrix has two linearly independent left null-vectors

$$\ell_1 = (1,0,0) , \qquad \ell_2 = (0,\nu_C,\nu_B) . \tag{6.13}$$

We also compute f_ε at $\varepsilon = 0$, which is given by

$$f_\varepsilon = \begin{pmatrix} 0 \\ 0 \\ -p \end{pmatrix} . \tag{6.14}$$

We can now use each of the vectors ℓ_1 and ℓ_2 given by (6.13) in (5.15), together with f_ε given by (6.14) and D given by (6.4). By using ℓ_1 in (5.15), we obtain

$$\alpha^0_\tau = D_A \alpha^0_{xx} . \tag{6.15}$$

This is just (4.1) of Section 4, with a replaced by α^0 and t replaced by τ. Then by using ℓ_2 in (5.15) we get

$$\nu_C \beta^0_\tau + \nu_B \gamma^0_\tau = \nu_C D_B \beta^0_{xx} + \nu_B D_C \gamma^0_{xx} - p(\gamma^0,d) . \tag{6.16}$$

This is exactly (4.2) with b, c and t replaced by β^0, γ^0 and τ respectively.

We have now shown how the simplified equations of Section 4 follow from the equations for the outer asymptotic expansion derived in Section 5. The corresponding initial conditions are obtained by using (6.9) in (5.12), and they are just those used in Section 4. An extension of the theory of Section 5 is needed to derive the boundary conditions (4.4) and $\alpha^0(0,\tau) = a_o$, but we shall not present it here.

This completes the application of the theory of fast reaction and slow diffusion to the equations of Section 3, when they are written in the form (6.1)-(6.3). Of course that way of writing the equations depends upon the parameters in the problem. They must be such that the terms in these equations have the indicated relative magnitudes.

REFERENCES

1. R. E. Liesegang, Phot. Archiv. 21, 221 (1896); Naturw. Wochschr. 11, 353 (1896).
2. W. Ostwald, Lehrbuch der allgemeinen Chemie, p. 778, Engelman, Leipzig (1897).
3. J. B. Keller and S. I. Rubinow, Recurrent precipitation and Liesegang rings, to be published.
4. J. B. Keller and P. S. Hagan, Fast reaction and slow diffusion, to be published.
5. C. Wagner, Mathematical analysis of the formation of periodic precipitations, J. Colloid Sci. 5, 85-97 (1950).
6. S. Prager, Periodic precipitation, J. Chem. Phys. 25, 279-283 (1956).
7. Ya. B. Zeldovich, G. I. Barenblatt and R. S. Salganik, Soviet Phys. Doklady 6, 869-871 (1962).
8. J.-M. Vanden-Broeck and J. B. Keller, Numerical calculation of Liesegang rings, to be published.

Supported by the Army Research Office, the Office of Naval Research, the Air Force Office of Scientific Research and the National Science Foundation.

Departments of Mathematics and Mechanical Engineering, Stanford University, Stanford, CA 94305.

Dynamics of Aerosols

John H. Seinfeld

1. INTRODUCTION

This paper is a survey of aspects important in simulating the dynamics of aerosols. Aerosols, or atmospheric particles, may vary in size from collections of a few molecules (diameter the order of 10Å or 0.001 μm) to cloud droplets (diameter the order of 10 μm). The evolution of these particles occurs as a result of particle-particle collisions (referred to as coagulation, agglomeration, or coalescence) and individual particle growth due to accretion of vapor molecules. The process of individual particle growth may involve chemical reactions occurring in the particles.

In Section 2 we present the basic conservation equations for the aerosol number concentration as a function of particle size. Then in Section 3 we discuss the kinetic coefficients that appear in the basic equations. Section 4 is devoted to non-dimensionalization of the equations and introduction of dimensionless groups that characterize the aerosol. In Section 5 we present the results of certain interesting cases of aerosol evolution during pure growth.

2. GENERAL DYNAMIC EQUATION FOR AEROSOLS

The dynamic behavior of an aerosol is described by a population balance equation which can be termed the General Dynamic Equation (GDE). In the most general form of this equation the independent variables are particle size and

ISBN 0-12-669550-4

composition [1], although in virtually all applications size is the only variable characterizing the aerosol.

In the most fundamental approach to deriving the GDE, particles are represented as consisting of integer multiples of a single structural unit, typically a molecule. In these discrete equations particles differ only in the number of monomers they contain. The GDE consists then of an infinite set of nonlinear ordinary differential equations for the number densities of all particles. The discrete GDE, while rigorously valid, is impractical for simulation of aerosol behavior because of the typical wide range in particle size. A popular alternative to the discrete GDE is the continuous GDE, in which the particle size spectrum is taken to be continuous rather than discrete. Whereas the continuous GDE is more tractable than the discrete version, it suffers from the disadvantage of inaccurately representing processes occurring among the very smallest particles.

A new form of the aerosol GDE which can be termed the discrete-continuous GDE has been derived by Gelbard and Seinfeld [2]. In this form the discrete representation is used up to a certain multiplet number past which the particle size distribution is represented as continuous. The discrete-continuous GDE has the capability of representing the entire aerosol size distribution from single molecules to the largest particles, including all relevant physical phenomena occurring.

2.1 Discrete-Continuous General Dynamic Equation

We consider the following phenomena to be occurring: (1) agglomeration of two particles, (2) evaporation or escape of a monomer from a particle, and (3) homogeneous particle generation or removal processes apart from those that occur as a result of evaporation and agglomeration. We restrict our attention to size distribution dynamics and do not consider particle composition as an independent variable. Thus, the aerosol may be considered as chemically homogeneous for the purposes of deriving the governing dynamic equation.

For a spatially homogeneous aerosol the quantity of interest is the concentration of particles containing i monomers, where $i \geq 1$. Assuming an i-mer has a volume v_i, the concentration of i-mers, $N(v_i,t)$, will vary with time due to agglomeration, evaporation, generation and removal processes. The rate of agglomeration of i-mers with j-mers is equal to the rate of formation of (i+j)-mers, and is given by

$$\frac{\beta(v_i,v_j)N(v_i,t)N(v_j,t)}{1+\delta_{i,j}} \qquad i,j \geq 1 \tag{2.1}$$

where $\delta_{i,j}$ is the Kronecker delta and $\beta(v_i,v_j)$ is the kinetic coefficient of agglomeration of two particles of volumes v_i and v_j. The functional form of $\beta(v_i,v_j)$ will be discussed later. If i is equal to j, we must divide by 2, so as not to count the agglomeration twice. The rate of evaporation of i-mers is $E(v_i)N(v_i,t)$, $i \geq 2$, where $E(v_i)$ is the evaporation coefficient. The rate of formation of i-mers by agglomeration is the sum of all agglomerations resulting in an i-mer and is given by

$$\frac{1}{2}\sum_{j=1}^{i-1} \beta(v_{i-j},v_j)N(v_{i-j},t)N(v_j,t) \qquad i \geq 2 \tag{2.2}$$

The factor of one-half is introduced because β is a symmetric function of its arguments and therefore the summation counts each agglomeration twice for i-j not equal to i. However, if i is an even integer, the term $\beta(v_{i/2},v_{i/2})N(v_{i/2},t)N(v_{i/2},t)$ is only counted once in the summation, but the factor of one-half is still required as given in (2.1). The rate of depletion of i-mers by agglomeration with all other particles is given by

$$N(v_i,t)\sum_{j=1}^{\infty} \beta(v_i,v_j)N(v_j,t) \qquad i \geq 1 \tag{2.3}$$

For j equal to i, the agglomeration rate is divided by 2 as given in (2.1), but because each agglomeration removes two i-mers, the rate is also multiplied by two, thereby cancelling the factor of one-half. The rate of formation of i-mers by evaporation from (i+1)-mers is $(1+\delta_{1,i})E(v_{i+1})N(v_{i+1},t)$, $i \geq 1$. The rate of depletion of i-mers due to evaporation is given by $E(v_i)N(v_i,t)$, $i \geq 2$.

The net rate of formation of monomers is thus

$$\frac{dN(v_1,t)}{dt} = -\,N(v_1,t)\sum_{j=1}^{\infty} \beta(v_1,v_j)N(v_j,t)$$

$$+ \sum_{j=2}^{\infty} (1+\delta_{2,j}) E(v_j) N(v_j,t) + S_0(v_1,t)$$

$$- S_1[v_1,t,N(v_1,t)] \qquad (2.4)$$

and the net rate of formation of i-mers for $i \geq 2$ is

$$\frac{dN(v_i,t)}{dt} = \frac{1}{2} \sum_{j=1}^{i-1} \beta(v_{i-j},v_j) N(v_{i-j},t) N(v_j,t)$$

$$- N(v_i,t) \sum_{j=1}^{\infty} \beta(v_i,v_j) N(v_j,t)$$

$$+ E(v_{i+1}) N(v_{i+1},t) - E(v_i) N(v_i,t)$$

$$+ S_0(v_i,t) - S_1[v_i,t,N(v_i,t)] \qquad (2.5)$$

where S_0 and S_1 represent all homogeneous generation and removal processes, respectively. Combined with the appropriate initial conditions, (i.e. $N(v_i,0)$, $i \geq 1$), (2.4) and (2.5) constitute the discrete GDE for a spatially homogeneous aerosol. Because agglomeration constantly produces larger particles, (2.4) and (2.5) are an infinite set of coupled ordinary differential equations.

Although the discrete GDE is an accurate description of aerosol dynamics, the number of equations needed to simulate actual aerosols can be immense. For large particles the difference in size between an i-mer and an (i+1)-mer is relatively small. Thus, for particles that contain k+1 or more monomers, where k>>1, the discrete concentrations can be represented by $n(v,t)$, which is a continuous function in the limit as $v_1/v_k \rightarrow 0$, defined by

$$N(v_i,t) = \int_{v_i - v_1/2}^{v_i + v_1/2} n(v,t)\, dv \qquad i \geq k + 1 \qquad (2.6)$$

We now divide the distribution into two regimes. For particle volumes smaller than or equal to v_k the discrete representation is used, and for particle volumes greater than or equal to v_{k+1} a continuous representation is used. Using (2.6), (2.4) becomes

$$\frac{dN(v_1,t)}{dt} = -N(v_1,t)\left[\sum_{j=1}^{k}\beta(v_1,v_j)N(v_j,t) + \int_{v_{k+1}-v_1/2}^{\infty}\beta(v_1,v)n(v,t)\,dv\right]$$

$$+\sum_{j=2}^{k}(1+\delta_{2,j})E(v_j)N(v_j,t) + \int_{v_{k+1}-v_1/2}^{\infty}E(v)n(v,t)\,dv$$

$$+ S_0(v_1,t) - S_1[v_1,t,N(v_1,t)], \tag{2.7}$$

and for $2 \leq i \leq k$, (2.5) becomes

$$\frac{dN(v_i,t)}{dt} = \frac{1}{2}\sum_{j=1}^{i-1}\beta(v_{i-j},v_j)N(v_{i-j},t)N(v_j,t)$$

$$-N(v_i,t)\left[\sum_{j=1}^{k}\beta(v_i,v_j)N(v_j,t) + \int_{v_{k+1}-v_1/2}^{\infty}\beta(v_i,v)n(v,t)\,dv\right]$$

$$+\begin{cases} E(v_{i+1})N(v_{i+1},t) & 2 \leq i \leq k-1 \\ \displaystyle\int_{v_{k+1}-v_1/2}^{v_{k+1}+v_1/2} E(v)n(v,t)\,dv & i = k \end{cases}$$

$$- E(v_i)N(v_i,t) + S_0(v_i,t) - S_1[v_i,t,N(v_i,t)]. \tag{2.8}$$

To derive the governing equation for $n(v,t)$ in the continuous regime, substitute $n(v,t)dv$ for $N(v,t)$ in (2.5) for $v \geq v_{k+1} - v_1/2$. Assuming conservation of volume, such that $v_j = jv_1$, for $v_{k+1} \leq v_i \leq v_{2k}$,

$$\frac{dn(v_i,t)}{dt} = \sum_{j=1}^{i-k-1}\beta(v_i-v_j,v_j)N(v_j,t)n(v_i-v_j,t) \qquad i \geq k+2$$

$$+\frac{1}{2}\sum_{j=i-k}^{k}\frac{\beta(v_i-v_j,v_j)N(v_i-v_j,t)N(v_j,t)}{v_1}$$

$$- n(v_i,t)\left[\sum_{j=1}^{k} \beta(v_i,v_j)N(v_j,t) + \int_{v_{k+1}-v_1/2}^{\infty} \beta(v_i,u)n(u,t)\,du \right]$$

$$+ E(v_i+v_1)n(v_i+v_1,t) - E(v_i)n(v_i,t)$$

$$+ \bar{S}_0(v_i,t) - \bar{S}_1[v_i,t,n(v_i,t)] \qquad (2.9)$$

for $v_{k+1} \leq v_i \leq v_{2k}$, $\bar{S}_0 = S_0/v_1$, and $\bar{S}_1 = S_1/v_1$. For $v > v_{2k} + v_1/2$

$$\frac{\partial n(v,t)}{\partial t} = \sum_{j=1}^{k} \beta(v-v_j,v_j)N(v_j,t)n(v-v_j,t)$$

$$+ \frac{1}{2}\int_{v_{k+1}-v_1/2}^{v-v_{k+1}+v_1/2} \beta(v-u,u)n(v-u,t)n(u,t)\,du$$

$$- n(v,t) \sum_{j=1}^{k} \beta(v,v_j)N(v_j,t)$$

$$- n(v,t)\int_{v_{k+1}-v_1/2}^{\infty} \beta(v,u)n(u,t)\,du + E(v+v_1)n(v+v_1,t)$$

$$- E(v)n(v,t) + \bar{S}_0(v,t) - \bar{S}_1[v,t,n(v,t)] \qquad (2.10)$$

Because agglomeration of two particles from the discrete regime can only introduce new particles in the continuous regime smaller than or equal to v_{2k}, the continuous regime is divided into two sections. The first section contains particles of volume v, where $v_{k+1}-v_1/2 \leq v \leq v_{2k} + v_1/2$, and the second section contains particles larger than $v_{2k} + v_1/2$. In the first section n(v,t) is governed by (2.9), which is at most a system of k equations. Note that the second term of (2.10) is evaluated only for $v \geq 2v_{k+1} - v_1/2$. Combined with the appropriate initial conditions, (2.7) - (2.10) constitute the discrete-continuous GDE.

We can define <u>clusters</u> as particles containing i monomers, where $2 \leq i \leq k$, and aerosol <u>particles</u> as those containing k+1 or more monomers. We can consider a k-mer to be the critical size particle for nucleation or that just below the lower size limit of detection by available instrumentation. In either case the discrete-continuous GDE can be used to determine the net rate of formation of particles smaller and larger than a k-mer. Thus, condensation will be defined as the net agglomeration rate of monomer and clusters with aerosol particles. Nucleation will be defined as the net formation from the discrete regime. Classical nucleation is then a special case of nucleation, in which the k-mer is the critical size particle. (However, classical nucleation theory neglects cluster-cluster agglomerations in determining the nucleation rate.) When we speak of the scavenging of an i-mer by a j-mer, we mean the net agglomeration rate of the two particles where $j >> i$.

The general aspects of the gas-to-particle conversion process are depicted in Figure 1. Monomer, the gaseous precursor, may either condense on aerosol particles or agglomerate with another monomer to enter the path involving molecular cluster dynamics. The molecular clusters may themselves be scavenged by aerosol particles or continue to grow to a point at which an aerosol particle is formed, the nucleation route (Figure 2).

2.2 Continuous General Dynamic Equation

If $N(v_i,t)$ is neglected for $2 \leq i \leq k$, (2.9) and (2.10) become

$$\frac{\partial n(v,t)}{\partial t} = \frac{1}{2}\int_{v_{k+1}-v_1/2}^{v-v_{k+1}+v_1/2} \beta(v-u,u)\,n(v-u,t)\,n(u,t)\,du$$

$$- n(v,t)\int_{v_{k+1}-v_1/2}^{\infty} \beta(v,u)\,n(u,t)\,du + \bar{S}_0(v,t) - \bar{S}_1[v,t,n(v,t)]$$

$$+ n(v-v_1,t)\,\beta(v-v_1,v_1)\,N(v_1,t)$$

$$- n(v,t)[\beta(v,v_1)N(v_1,t) + E(v)]$$

$$+ E(v+v_1)\,n(v+v_1,t). \qquad (2.11)$$

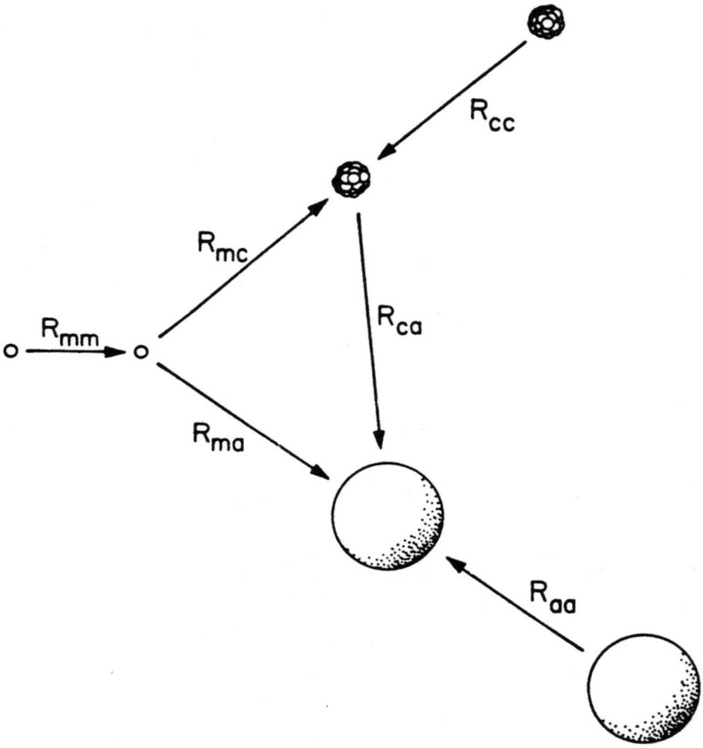

Figure 1. Formation and growth of aerosols by coagulation. R_{ij} is the net agglomeration rate of particle type i with particle type j, where m, c, and a refer to monomer, cluster and aerosol, respectively.

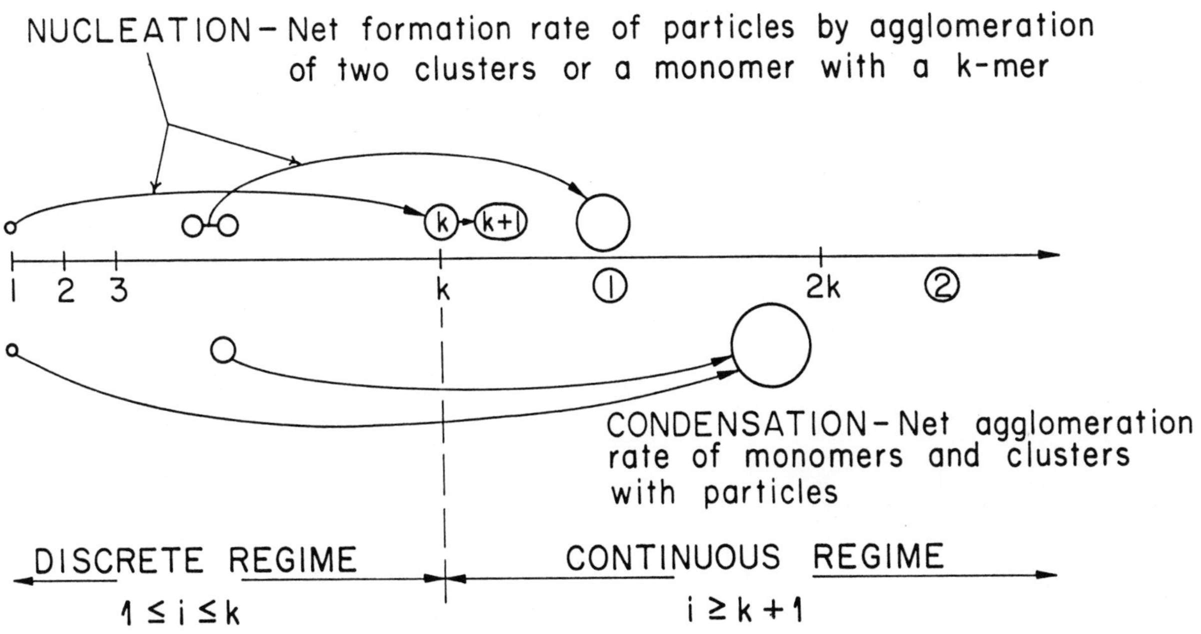

Figure 2. Discrete-continuous representation of the aerosol size spectrum with associated definitions of nucleation and condensation.

In the limit as $v_1/v \to 0$, the last three terms of (2.11) reduce to

$$- \frac{\partial}{\partial v}\left\{[\beta(v,v_1)N(v_1,t) - E(v)]v_1 n(v,t)\right\} + \frac{\partial^2}{\partial v^2}\left\{[\beta(v,v_1)N(v_1,t) + E(v)]\frac{v_1^2 n(v,t)}{2}\right\} \tag{2.12}$$

For most aerosols it has been shown that the second term of (2.12) can be neglected [3]. Therefore (2.11) becomes

$$\frac{\partial n(v,t)}{\partial t} = \frac{1}{2}\int_{v_{k+1}-v_1/2}^{v-v_{k+1}+v_1/2} \beta(v-u,u)n(v-u,t)n(u,t)du - n(v,t)\int_{v_{k+1}-v_1/2}^{\infty} \beta(v,u)n(u,t)du + \bar{S}_0(v,t) - \bar{S}_1[v,t,n(v,t)] - \frac{\partial[I(v,t)n(v,t)]}{\partial v} \tag{2.13}$$

where $I(v,t) = [\beta(v,v_1)N(v_1,t) - E(v)]v_1$. $I(v,t)$ is the net growth rate of a particle of volume v due to condensation and evaporation of monomers, and is commonly called the condensation growth rate or the growth law. Notice that (2.13) is defined only over the domain $v \geq v_{k+1}-v_1/2$, and all information on the distribution below this size is lost.

Notice that the first integral on the right hand side of (2.13) is not evaluated in the region $[v_{k+1}-v_1/2,\ 2v_{k+1}-v_1/2]$. This is because agglomerations that form new particles in this region are neglected, (except for the agglomeration of monomers with k-mers, which is accounted for in the boundary condition.)

To determine the boundary condition at $v - v_{k+1}-v_1/2$, we note that implicit in (2.6) is constancy of $n(v,t)$ in the regions $[v_{k+1}-v_1/2 + \ell v_1,\ v_{k+1} + v_1/2 + \ell v_1]$, where ℓ is a positive integer or zero. (Although $n(v,t)$ is actually a series of step functions, because the spacing of the steps is so small relative to v, $n(v,t)$ can be used as a continuous function.) Thus, at the boundary, $n(v_{k+1}-v_1/2,t)$ is given by

$N(v_{k+1},t)/v_1$. Using (2.9) and neglecting the clusters we have the boundary condition,

$$\frac{dn(v_{k+1}-v_1/2,t)}{dt} = \frac{J}{v_1} - n(v_{k+1},t)\left[E(v_{k+1}) + \beta(v_{k+1},v_1)N(v_1,t)\right]$$
$$+ E(v_{k+2})n(v_{k+2},t)$$
$$- n(v_{k+1},t)\int_{v_{k+1}-v_1/2}^{\infty} \beta(v_{k+1},u)n(u,t)\,du \qquad (2.14)$$

where J is equal to $\beta(v_1,v_k)N(v_1,t)N(v_k,t)$. There are difficulties associated with determining J and hence the boundary condition. Due to the scavenging by particles in the continuous regime, the discrete profile, and hence the nucleation rate, is dependent on the continuous distribution. Therefore, if the profile in the discrete regime is affected by scavenging, J will vary as the continuous regime evolves and can not be determined in the absence of knowledge of the dynamics in the continuous regime.

If particle diameter is chosen as the size variable of interest, the continuous GDE assumes the form

$$\frac{\partial n_D(D,t)}{\partial t} = \frac{-\partial[I_D(D,t)n_D(D,t)]}{\partial D}$$
$$+ D^2\int_{D_a}^{D/2^{1/3}} \frac{\beta_D[(D^3-\tilde{D}^3)^{1/3},\tilde{D}]n_D[(D^3-\tilde{D}^3)^{1/3},t]n_D(\tilde{D},t)d\tilde{D}}{(D^3-\tilde{D}^3)^{2/3}}$$
$$- n_D(D,t)\int_{D_a}^{D_b} \beta_D(D,\tilde{D})n_D(\tilde{D},t)d\tilde{D}$$
$$+ S_D[n_D(D,t),D,t] \qquad (2.15)$$

where $n_D(D,t)$ is the size distribution density function at time t, D is the particle diameter, $\beta_D(D,\tilde{D})$ is the coagulation coefficient for particles with diameters D and $\tilde{D}$, $I_D(D,t) = dD/dt$, the rate of particle growth from condensation, $S_D[n_D(D,t),D,t]$ represents the net influx of particles in the size range [D,D+dD] from all homogeneous sources and sinks, and D_b and D_a are the upper and lower bounds of the size domain, respectively.

3. KINETIC COEFFICIENTS

3.1 Agglomeration Coefficient

For typical atmospheric aerosols the dominant mechanism for agglomeration is Brownian motion. The functional form of β for Brownian coagulation is highly dependent on the sizes of the coagulating particles relative to the mean free path of the medium, λ. The Knudsen number is defined as $Kn = 2\lambda/D$, where D is the particle diameter.* For particles with $Kn \gg 1$ or $Kn \ll 1$ the coagulation processes is said to be in the free molecule or continuum regime, respectively. If, for one particle $Kn \ll 1$ and the other $Kn \gg 1$, the coagulation process is essentially classical diffusion or condensation. For these three limiting cases the functional form of β is well known and is given in Table 1 [4], where $\beta_D(D_i, D_j) = \beta(v_i, v_j)$, $\mathcal{D}$ is the particle diffusivity usually determined from the Stokes-Einstein expression, k is the Boltzmann constant, T is the absolute temperature, η is the viscosity of the medium and m is the mass of the particle. In the transition regime, where Kn is of order one, the slip correction for the particle diffusivity can be based on either Millikan's or Phillips' correction. For determining β in the transition regime the interpolation formula of either Fuchs [5] or Sitarski and Seinfeld [4] can be used.

3.2 Growth Laws

For most atmospheric aerosols, gas-phase diffusion is believed to be the rate determining step for particle growth. However, there is also evidence that heterogeneous chemical reactions involving gaseous species and the surface or interior of particles may serve as the rate determining step in gas-to-particle conversion. The particular mechanism of growth has an important influence on the rate at which particles of a certain size grow. Since temperature changes resulting from condensational growth of atmospheric aerosols are considered to be negligible [6], it will be assumed that isothermal conditions prevail. In this section, growth law expressions for I(v,t) in the cases of diffusion-, surface-, and volume reaction-controlled growth are presented [7].

---

*It is usually assumed that the particles are spherical.

Table 1. Brownian Coagulation Coefficients, $\beta_D(D_i, D_j)$*

Continuum $Kn_i \ll 1$, $Kn_j \ll 1$	Transition $Kn_i, Kn_j \sim O(1)$, $Kn_i = 2\lambda/D_i$	Free Molecule $Kn_i, Kn_j \gg 1$
$2\pi(\mathcal{D}_i+\mathcal{D}_j)(D_i+D_j)$	$2\pi(\mathcal{D}_i+\mathcal{D}_j)(D_i+D_j)\left[\frac{D_i+D_j}{D_i+D_j+2g_{ij}} + \frac{8(\mathcal{D}_i+\mathcal{D}_j)}{\bar{v}_{ij}(D_i+D_j)}\right]^{-1}$	$\frac{\pi}{4}(D_i+D_j)^2\,\bar{v}_{ij}$
$\mathcal{D}_i = \frac{kT}{3\pi D_i \eta}$	$\mathcal{D}_i = \begin{cases} \frac{kT}{3\pi D_i \eta}\left[1+Kn_i(1.257 + 0.4\exp(-1.1/Kn_i))\right] & \text{(Millikan) [9]} \\ \frac{kT}{3\pi D_i \eta}\left[\frac{5 + 4Kn_i + 6Kn_i^2 + 18Kn_i^3}{5 - Kn_i + (8+\pi)Kn_i^2}\right] & \text{(Phillips)[10]} \end{cases}$	$\bar{v}_{ij} = (\bar{v}_i^2+\bar{v}_j^2)^{1/2}$; $\bar{v}_i = (8kT/(\pi m_i))^{1/2}$
	$g_{ij} = (g_i^2 + g_j^2)^{1/2}$	
	$g_i = \frac{1}{3D_i \ell_i}\left\{(D_i+\ell_i)^3 - (D_i^2 + \ell_i^2)^{3/2}\right\} - D_i$	
	$\ell_i = 8\mathcal{D}_i/(\pi\bar{v}_i)$	

*In using the expressions in this table, one assumes that each collision results in coalescence.

We consider the growth of a particle resulting from condensation of a single species A on the particle. In the case of diffusion-controlled growth, we restrict ourselves to the case in which A is condensing on a particle of pure A or on a particle consisting of several species but for which the vapor pressure of A is independent of the particle's composition. In the cases of surface reaction- and volume reaction-controlled growth, A is presumably converted to another species B. It is assumed that in these two cases the equilibrium concentration of A above the particle surface is linearly related to the concentration of A in the particle. This condition can probably be expected to be valid as long as the concentration of A in the particle is small.

3.2.1 Diffusion-Controlled Growth

As shown in section 2.2 the growth law, $I(v,t)$, for diffusion-controlled growth can be determined from the agglomeration and evaporation coefficients. Just as for the agglomeration coefficient, the functional form of $I(v,t)$ is highly dependent on the Knudsen number of the particle. For particles in the continuum and free molecule regime, $I(v,t)$ is given in Table 2 in terms of the particle diameter. Although in the transition regime, one can use the interpolation formula for $\beta(v,v_1)$ given in Table 1, it is more convenient to use one of two interpolation formulae for $I(v,t)$ as given by Fuchs' flux matching formula or Sahni's formula [8].

Although both expressions are valid in the limits of Kn large and small, difficulties arise when applying the formulae to situations in which the monomer is condensing on a particle not much larger than the monomer itself. The growth laws can be corrected for monomer condensation on an i-mer by multiplying the expressions by the terms, $\{(1+i^{1/3})/i^{1/3}\}^2\ (1+i)/i^{\ 1/2}$. With this correction, $I(v,t)/N(v_1,t)$ approaches $\beta(v,v_1)$ for $E(v)=0$.

3.2.2 Surface Reaction-Controlled Growth

If surface chemical reaction is the rate-determining step for particle growth, then the rate at which reactants diffuse to the particle must equal the rate of reaction on the particle surface. Therefore, assuming the molecular volumes of the reactant and product to be identical,

Table 2. Diffusion-Controlled Growth Laws, I(v,t)*

Continuum $Kn \ll 1$	Transition $Kn \sim 0(1)$		Free Molecule $Kn \gg 1$
$2\pi D\mathcal{D}\Delta v_1$	$\dfrac{2\pi(D+2\alpha\lambda)\mathcal{D}\Delta v_1 D^2}{D^2+2b\lambda D+4a\lambda^2}$		$\pi D^2\left[\dfrac{kT}{2\pi m}\right]^{1/2} v_1\Delta$
	Fuchs	Fuchs and Sutugin	
	$a = \dfrac{\alpha\mathcal{D}}{\lambda}\left[\dfrac{2\pi m}{kT}\right]^{1/2}$	1.33	
	$b = a/\alpha$	1.71	
	$\alpha = \sim 0(1)$	1.0	

*$Kn = 2\lambda/D$, $\Delta = N_\infty - N_+\exp\left[(4\sigma v_1)/(DkT)\right]$, $v = \frac{\pi}{6}D^3$, $N_\infty \approx N(v_1,t)$

$$\pi D^2 k_s N_s^\gamma = \frac{2\pi\mathcal{D}(D^3+2\alpha\lambda D^2)}{(D^2+2b\lambda D+4a\lambda^2)}\left[N_\infty - N_+\exp\left(\frac{4\sigma v_1}{DkT}\right)\right] \tag{3.1}$$

where k_s is the surface reaction rate constant, N_s is the concentration of reactant on the surface of the particle, and γ is the order of the surface reaction. If

$$N_+ = H_s N_s \tag{3.2}$$

where H_s is constant, then by substituting (3.2) into (3.1) and rearranging, we obtain the following equation for N_s,

$$N_s^\gamma + \frac{2N_s H_s(D+2\alpha\lambda)\mathcal{D}\exp\left(\frac{4\sigma v_1}{DkT}\right)}{k_s(D^2+2b\lambda D+4a\lambda^2)} - \frac{2N_\infty(D+2\alpha\lambda)\mathcal{D}}{k_s(D^2+2b\lambda D+4a\lambda^2)} = 0 \tag{3.3}$$

For surface reactions of order γ,

$$I(v,t) = \pi D^2 v_1 k_s N_s^\gamma \tag{3.4}$$

The growth laws are given in terms of the particle diameter for $\gamma = 1/2$, 1, and 2 in Table 3.

3.2.3 Volume Reaction-Controlled Growth

If chemical reaction occurring throughout the volume of particle is the rate-determining step for growth, then the equating of diffusion and reaction rates gives

$$\frac{\pi D^3}{6} k_v N_v^\gamma = \frac{2\pi \mathcal{D}(D^3+2\alpha\lambda D^2)}{(D^2+2b\lambda D+4a\lambda^2)}\left[N_\infty - N_+ \exp\left(\frac{4\sigma v_1}{DkT}\right)\right] \tag{3.5}$$

where k_v is the volume reaction rate constant, and N_v is the concentration of reactant in the particle. Assuming

$$N_+ = H_v N_v \tag{3.6}$$

where H_v is a constant, then by substituting (3.6) into (3.5) and rearranging, we obtain the following equation for N_v,

$$N_v^\gamma + \frac{12H_v(D+2\alpha\lambda)\mathcal{D}N_v \exp\left(\frac{4\sigma v_1}{DkT}\right)}{k_v D(D^2+2b\lambda D+4a\lambda^2)} - \frac{12N_\infty(D+2\alpha\lambda)\mathcal{D}}{k_v D(D^2+2b\lambda D+4a\lambda^2)} = 0 \tag{3.7}$$

For volume reactions of order γ the growth law is then

$$I(v,t) = \frac{\pi D^3}{6} v_1 k_v N_v^\gamma \tag{3.8}$$

The growth laws for $\gamma = 1/2$, 1, and 2 are given in Table 3.

4. DIMENSIONLESS FORM OF THE CONTINUOUS GDE

In this section we will present dimensionless forms of the growth law, $I_D(D,t)$, based on diffusion-, surface reaction-, and volume reaction-controlled mechanisms. We also introduce a dimensionless form of the coagulation coefficient $\beta_D(D,\tilde{D})$ for Brownian coagulation.

For a particle in a fluid the characteristic length scale that determines many of the transport properties of the particle is the mean free path of the medium λ. Thus, all aerosol kinetic coefficients will be expressed in terms of the Knudsen number, defined as $Kn = 2\lambda/D$.

4.1 Dimensionless Growth Laws

The dimensionless form of the growth laws given in Tables 2 and 3 are summarized in Table 4. The dimensionless groups C_i and E_i, (i=0,1,2) are the ratios of characteristic rates of diffusion to the rate of reaction. For reaction-limited growth, which depends on the concentration of the reactant, the maximum growth rate occurs when the reaction rate is so fast relative to diffusion that all of the reactant that diffuses to the particle reacts immediately. Hence, the maximum growth

Table 3. Reaction-Controlled Growth Laws, I(v,t)*

Reaction Order	Surface Reaction	Volume Reaction
$\gamma = 1/2$	$\phi_s \left[\frac{1+2\phi_1\phi_2 - (1+4\phi_1\phi_2)^{1/2}}{2\phi_1^2} \right]^{1/2}$	$\phi_v \left[\frac{1+2\phi_3\phi_4 - (1+4\phi_3\phi_4)^{1/2}}{2\phi_3^2} \right]^{1/2}$
$\gamma = 1$	$\frac{\phi_s\phi_2}{1+\phi_1}$	$\frac{\phi_v\phi_4}{1+\phi_3}$
$\gamma = 2$	$\frac{\phi_s}{4} \left[-\phi_1 + (\phi_1^2+4\phi_2)^{1/2} \right]^2$	$\frac{\phi_v}{4} \left[-\phi_3 + (\phi_3^2+4\phi_4)^{1/2} \right]^2$

*$\phi_s = \pi D^2 v_1 k_s$ $\qquad \phi_1 = \frac{H_s\phi}{k_s} \exp\{(4\sigma v_1)/(DkT)\}$ $\qquad \phi_4 = \frac{6N_\infty\phi}{k_v D}$

$\phi_v = \frac{\pi}{6} D^3 v_1 k_v$ $\qquad \phi_2 = N_\infty\phi/k_s$

$\phi = \frac{2(D+2\alpha\lambda)\mathcal{D}}{D^2 2b\lambda D+4a\lambda^2}$ $\qquad \phi_3 = \frac{6H_v\phi}{k_v D} \exp\{(4\sigma v_1)/(DkT)\}$

Table 4. Dimensionless Growth Laws, $I_D(D,t) = 2\phi\bar{I}(Kn)$*

Controlling Mechanisms	$\bar{I}(Kn)$
Gas phase diffusion	$[1-\bar{D}\exp(BKn)]f$
Surface reaction	
half order ($\gamma=1/2$)	$\frac{\left[2+2C_0\exp(BKn)f^2-2(1+2C_0\exp(BKn)f^2)^{1/2}\right]^{1/2}}{C_0\exp(BKn)f}$
first order ($\gamma=1$)	$\frac{Kn+\alpha Kn^2}{aKn^2+bKn+1+C_1(Kn+\alpha Kn^2)\exp(BKn)}$
second order ($\gamma=2$)	$C_2\exp(2BKn)f^2\left[-1+(1+1/(C_2 f\exp(2BKn)))^{1/2}\right]^2$
Volume reaction	
half order ($\gamma=1/2$)	$\frac{\left[2+2E_0\exp(BKn)f^2Kn^2-2(1+2E_0\exp(BKn)f^2Kn^2)^{1/2}\right]^{1/2}}{E_0\exp(BKn)fKn^2}$
first order ($\gamma=1$)	$\frac{Kn+\alpha Kn^2}{aKn^2+bKn+1+E_1(Kn^2+\alpha Kn^3)\exp(BKn)}$
second order ($\gamma=2$)	$E_2Knf^2\exp(2BKn)\left[-1+(1+1/(E_2 f\exp(2BKn)Kn))^{1/2}\right]^2$

*See footnote on following page.

*Footnote from previous page (Table 4).

$$B = \frac{2\sigma v_1}{\lambda kT}, \quad \bar{D} = \frac{N_+}{N_\infty}, \quad f = (Kn+\alpha Kn^2)/(aKn^2+bKn+1)$$

$$C_0 = \frac{2H_s \mathcal{D}^2 N_\infty}{\lambda^2 k_s^2} \qquad C_1 = \frac{H_s \mathcal{D}}{k_s \lambda} \qquad C_2 = \frac{H_s^2 \mathcal{D}}{4k_s N_\infty \lambda}$$

$$E_0 = \frac{18H_v \mathcal{D}^2 N_\infty}{k_v^2 \lambda^4} \qquad E_1 = \frac{3H_v \mathcal{D}}{k_v \lambda^2} \qquad E_2 = \frac{3H_v^2 \mathcal{D}}{4N_\infty k_v \lambda^2}$$

$$\phi = \frac{N_\infty v_1 \mathcal{D}}{\lambda}$$

---

rate must be given by the diffusion-limited growth law with no particle vapor pressure. Therefore, as C_i or $E_i \to 0$, $(i=0, 1,2)$ for $\gamma = 1/2$, 1 and 2, $I_D(D,t)$ approaches the gas phase diffusion-limited growth law with $\bar{D} = 0$.* Conversely, as C_1 or $E_i \to \infty$, the reaction rate is much slower than the diffusion rate. Since growth is reaction limited, $I_D \to 0$ in this case.

4.2 Dimensionless Coagulation Coefficient

In dimensionless form $\beta_D(D_i,D_j)\eta/kT$ given by Fuch's formula in Table 1 is equal to,

$$\frac{\beta_{Kn}(Kn_i,Kn_j)\eta}{kT} = \frac{2Q(p_iKn_i+p_jKn_j)/3}{\dfrac{Q}{A+(H_i^2+H_j^2)^{1/2}} + \dfrac{\pi A(p_iKn_i+p_jKn_j)}{2Q(Kn_i^3+Kn_j^3)^{1/2}}} \tag{4.1}$$

where

$$Q = \frac{1}{Kn_i} + \frac{1}{Kn_j}$$

$$H_i = \frac{Kn_i^{3/2}}{6Ap_i}\left[\left(\frac{2}{Kn_i} + \frac{Ap_i}{Kn_i^{1/2}}\right)^3 - \left(\frac{4}{Kn_i^2} + \frac{A^2p_i^2}{Kn_i}\right)^{3/2}\right] - \frac{2}{Kn_i}$$

$$A = \left[\frac{8\rho kT}{27\pi^2\eta^2\lambda}\right]^{1/2}$$

---

*If however, the molecular volume of the product is greater than that for the diffusing species, reaction-limited growth for $\gamma>0$ may exceed the growth rate for diffusion-limited growth.

$$p_i = \begin{cases} 1 + Kn_i[1.257 + 0.4\exp(-1.1/Kn_i)] & \text{(Millikan)} \\ \dfrac{5+4Kn_i+6Kn_i^2+18Kn_i^3}{5-Kn_i+(8+\pi)Kn_i^2} & \text{(Phillips)} \end{cases}$$

$$Kn_i = 2\lambda/D_i$$

Notice that the only parameter which is dependent on the physical properties of the medium or the particle is A. For typical atmospheric aerosols, A is approximately 0.075.

4.3 Dimensionless Form of the Continuous GDE

To nondimensionalize (2.15) characteristic length and time scales must be chosen. As discussed previously, the mean free path of the medium, λ, is the appropriate length scale. The characteristic time scale can be based on coagulation, growth, source or sink mechanisms. When growth occurs, since the mechanism for transport to the particle is diffusion, $\lambda^2/\mathcal{D}$ can be chosen as the characteristic time scale. Equation (2.15) contains a general term $S_D[n(D,t),D,t]$ that represents the net influx of particles in the size range [D,D+dD] from all homogeneous sources and sinks. For the purpose of non-dimensionalization it is useful to specify the form of S_D somewhat more explicity. A common situation is that in which both a homogeneous particle source, perhaps from nucleation, and a first order removal process exist. Thus, we let $S_D[n(D,t),D,t] = S_0(D,t) - S_1(D,t)n(D,t)$. Defining a dimensionless time, $\tau = \mathcal{D}t/\lambda^2$, (2.15) can be transformed into the dimensionless form

$$\frac{\partial \bar{n}(Kn,\tau)}{\partial \tau} + \frac{\Lambda\zeta(\tau)\partial[\bar{n}(Kn,\tau)Kn^2\bar{I}]}{\partial Kn} + \xi\int_{Kn_b}^{2^{1/3}Kn} \bar{\beta}(Kn'',Kn')\bar{n}(Kn'',\tau)\bar{n}(Kn',\tau)\left(\frac{Kn''}{Kn}\right)^4 dKn'$$
$$- \xi\int_{Kn_b}^{Kn_a} \bar{\beta}(Kn,Kn')\bar{n}(Kn',\tau)\bar{n}(Kn,\tau)\, dKn'$$
$$- \bar{S}_0(Kn,\tau) - \bar{S}_1(Kn,\tau)\bar{n}(Kn,\tau) \qquad (4.2)$$

where

$$\bar{n}(Kn,\tau) = \frac{-2\lambda n(2\lambda/Kn,\tau\lambda^2/\mathcal{D})}{N_\infty Kn^2}$$

$$\bar{\beta}(Kn,Kn') = \beta_{Kn}(Kn,Kn')\eta/kT$$

$$Kn'' = \left[\frac{1}{1/Kn^3-1/Kn'^3}\right]^{1/3}$$

$$\bar{S}_0(Kn,\tau) = \frac{2S_0(2\lambda/Kn,\tau\lambda^2/\mathcal{D})\lambda^3}{\mathcal{D}N_\infty Kn^2}$$

$$\bar{S}_1(Kn,\tau) = \frac{S_1(2\lambda/Kn,\tau\lambda^2/\mathcal{D})\lambda^2}{\mathcal{D}}$$

$$\Lambda = \frac{\lambda\phi}{\mathcal{D}} \qquad \xi = \frac{N_\infty kT\lambda^2}{\mathcal{D}\eta}$$

$$Kn_a = 2\lambda/D_b \qquad Kn_b = 2\lambda/D_a$$

To account for any temporal variations in I_D, $\zeta(\tau)$ is introduced as an arbitrary positive function of τ. Notice that the definition of Λ will vary depending on the specific growth mechanism, and ϕ is defined in Table 4 for each mechanism.

5. AEROSOL SIZE DISTRIBUTION EVOLUTION UNDER PURE GROWTH

At relatively small aerosol number densities, such as those frequently prevalent in ambient atmospheres, coagulation can often be neglected relative to condensation in influencing the temporal variations in $\bar{n}$. Neglecting the coagulation integrals in (4.2), we obtain

$$\frac{\partial\bar{n}(Kn,\tau)}{\partial\tau} = \Lambda\zeta(\tau)\frac{\partial}{\partial Kn}[\bar{n}(Kn,\tau)Kn^2\bar{I}] - \bar{S}_0(Kn,\tau) - \bar{S}_1(Kn,\tau)\bar{n}(Kn,\tau) \tag{5.1}$$

The characteristic equations for (5.1) are

$$\frac{dKn}{d\tau} = -\Lambda Kn^2\zeta(\tau)\bar{I} \tag{5.2}$$

$$\frac{d\bar{n}(Kn,\tau)}{d\tau} = \left[\Lambda\zeta(\tau)\frac{d(Kn^2\bar{I})}{dKn} - \bar{S}_1(Kn,\tau)\right]\bar{n}(Kn,\tau) - \bar{S}_0(Kn,\tau) \tag{5.3}$$

We note that (5.2) is simply the dimensionless growth law. Integrating (5.2) for some of the growth laws given in Table 4, we have the characteristic growth curves as given in Table 5.

For gas phase diffusion-limited growth a two term expansion of the exponential term for BKn<<1 is required, however

Table 5. Characteristic Growth Curves*

Controlling Mechanism	$\int \frac{dKn}{Kn^2 \bar{I}(Kn)}$
Gas phase diffusion	$\psi_1 \ln Kn + \psi_2 Kn^{-1} + \psi_3 Kn^{-2} + \psi_4 \ln(1+\alpha Kn) + \psi_5 \ln(1-\bar{D}-\bar{D}BKn)$
Surface reaction	
first order ($\gamma=1$)	$\frac{\alpha-b}{Kn} - \frac{1}{2} Kn^{-2} + (\alpha^2-b\alpha+a)\ln\left(\frac{Kn}{1+\alpha Kn}\right) - C_1 Kn^{-1} e^{BKn} + C_1 B\left[\ln Kn + \sum_{j=1}^{\infty} \frac{(BKn)^j}{jj!}\right]$
Volume reaction	
first order ($\gamma=1$)	$\frac{\alpha-b}{Kn} - \frac{1}{2} Kn^{-2} + (\alpha^2-b\alpha+a)\ln\left(\frac{Kn}{1+\alpha Kn}\right) + E_1\left[\ln Kn + \sum_{j=1}^{\infty} \frac{(BKn)^j}{jj!}\right]$

$$^*\psi_1 = \frac{[\bar{D}B+\alpha(\bar{D}-1)][\bar{D}B+b(1-\bar{D})]+(1-\bar{D})^2(\alpha^2+a)}{(1-\bar{D})^3}$$

$$\psi_2 = \frac{(1-\bar{D})(\alpha-b)-\bar{D}B}{(1-\bar{D})^2}$$

$$\psi_3 = \frac{1}{2(\bar{D}-1)}$$

$$\psi_4 = \frac{\alpha(a-b\alpha+\alpha^2)}{\alpha(\bar{D}-1)-\bar{D}B}$$

$$\psi_5 = \frac{\bar{D}B\{\bar{D}B[b(\bar{D}-1)-\bar{D}B]-a(\bar{D}-1)^2\}}{(1-\bar{D})^3[\bar{D}B+\alpha(1-\bar{D})]}$$

if $\bar{D} = 0$ or $B = 0$, entry 1 of Table 5 is exact. For all other solutions given in Table 5 no approximations were required to evaluate the integral. Physically, the characteristics given in Table 5 represent the change in Knudsen number of a particle as a function of $\Lambda \int_{\tau_0}^{\tau} \zeta(\bar{\tau})d\bar{\tau}$. Given the Knudsen number, Kn_0 of a particle at $\tau_0 = 0$, the dimensionless time for a particle to grow to $Kn_1 < Kn_0$ can be determined for different mechanisms from Table 5.

For positive values of $\bar{D}$, a critical particle size, given by

$$Kn_{crit} = -\frac{\ell n(\bar{D})}{B} \quad (5.4)$$

exists such that no growth occurs, i.e. $dKn/d\tau = 0$. For $Kn > Kn_{crit}$ the particle vaporizes and for $Kn < Kn_{crit}$ the particle grows.

Let us consider the cases of diffusion-, surface reaction-, and volume reaction-controlled growth. We wish to study the evolution of an aerosol from a prescribed initial distribution under the three forms of growth. The initial aerosol size distribution chosen is log normal with parameters, $D_g = 0.026$ μm and $\sigma_g = 1.98$. The distribution is shown in dimensionless form by the solid curves in Figures 3-5.

In addition to the number distribution, as shown in Figures 3-5, data are frequently represented in terms of the volume distribution, which, in dimensionless form, is related to the number distribution by

$$\bar{V}\left[\log_{10} Kn^{-1},\tau\right] = \frac{4\pi}{3Kn^3}\,\bar{n}\left[\log_{10} Kn^{-1},\tau\right] \quad (5.5)$$

Figures 3-8 show the evolution of the dimensionless number and volume distributions for diffusion-, first order surface reaction-, and first order volume reaction-controlled growth. The parameter values used in the cases shown in Figures 3-8 are:

$a = 1.33$	$b = 1.71$	$\alpha = 1.0$
$\lambda = 0.066$ μm	$T = 298^{\circ}K$	$\mathcal{D} = 0.05\ cm^2\ sec^{-1}$
$\bar{D} = 0.1$	$B = 0.01$	
$C_1 = 10$	$E_1 = 10$	

C_1 and E_1 are the dimensionless groups that represent ratios of the characteristic rate of diffusion of species to the particle to the characteristic rates of surface and volume reaction. The dimensionless size distributions in Figures 3-8 are shown at values of $\Lambda\tau = 0, 2,$ and 5. The values of t corresponding to values of $\Lambda\tau$ can be estimated for typical situations. We note that $\Lambda\tau = N_\infty v_1 \mathcal{D} t/\lambda^2$. If the condensing species ambient concentration N_∞ is one part-per-billion by volume, then $N_\infty = 2.46\times10^{10}$ molecules cm^{-3}. A typical molecular diameter for the condensing species is 4Å, and a typical surface tension is

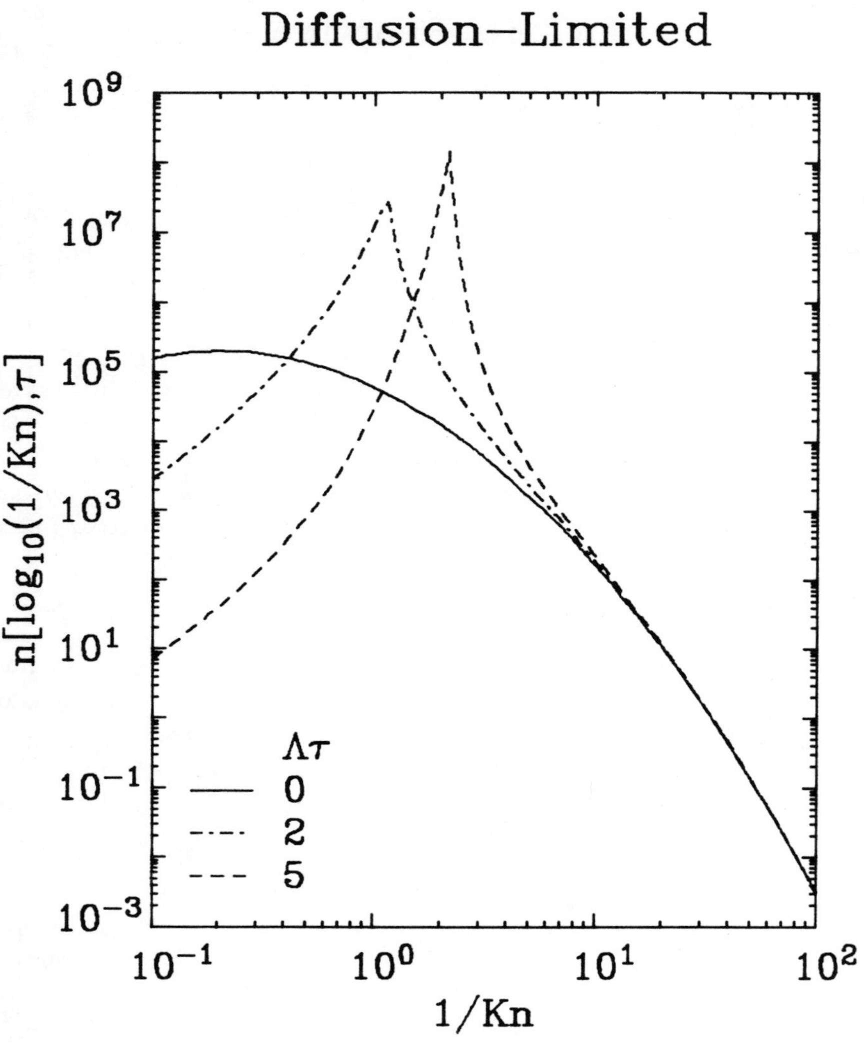

Figure 3. Evolution of the dimensionless number distribution by pure growth under diffusion-controlled conditions

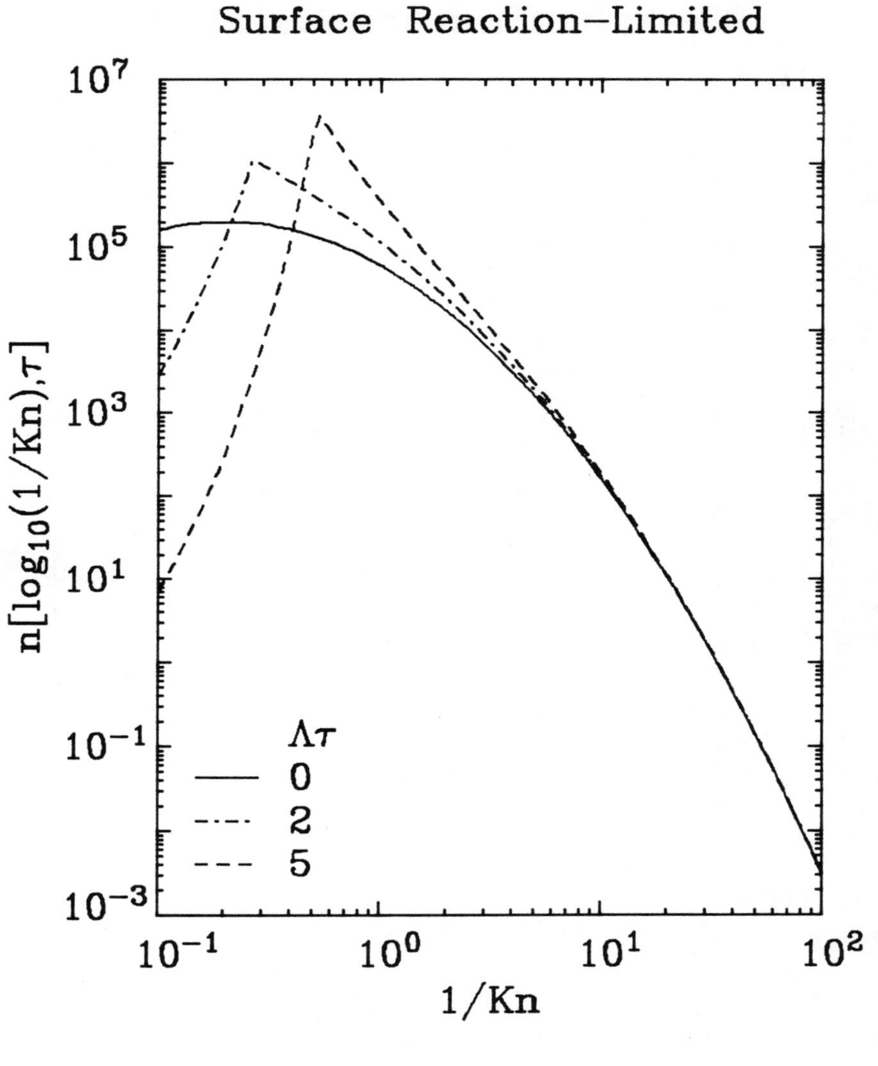

Figure 4. Evolution of the dimensionless number distribution by pure growth under surface-reaction controlled conditions

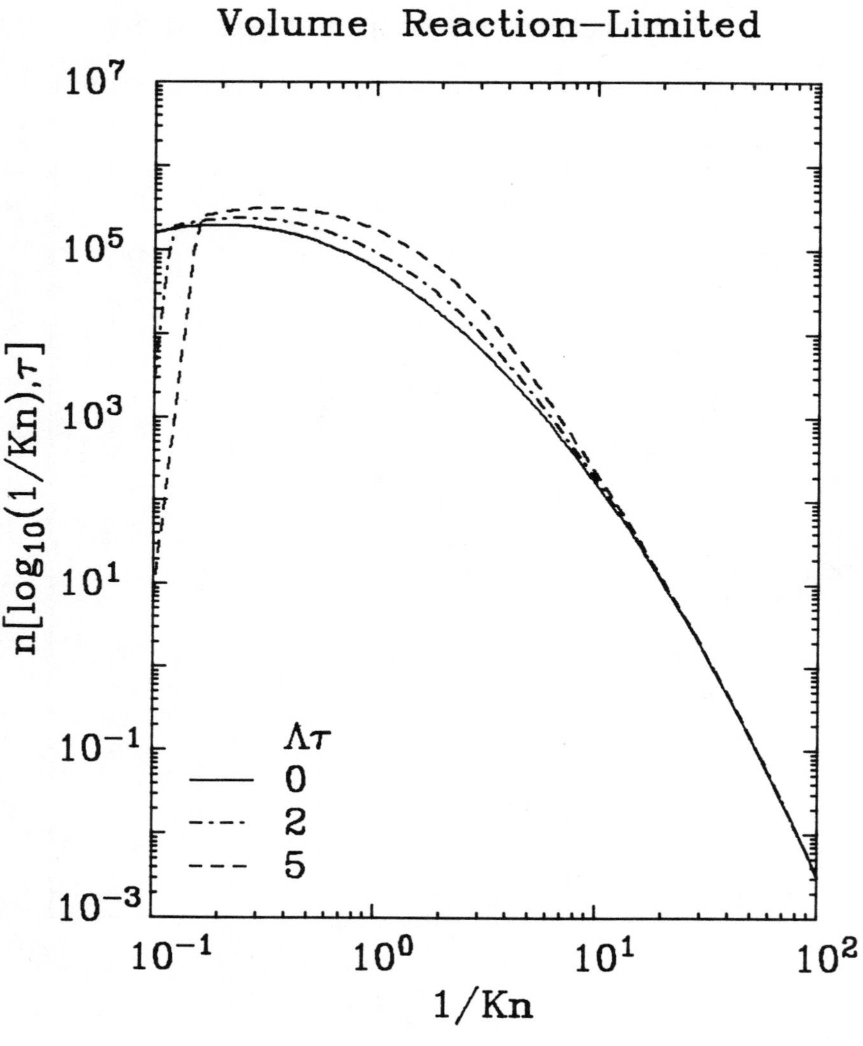

Figure 5. Evolution of the dimensionless number distribution by pure growth under volume reaction-controlled conditions

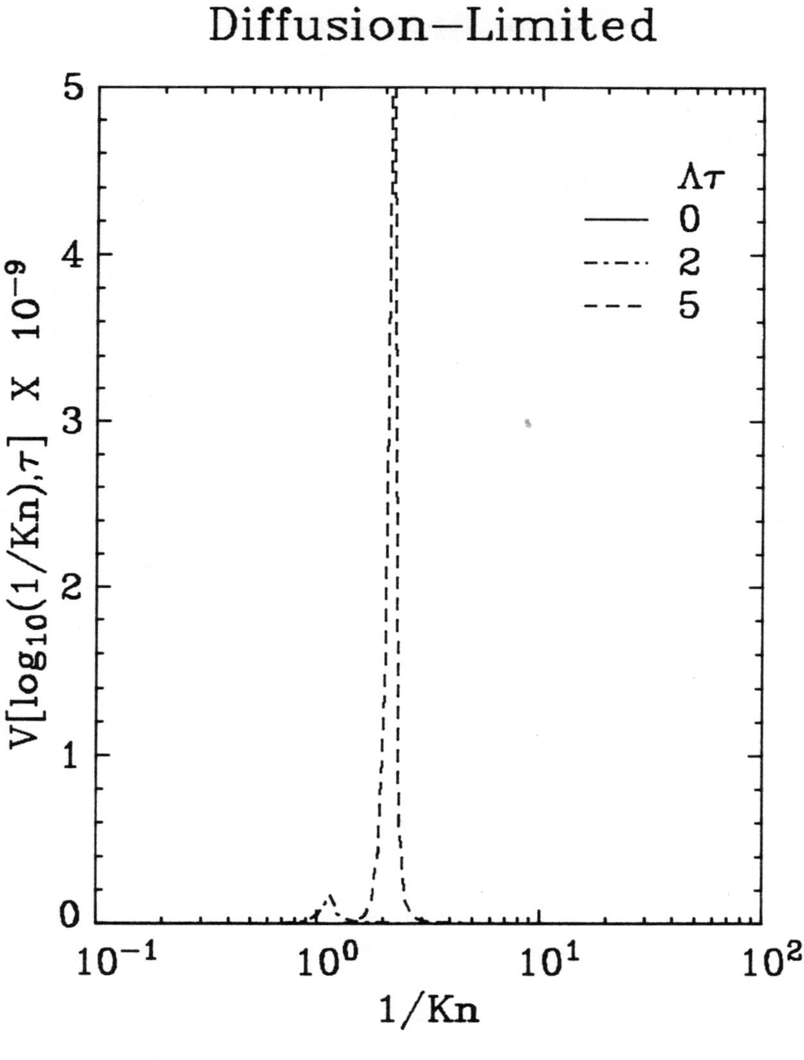

Figure 6. Evolution of the dimensionless volume distribution by pure growth under diffusion-controlled conditions

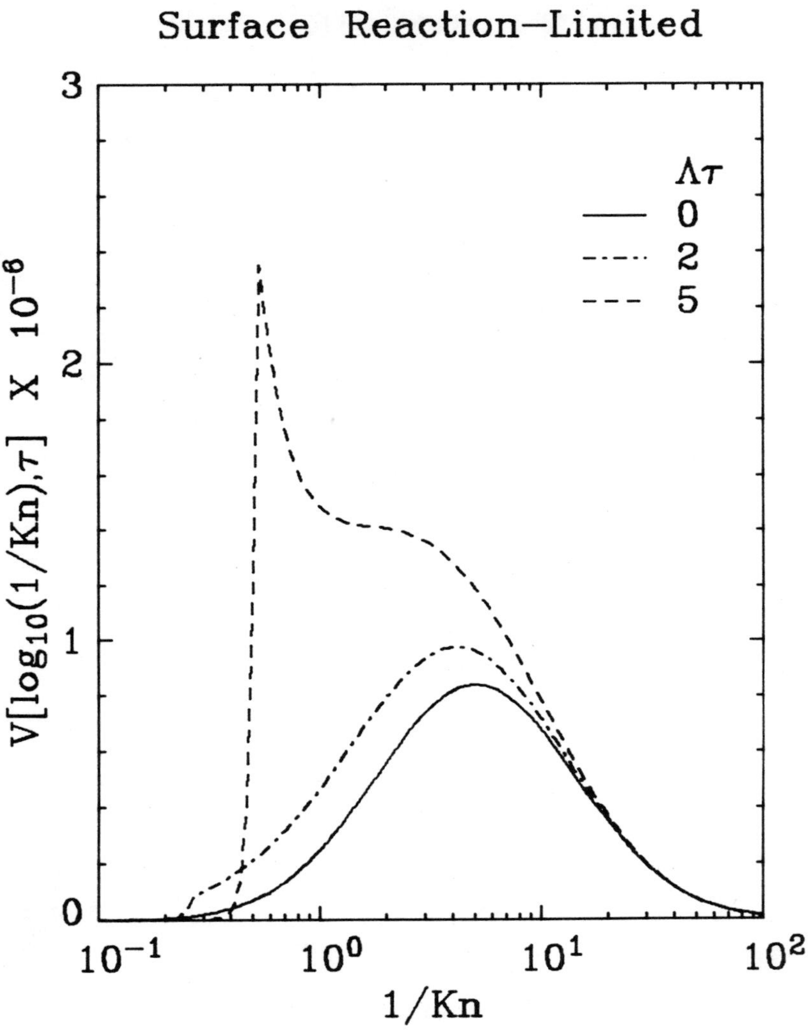

Figure 7. Evolution of the dimensionless volume distribution by pure growth under surface reaction-controlled conditions

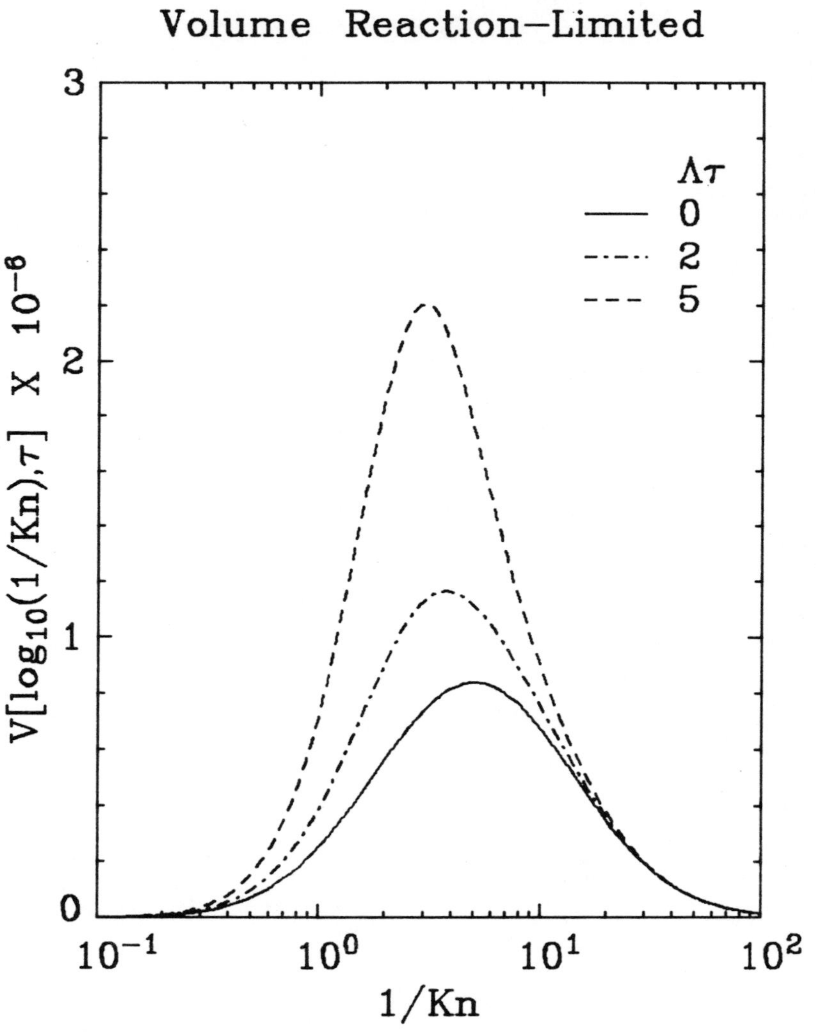

Figure 8. Evolution of the dimensionless volume distribution by pure growth under volume reaction-controlled conditions

σ = 40 dynes cm^{-3}. Using these values, $\Lambda\tau$ = 5 corresponds to a time t of approximately 90 minutes.

It is necessary to specify a boundary condition on (5.1) at Kn = Kn_o. We will set the boundary condition as

$$\bar{n}(Kn_o,\tau) = e^{-2\Lambda\tau}\bar{n}(Kn_o,0) \qquad (5.6)$$

where the upper limit on the Knudsen number Kn_o is chosen as 10. This boundary condition has been selected arbitrarily to reflect the fact that the influx of fresh particles across the boundary at Kn_o must decrease as more aerosol surface area is produced. As noted previously, it is not possible to specify the boundary condition rigorously without the considerations inherent in (2.14). We choose (5.6) only for the convenience of illustrating the nature of the solution.

For the particular initial distribution chosen, the size spectra in Figures 3-5 have the property that the point of discontinuity in the first derivative of $\bar{n}$ indicates the division between the two regions of the Kn-τ plane. The distribution for Kn values larger than that at the point is influenced by the boundary condition (5.6); that to the right of the point is influenced only by the growth of particles initially present.

In the three cases, the dimensionless particle growth laws $\bar{I}(Kn)$ of Table 4 become:

Diffusion $$\frac{(1-\bar{D}e^{BKn})(Kn+Kn^2)}{aKn^2+bKn+1}$$

Surface reaction (γ=1) $$\frac{Kn+Kn^2}{aKn^2+bKn+1+C_1(Kn+Kn^2)\exp(BKn)}$$

Volume reaction (γ=1) $$\frac{Kn+Kn^2}{aKn^2+bKn+1+E_1(Kn^2+Kn^3)\exp(BKn)}$$

The relative behavior of the size distributions in Figures 3-8 can be understood by reference to the above growth laws. First, as noted previously, the maximum growth rate of any particle occurs under diffusion-controlled conditions. Therefore, the shift in the size distribution, as evidenced by the movement with time of the peaks in the number distribution in Figures 3-5, is most pronounced for diffusion-controlled growth (Figure 3). The dimensionless groups C_1

and E_1, representing the relative importance of diffusion to surface and volume reaction, respectively, are both equal to 10, allowing direct comparison of surface reaction- and volume reaction-controlled growth. We note, from the above growth laws, that when Kn>1, the rate of growth by surface reaction exceeds that for volume reaction, whereas when Kn<1, the opposite is obtained. Physically, this behavior is due to the fact that when Kn>1, the particles are relatively smaller, and there is more surface area per unit volume than when Kn<1. Since the rate of surface reaction depends on the amount of surface area present, the rate of growth by surface reaction exceeds that for volume reaction when Kn>1. On the other hand, if Kn<1 particles are relatively larger, the surface area per unit volume is smaller, and volume reaction predominates.

The comparative aspects of the three growth mechanisms are demonstrated even more vividly in the volume distributions of Figures 6-8. Because of the rapid growth by diffusion, the scale of Figure 6 is such that the initial distribution is too close to the abscissa to be seen. The diffusional growth is characterized by a sharp peak in the volume distribution centered at Kn $\cong$ 0.5 at $\Lambda\tau = 5$. Under surface- and volume-reaction-limited conditions the rate of growth is substantially slower than for diffusion-controlled growth for the particular parameter values chosen. (Note the three orders of magnitude difference between the ordinates of Figure 6 and Figures 7 and 8.) With surface reaction growth, for those particles for which Kn>1, a peak develops in the volume distribution at Kn $\cong$ 2 that is not present for volume reaction growth. This behavior is due to the preferential growth of relatively smaller particles by surface reaction as compared to volume reaction.

6. SUMMARY

In this paper we have developed the basic mathematical framework for studying the dynamic behavior of aerosols. First, the basic number conservation equation, the so-called General Dynamic Equation, was presented in a discrete-continuous form that allows simulation of a complete size spectrum. The frequently used continuous form of this equation was then derived from the discrete-continuous form, and the approximations inherent in the continuous equation were noted. Next, the kinetic coefficients that appear in the General Dynamic Equation for atmospheric situations were introduced, with particular attention being given to those coefficients that represent the rate of growth of a particle by condensation of vapor molecules. Three mechanisms were delineated for such growth: molecular diffusion of the vapor molecules to the particle surface and chemical reaction occurring either on the surface of the particle or throughout the particle volume. Expressions for the particle growth rate in terms of dD/dt, the rate of change of its diameter D, have been formulated. The full General Dynamic Equation was then placed in dimensionless form. Exact solutions to the dimensionless equation were presented for the case of pure growth, that is when each particle grows independently and no particle interactions (coagulation) occur. The evolving size spectra in the three cases of diffusion-, surface reaction-, and volume reaction-controlled growth were found to exhibit certain distinct features characteristic of the growth mechanism.

Several aspects important in simulating aerosol dynamics have not been considered in the present paper. In general, numerical techniques are required to solve the General Dynamic Equation when coagulation is occurring. Such techniques are discussed by Gelbard and Seinfeld [11] and Peterson et al. [12]. We have not considered here cases in which the multicomponent nature of the particles must be explicitly accounted for. In fact, simulation of the dynamics of multicomponent aerosols with particle-phase chemical reactions and particle-particle interactions represents a particularly challenging problem. (The case of pure growth has been formulated by Peterson and Seinfeld [6].) The behavior of aerosols in chemical reactors, such as the continuous stirred tank reactor and the tubular reactor, in which simultaneous mass and heat transfer take place, has not yet been adequately explored. Finally, some interesting cases of periodic behavior during aerosol formation by nucleation in systems with simultaneous diffusion have been observed that require mathematical analysis.

REFERENCES

1. Chu, K. J. and J. H. Seinfeld, Formulation and initial application of a dynamic model for urban aerosols, Atmos. Environ. 9, (1975) 375
2. Gelbard, F. and J. H. Seinfeld, The general dynamic equation for aerosols - Theory and application to aerosol formation and growth, J. Colloid Interface Sci. 68, (1979) 363.
3. Ramabhadran, T. E.,T. W. Peterson, and J. H. Seinfled, Dynamics of aerosol coagulation and condensation, A.I.Ch.E.J. 22 (1976) 840.
4. Sitarski, M. and J. H. Seinfeld, Brownian coagulation in the transition regime, J. Colloid Interface Sci. 61, (1977) 261.
5. Hidy, G. M. and J. R. Brock, The dynamics of aerocolloidal systems, Pergamon, Oxford, 1970.
6. Peterson, T. W. and J. H. Seinfeld, Heterogeneous condensation and chemical reaction in droplets - application to the heterogeneous atmospheric oxidation of SO_2, Adv. in Envrion. Sci. Technol. 10 (in press)
7. Gelbard, F. and J. H. Seinfeld, Exact solution of the General dynamic equation for aerosol growth by condensation, J. Colloid Interface Sci. 68 (1979) 173.
8. Fuchs, N. A. and A. G. Sutugin, in Topics in Current Aerosol Research (G. M. Hidy and J. R. Brock, Eds.) Pergamon, Oxford, 1971.
9. Millikan, R. A., The general law of fall of a small spherical body through a gas, and its bearing upon the nature of molecular reflection from surfaces, Phys. Rev. 22 (1923) 1.
10. Phillips, W. F., Drag on a small sphere moving through a gas, Phys. Fluids 18, (1975) 1089.
11. Gelbard, F. and J. H. Seinfeld, Numerical solution of the dynamic equation for particulate systems, J. Computational Physics 28, (1978) 357.
12. Peterson, T. W., F. Gelbard, and J. H. Seinfeld, Dynamics of source-reinforced, coagulating, and condensing aerosols, J. Colloid Interface Sci. 63 (1978) 426.

The author was partially supported by National Science Foundation Grant PFR 76-04179.

California Institute of Technology
Pasadena, CA 91125
and
Mathematics Research Center
University of Wisconsin-Madison
Madison, WI 53706

Impulse Propagation in Excitable Systems

John Rinzel

I. Introduction

The interaction of diffusion with reaction dynamics which exhibit multiple steady states or temporal oscillations can lead to a variety of spatio-temporal patterns. But this can also occur in cases where the reaction dynamics has a unique stable steady state as a global attractor. We will focus here on such cases for which the reaction-diffusion system is said to be excitable. A rigorous and comprehensive definition of excitability will not be attempted although the characteristics for the simplest examples are as follows; see also [15,19]. The reaction dynamics for $y(t) \in R^n$,

$$y_t = f(y) \tag{1}$$

exhibit a unique, stable, globally attracting, steady state: $y(t) \equiv \bar{y}$, constant. Any other t-independent steady states or oscillatory solutions (e.g., limit cycles) are unstable. From local stability it follows, for a sufficiently small displacement δy from $\bar{y}$, $y(0) = \bar{y} + \delta y$, that $y(t)$ returns abruptly to $\bar{y}$ without amplification. However, if δy is not too small then, in an excitable system, y undergoes a large excursion before eventually returning to $\bar{y}$. Thus y versus t will appear pulse-like. Under the action of diffusion, the pulse-like excursion may propagate as an unattenuated wave through the medium in response to a spatially localized disturbance of adequate size. Pulse propagation is not guaranteed just because (1) exhibits pulse-like behavior; the combined influence of

 ISBN 0-12-669550-4

reaction and diffusion must be considered. The term excitable is sometimes reserved only for those systems capable of pulse propagation [19].

Let us consider a few examples of dynamics from excitable systems. Perhaps the simplest is the "ring dynamics" $\phi_t = f(\phi)$, where the scalar ϕ describes the state or phase of the system (e.g., see [37]). Here, $f(\phi)$ is 2π-periodic with $f(0) = f(2\pi) = 0$, $f(\phi) < 0$ for $0 < \phi < \phi_T$ and $f(\phi) > 0$ for $\phi_T < \phi < 2\pi$. In this model, the singular points $2\pi m$, $m = 0, 1, \ldots$ are identified as equivalent rest states. For small displacements from rest, ϕ returns to rest. If, however, a positive displacement exceeds the threshold ϕ_T then ϕ increases to the next multiple of 2π; the system has executed a large excursion and returned to rest. To propagate, by diffusive coupling ($\phi_t = \phi_{xx} + f(\phi)$), the excitation transition from $\phi = 2\pi m$ to $\phi = 2\pi(m+1)$, it is necessary and sufficient that $\int_0^{2\pi} f(\phi)d\phi > 0$. This transition wave, although not strictly pulse-like, qualifies to admit this system as excitable since the rest state is recovered. Discrete analogs of one-dimensional dynamics have also been studied (see [15] and its references).

A two variable example, formally an electrical circuit, but which also serves as a conceptual model for ionic current flows across nerve membrane, is the FitzHugh-Nagumo (FHN) equation [12, 23]:

$$
\begin{aligned}
v_t &= -f(v) - w + I \\
w_t &= \varepsilon(v - \gamma w), \qquad \varepsilon, \gamma \geq 0.
\end{aligned}
\tag{2}
$$

Here, v is voltage across a three branch parallel circuit. One branch is capacitive. Another is an inductor in series with a resistor; w is the current through this branch. The third branch is a nonlinear resistor (e.g., tunnel diode) for which f(v) is the cubic-shaped current-voltage relation (e.g., negative of the dashed cubic in Fig. 1-A). Hence f(v) actually generates rather than dissipates current for v which satisfies $v_T < v < 1$, where $v_T > 0$. In (2), I corresponds to an externally applied current source. The phase plane for

(2) is shown in Fig. 1-A for I = 0 with the two nullclines shown dashed. Sample trajectories are shown for $\varepsilon = 0.02$ and 0.002.

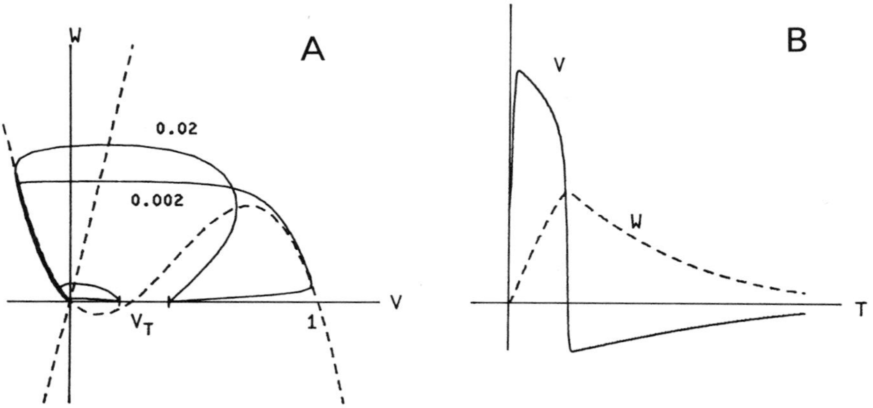

Fig. 1. A. Phase plane portrait for (2) with $f(v) = v(v_T - v)(1 - v)$, $v_T = 0.25$, $\gamma = 1.0$, $I = 0$, and ε as indicated. B. v and w versus t for the pulse trajectory in A for $\varepsilon = 0.002$.

For a small horizontal displacement from rest ($w(0) = 0$), v, w quickly return to the unique stable rest state (0,0). However, if $v(0) > v_T$ then v increases before returning to rest and v versus t is pulse-shaped as in Fig. 1-B for $\varepsilon = 0.002$. The value v_T distinguishes initial values (with $w(0) = 0$ for simplicity) for those solutions which show some v-amplification from those which do not. The threshold behavior exhibited here is not a singular point threshold phenomenon [13] as in the preceding example. However, if ε is much smaller (compare the two cases in Fig. 1-A), the flow is nearly horizontal except near the v-nullcline and the threshold behavior is sharper. The pulse shape is more distinct with a broad plateau during which the phase point slowly moves upward along the right branch of the cubic, the excited phase, followed by the return-to-rest or recovery or refractory phase with (v,w) moving slowly downward along the left branch of the cubic. In such models, w is often called the recovery variable. Observe that a brief stimulus I(t) applied during the recovery phase may lead to a second pulse if its intensity is sufficiently strong to carry (v,w) to the right of the middle branch. The larger is the value of w, the greater must be the stimulus intensity, and the level of excitability is then said to be lower.

Two variable systems with pulse-like dynamics in the biochemical context of enzyme mediated reactions have been studied by several investigators, e.g. [1, 14, 24]. For a purely chemical system, the three-variable Oregonator or Field-Noyes (FN) model [11] for the Belousov-Zhabotinskii reaction exhibits excitable behavior in some parameter range [26, 34]. An example of a single exothermic reaction for a continuous stirred chemical reactor is considered in [35]; see Fig. 6-C. Perhaps the earliest, quantitatively accurate model of excitable behavior was the four-variable Hodgkin-Huxley [17] model to describe membrane potential (like v, above) and ionic current flows (f(v) and w are analogs) across the membrane of the giant nerve fiber of squid.

In the above examples for spatially homogeneous dynamics, a pulse is not precisely defined; each different initial condition leads to a somewhat different excursion trajectory. When ε is sufficiently small however, the pulse-type trajectories have similar plateau and recovery phases. In contrast, the propagating pulse is well-defined. We next consider the effects of diffusive coupling. If FHN circuits are cascaded linearly with pure resistive coupling, one obtains the following equation

$$v_t = v_{xx} - f(v) - w + I(x,t) \qquad (3)$$
$$w_t = \varepsilon(v - \gamma w).$$

A similar reaction-diffusion structure arises in nerve conduction models [28]. There the single diffusion term corresponds to spread of current along the fiber's intracellular core of constant ohmic resistivity. Membrane potential is v and $v_{xx} + I(x,t) = v_t + I_{ion}$ where v_t and I_{ion} are the capacitive and ionic contributions to membrane current. Auxiliary ordinary differential equations in t describe I_{ion} and the dynamics of the membrane ionic channels whose descriptive variables are not modeled as diffusible. For more chemical relevance, Winfree [36] added w_{xx} to the second of (3).

Now suppose $I(x,t)$ is a spatially localized stimulus $I(x,t) = I_s(t)\,\delta(x)$. A brief and weak I_s produces only a local response which attenuates with x and t; the rest state is stable for (3). A sufficiently strong stimulus, however, excites the region near $x = 0$. The gradient v_x supplies current to neighboring regions which then also become excited. After each region is successively recruited, it passes through the excitation phase and eventually returns to rest. The pulse of excitation reaches a unique constant velocity and shape independent of the adequate I_s. In nerve, propagation speed is of the order mm/msec. For an excitable Belousov-Zhabotinskii mixture, pulse speed is like mm/min and a pulse may be initiated with a localized heat source (Winfree, private communication) or a silver electrode [31]. Corresponding to the above or similar experiments, numerical solutions of the FHN and HH (see [28] for references) and FN [26] models exhibit propagation of single pulses as well.

A second, brief but adequate, I_s may initiate a second propagating pulse. Because this second pulse will be advancing not into regions at rest but still returning to rest, its speed will not be the same as the first pulse. If it sees a state of less excitability (as for FHN with $\varepsilon << 1$) one expects intuitively, and observes, its speed to be slower than the first pulse. Suppose I_s is applied periodically. If this rate is not too high, a periodic state will be approached with each stimulus initiating a pulse. Here again these pulses will travel at a different speed than a solitary pulse. The speed will depend on the period, i.e. on how closely each pulse follows its predecessor.

In the following sections we will discuss pulse propagation in terms of particular solutions, traveling waves, of the reaction-diffusion equations. For periodic wave trains of uniformly spaced pulses we describe the dependence of speed on period, the dispersion relation. For nerve conduction models, we illustrate both monotone (e.g., speed decreasing with frequency) and non-monotone (e.g., maximum speed at some preferred non-zero frequency) dispersion relations for different parameter values. These different cases carry different implications for spatio-temporal patterning during propagation of pulses evoked by stimuli not necessarily periodic. Stimulus frequencies in the monotone range of the dispersion relation suggest loss of pattern while instantaneous frequencies which span the non-monotone range can lead to establishment of patterning in the pulse train during propagation. Computed solutions to the full reaction-diffusion equations (e.g., (3)) provide evidence of this. Moreover, individual pulse trajectories are predicted quantitatively by a kinematic approximation for these examples.

II. Traveling wave solutions and the dispersion relation

A train of one or more pulses which propagate with fixed shape and speed c corresponds to a traveling wave solution, a function of $z = x + ct$, of the reaction-diffusion equations. Such a solution satisfies a system of ordinary differential equations, 2n equations if all species diffuse, in which c is a parameter. Nerve conduction is a special case in which there are only n+1 equations. For excitable systems, more is presently known rigorously and formally for the (n+1)-case than for the general 2n-case (a restricted 2n-case is considered in [8]) and we will next briefly sketch some of this accumulated understanding. Additional details and references for nerve conduction may be found in [16, 28]; for an (n+1)-order treatment of the Oregonator model see [34]. The reader may consider the FHN equation:

$$c\, v' = v'' - f(v) - w \qquad (4)$$

$$c\, w' = \varepsilon(v - \gamma w)$$

as a model system for many purposes.

In the (n+1)-case, the unique singular point is $(0, \bar{y}_1, \ldots, \bar{y}_n)^T$; for (4), it is $(0,0,0)^T$. By a linear stability analysis, this singular point (the rest state) has a hyperbolic structure: a one-dimensional unstable manifold, tangent near the rest state to a vector U, and an n-dimensional stable manifold, locally a hyperplane S. For a solitary pulse, one seeks a value of c such that the trajectory of the unstable manifold is bounded with its large excursion and eventual return to rest along the stable manifold, i.e. the singular point is entered as $z \to \pm \infty$; see Fig. 2-A for a schematic illustration. Such a homoclinic orbit occurs only for an appropriate speed c_0. If c is slightly different, the unstable manifold passes above S (c too low) or below S (c too high). Shooting techniques, numerical and analytical, have been applied to calculate c or prove its existence (see [16, 28] and their references). In addition to this solitary pulse which corresponds to the one observed

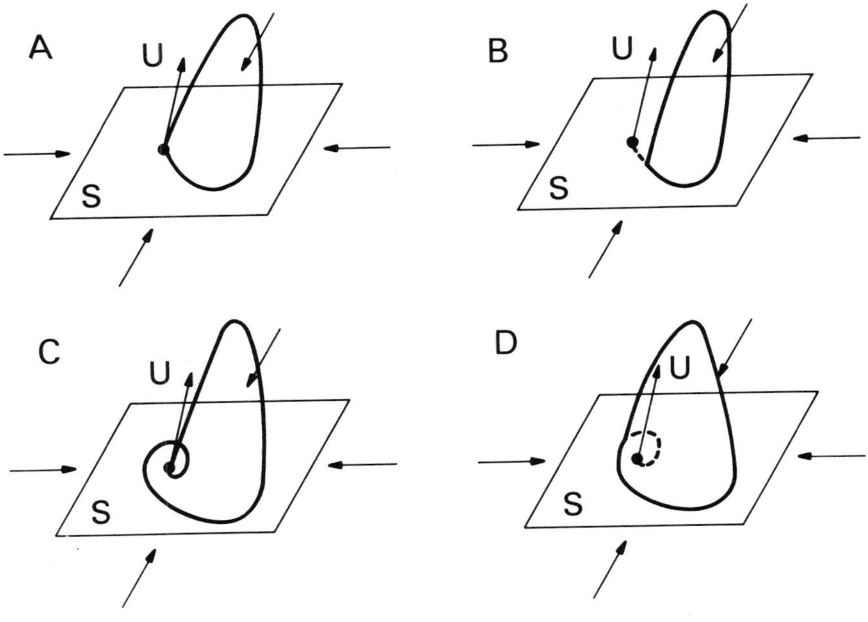

Fig. 2. Schematics of phase space for traveling wave equation, (n+1)-case. Unstable and stable manifolds of rest point are represented locally by U and S. Cases A (solitary pulse) and B (periodic train) are for dynamics with monotone return to rest in S. C and D are for damped oscillatory return in S.

experimentally or numerically in response to stimuli, there is typically a second, smaller amplitude, solitary pulse solution with a different speed, typically slower. This one is conjectured, proven in some cases [9, 10, 27], to be an unstable solution to the reaction-diffusion equations. For increasing values of a parameter in the model which reduces excitability, e.g., ε or v_T in FHN, these two solutions converge and, for a critical value, they coalesce and disappear; ability to steadily propagate a pulse is lost. For FHN when $f(v) = v(v - v_T)(v - 1)$ and $\varepsilon << 1$ such a restriction is $v_T < 1/2$.

For steady repetitive pulse propagation one seeks a periodic traveling wave solution. If P denotes period in z then $P = \infty$ corresponds to the solitary pulse with speed c_o. One can expect, for large but finite P and an appropriate c close to c_o, that the solitary pulse homoclinic has disengaged itself from the singular point to become a periodic orbit (Fig. 2-B). For such a closed orbit the return to rest or recovery phase takes up most of the period as the trajectory passes close to the singular point. This portion of the trajectory is nearly identical to that of the solitary pulse. The upstroke for the periodic pulse (locally, parallel to U) takes off from this recovery trajectory and the degree of excitability at take-off determines the speed c. Hence, there is a family of periodic wave trains parametrized by P with a dispersion relation $c = c(P)$ and with the solitary pulse as a limit for $P = \infty$. Similarly, there is a family of periodic waves with the slow, smaller amplitude, solitary pulse as a limit.

This one parameter family of periodic wave trains may be solved for analytically if one chooses for the FHN equation the piecewise linear $f(v) = v - H(v-a)$ where $H(\cdot)$ is the Heaviside step function; c is calculated numerically by solving one or two transcendental equations [27]. The dispersion relation may be used to express c versus spatial period or wavelength P, or equivalently, c versus temporal frequency ω, $\omega = Pc$, or just as well, ω versus wavenumber 1/P. Figs. 3-A,B,C provide computed examples for the case $\gamma = 0$, $v_T = 0.1$ and three values of ε. Observe that the fast wave trains and the slow trains correspond to different branches of the same family. The limit as $P \to \infty$ or $\omega \to 0$ yields the respective solitary pulse speeds. For this solvable case, stability of the fast pulse [10] and instability of the slow pulse along with certain of the wave trains, continuing inward along the slow branch, [27, 28] has been demonstrated. Here we will concern ourselves with the fast waves, the upper branch in the representation of Fig. 3-B.

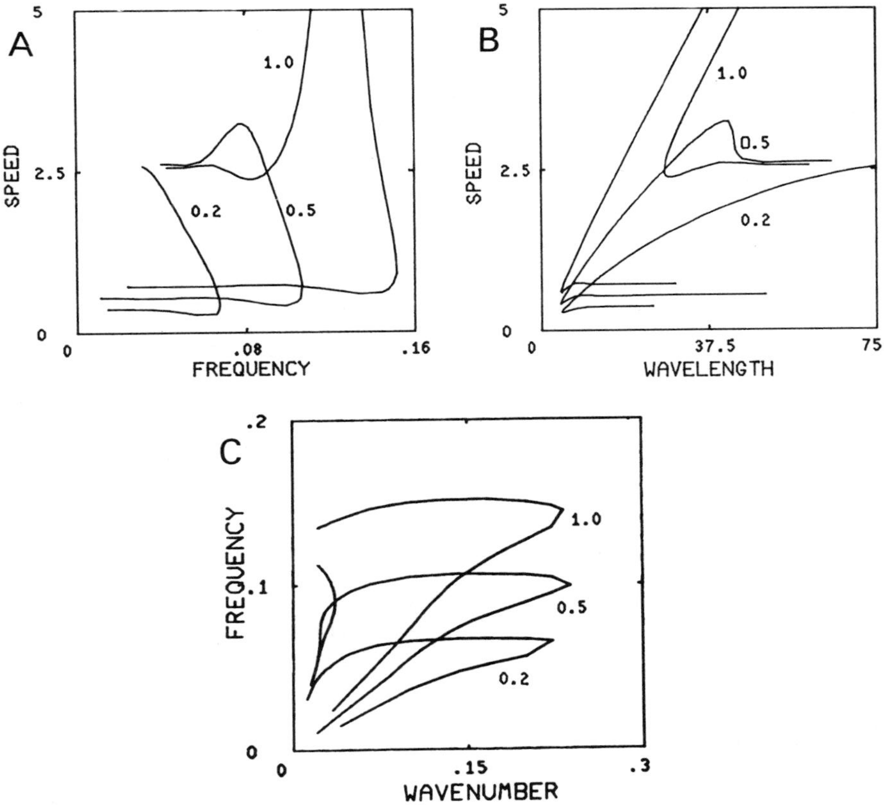

Fig. 3. Three different, but equivalent, representations (A,B,C) of the dispersion relation for periodic (uniform spacing) traveling wave solutions of piecewise linear FHN equation (4) for $v_T = 0.1$, $\gamma = 0$ and three values of ε.

In the case of ε small, $\varepsilon = 0.2$ in Fig. 3, c decreases monotonically with ω or $1/P$. Correspondingly, the solitary pulse has a monotonic return-to-rest phase (v increasing, w decreasing) with a well-defined asymptotic direction in S as suggested in Fig. 2-A. Hence for the periodic waves as P decreases from ∞, the pulse upstroke occurs for a state of lesser excitability so the speed decreases. This monotonic character may be exposed analytically for the general FHN equation (4) by a singular perturbation construction of the wave trains in the case $0 < \varepsilon << 1$ [5, 7, 20].

In contrast to this monotonic approach $c \nearrow c_o$ as $\omega \to 0$ for ε small, larger values of ε evidence a damped oscillatory ultimate approach (Fig. 3). Correspondingly, for such ε, the solitary pulse exhibits a return-to-rest which is a damped oscillation (schematic of Fig. 2-C). Hence for a periodic train of a given large period the pulse upstroke may occur during a phase of either sub- or super-excitability (Fig. 2-D) and therefore the wave train speed will be greater than or less than c_o; as $\omega \to 0$, c approaches c_o with a damped oscillatory variation. From Fig. 2 and from this heuristic argument, which although lacks precise definition of sub- or super-excitability and take-off, we expect in general the asymptotic approach, $c \to c_o$ for $\omega \to 0$, to be determined by the ultimate decay of the solitary pulse in S: either monotone, i.e., nodal structure in S, or oscillatory, i.e., spiral structure in S. We remark however, that the nodal case does not imply that c must necessarily <u>increase</u> monotonically to c_o as $P \to \infty$. Examples, qualitatively of HH form with more than one slow variable, are shown by singular perturbation arguments to exhibit a monotone approach with $c \searrow c_o$ (see [6] which also describes additional classes of traveling waves, e.g. steadily propagating periodic pulse bursts). It should further not be assumed, if recovery processes take place on a very slow time scale, that an oscillatory approach is precluded; a case studied in [4] provides a counterexample. For either a decreasing monotone approach $c \searrow c_o$ or an oscillatory approach as $\omega \to 0$, there will be frequency ranges where c exceeds c_o, supernormal speeds. Either possibility provides a mechanism for non-monotone c versus ω behavior at intermediate frequency ranges.

An additional and somewhat distinct factor which may contribute to non-monotone behavior (e.g., local maximum in c) for frequencies not in the asymptotic range $\omega \to 0$ is a tendency of the reaction dynamics (1) to exhibit a large amplitude limit cycle. In several models, e.g., HH [29], cubic FHN [34], and Oregonator (Field, private communication), for a certain range of parameter values the spatially homogeneous dynamics may exhibit a stable singular point in addition to a coexisting stable, large amplitude, limit cycle. Similarly, for the piecewise linear FHN equation with fixed $v_T > 0$ the rest state is stable and if ε is large enough (larger than approximately 0.969 if $v_T = 0.1$) there is also a stable limit cycle [2]. In these cases, as a parameter approaches such a range, a critical value is reached at which a large amplitude limit cycle suddenly appears and splits into a pair of limit cycles, one stable and the other unstable. One therefore anticipates, as the parameter is set close to but not beyond its critical value, that the c versus ω curve develops a local maximum around the frequency of the impending limit cycle (e.g., Fig. 3, $\varepsilon = 0.5$). This is non-monotone behavior away from the $\omega = 0$ limit and not necessarily (or obviously) attributable to the character of the return-to-rest phase of the solitary pulse. Upon passage through the critical parameter value, this hump should reach to $c = \infty$ as the case $\varepsilon = 1.0$ in Fig. 3 suggests. We interpret the $c = \infty$ limit as corresponding to the spatially homogeneous limit cycle, i.e. with $1/P = 0$. This observation is consistent with a theorem of Kopell and Howard [21] for the 2n-case on existence of a family of wave trains which emerge with $P = c = \infty$ from a spatially uniform limit cycle. A parameter variation as above provides an opportunity to study the transition from excitable to oscillatory dynamics but here we consider only the excitable case (subcritical parameter values). We mention it only to illustrate an additional mechanism for non-monotonicity in the dispersion relation.

We have used or referenced examples of dispersion relations and periodic traveling wave solutions for the solvable piecewise linear FHN model or models solvable by singular perturbation methods. Beyond these special cases, solutions must generally be computed numerically. An approach we have used, which avoids some pitfalls a shooting method can exhibit for these problems, is described in [22]; similar alternatives to shooting for the solitary pulse case have been described by other investigators (see [22] for references).

We have illustrated different possibilities in the speed-frequency relation and these carry different implications for spatio-temporal patterning during pulse propagation. Suppose pulses are initiated by a stimulus at $x = 0$ which is not necessarily time periodic. For a system in which c decreases with ω, one expects each pulse except the first to travel more slowly than a solitary pulse. The shorter the time interval since the preceeding pulse (the higher the instantaneous frequency) or the closer it follows its predecessor, the slower will be the speed of a pulse. The tendency to slow and spread apart is greatest for the closest spaced pulses. Hence one expects continual widening of interpulse intervals and therefore loss of temporal pattern as generated by the stimulus. On the other hand, for a system with a local maximum in its $c - \omega$ relation, pulses with interpulse times through some frequency ranges will have a tendency to speed up and shorten interpulse intervals. During propagation, some intervals may decrease while others increase and a pattern of preferred uneven spacings may develop. For the above argument we have formally applied the dispersion relation for periodic wave trains to individual pulses in an arbitrary train. In the next two sections we will describe numerical calculations and kinematic approximations for some signaling problems for which such conclusions are reasonable.

III. Pulse dispersion during propagation

In this section we will depart from the idealized situation of a steadily propagating wave train. Our goal is to consider how the speed of an individual pulse changes during propagation as it is influenced by preceding pulses. For our demonstrations, pulses will be initiated by a spatially localized (point) source, $I_s(t)$ at $x = 0$, which delivers a sequence of brief stimuli. One can imagine a nerve being stimulated by an experimenter's electrode since the equations we have solved numerically are of the nerve conduction type. Also in this context, we ask how interpulse temporal intervals may be altered during propagation along a uniform nerve fiber. According to some physiological forklore, successive pulses propagate with approximately the same constant speed. Therefore if one assumes that neural information is coded by the temporal sequence of pulses, this information would be maintained during propagation. As we have suggested, the speed of an individual pulse in a sequence should not be expected to remain constant during propagation and therefore temporal information may be altered. Certainly the direction, and hopefully an estimate of the degree, of alteration should be predictable from the dispersion relation. Analogous questions may be addressed for pulse propagation in chemical or biochemical systems.

First we consider a signaling problem in which the stimulating frequencies are restricted to the monotone decreasing range of the c - ω curve. This example for the HH model at 6.3°C has the dispersion relation of Fig. 4. Maximum

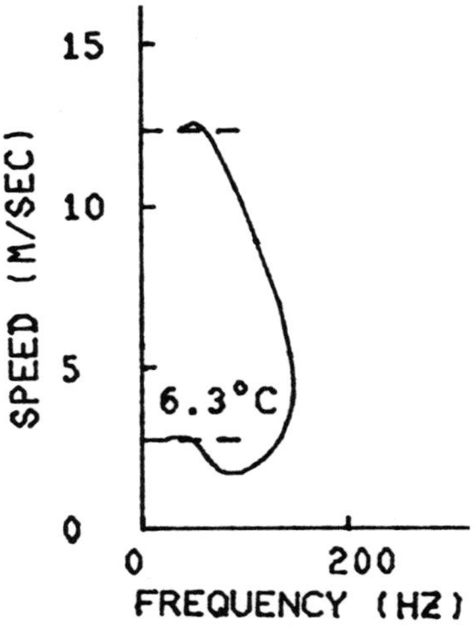

Fig. 4. Dispersion relation for HH model [22] for 6.3°C; c versus ω has prominent monotone drop.

frequency here is about 147 Hz at which c has dropped by nearly a factor of three from c_o. The problem is formulated as follows

$$v_t = v_{xx} - I_{ion}(v,\underline{w}) + I_s(t)\delta(x)$$
$$\underline{w}_t = \underline{g}(v, \underline{w}) \quad , \underline{w} \in R^3 \qquad (5)$$

with boundary conditions

$$v(-x,t) = v(x,t), \qquad v_x(L,t) = 0$$

and initial conditions which specify $v(x,0)$, $\underline{w}(x,0)$ at their uniform resting values. Because of imposed even symmetry about $x = 0$, we consider only $0 < x < L$; here $L = 7.5$ cm. This problem was solved numerically [22] by the Crank-Nicholson finite difference method with $\Delta x = 0.05$ cm, $\Delta t = 0.01$ msec. To present our simulation results we specify a value of $v = v^*$ (say, 20 mV) on the pulse upstroke and then locate the upstroke position as a function of time $x^*(t)$ according to $v(x^*(t),t) = v^*$. This trajectory is plotted in the $x - t$ plane. We have also computed, by numerical differencing, the instantaneous velocity of this upstroke trajectory.

Our first case is for regular periodic stimulus pulses at 125 Hz (Fig. 5-A,B). Vertical bars just left of the ordinate axis indicate duration of the square pulses of I_s. For convenience we will refer to the n-th pulse as P_n. The first pulse P_1 settles into the constant speed c_o of the solitary pulse traveling wave solution. P_2 starts slowly but accelerates during propagation because its predecessor, always traveling faster, is pulling away and therefore the interval between P_1 and P_2 is ever increasing (Fig. 5-A). Eventually, by the fifth I_s pulse, a regular periodic state is virtually reached with P_5 moving at the constant speed (Fig. 5-B) predicted by the dispersion relation at $\omega = 125$ Hz. During the transient phase, at each x a pulse travels more slowly than its predecessor. As each pulse approaches the sealed end at $x = L$, its speed (according to our definition) increases without bound. Because of even symmetry at $x = L$, one may imagine an opposing pulse which approaches from $x > L$. As these pulses collide they both supply current to the region between themselves and rapidly excite it. This boundary effect is local and restricted to within 1 cm of $x = L$, a small distance compared, for example, to the wavelength of about 6 cm at 125 Hz.

Our next simulation (Fig. 5-C,D) for this HH model demonstrates that significant changes in temporal patterning can occur. We stimulate at a frequency of 140 Hz but with a two-skip-one stimulus pattern. Here again, P_1 behaves as a solitary pulse while P_2 starts slow and accelerates by the same mechanism as above. P_3 starts slower than P_1 but faster than P_2 since, near $x = 0$, the interpulse time interval between P_3 and P_2 is twice that between P_2 and P_1. During propagation, P_3 slows down since it advances into the recovery wake of P_2 faster than P_2 pulls away. The periodic stimulus quite rapidly induces a periodic response with an alternating pattern of P_{2n} accelerating and P_{2n+1} decelerating (Fig. 5-D) and with the interpulse time interval between P_{2n+1} and P_{2n} decreasing and that between P_{2n+2} and P_{2n+1} increasing (Fig. 5-C). The two-skip-one pattern is not preserved during propagation. The interstimulus times are in the ratio 2:1 but during propagation the interpulse times have tended toward uniformity and near $x = L$ the ratio is approximately 1.2:1. For a much longer fiber one expects the initial pattern to be almost completely lost during transmission. A corresponding example for the piecewise linear FHN model (parameter choices for a strictly monotone dispersion relation) exhibits a similar loss of input pattern (E. Pate and G. Odell, private communication). The physiological consequences for squid per se of our observations are unclear since it is apparently not known if precise timing of pulses in a train is functionally important. Nevertheless, these theoretical results (see [22] for details and further HH examples) should be testable for the squid nerve viewed as a convenient experimental model system. We remark that experimental data on pulse propagation in signaling problems for excitable Belousov-Zhabotinskii chemical systems suggests monotone decreasing $c - \omega$ relations for those mixtures tested (Winfree, private communication, and discussion in [31]).[1]

---

[1]Experimental dispersion data for an oscillatory mixture indicate that c versus ω is monotone, but with slight <u>increase</u>, over the limited frequency range considered [33].

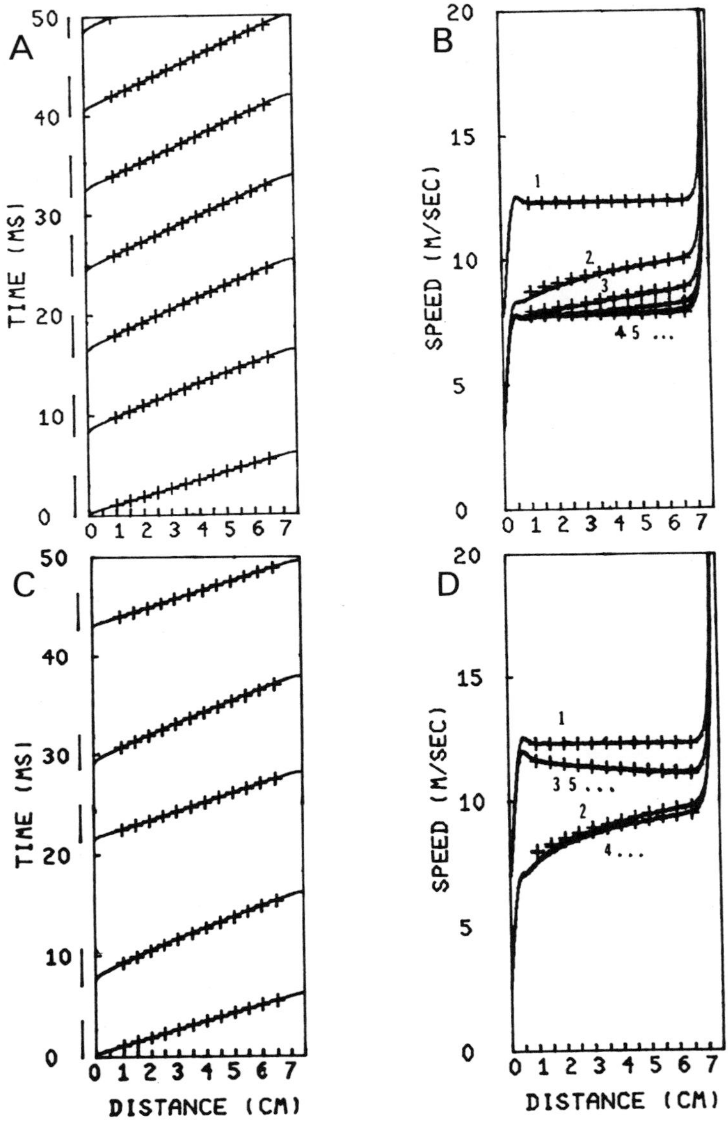

Fig. 5. Numerical solution of two HH signaling problems (5); parameters as in Fig. 4. A and B show x-t and c-x plots of pulses for periodic stimulus of 125 Hz. C and D are for two-skip-one stimulus pattern of 140 Hz. This case illustrates tendency toward uniform interpulse times during propagation. Crosses are for kinematic approximation.

Next we consider an example for which the c - ω curve has a prominent supernormal hump. As Fig. 3 evidences, the piecewise linear FHN equation offers such an opportunity. For $\varepsilon = 0.5$, $v_T = 0.05$ and $\gamma = 0$, a parameter set also used in [36], the dispersion relation is shown in Fig. 6. Here,

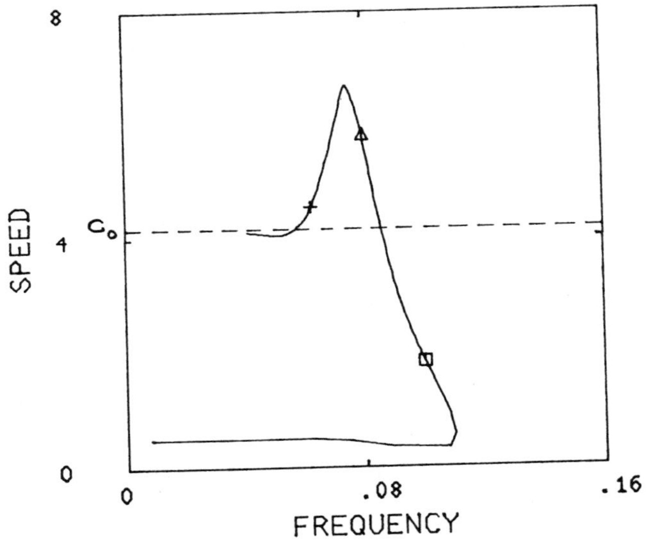

Fig. 6. Dispersion relation for piecewise linear FHN model with $\varepsilon = 0.5$, $v_T = 0.05$, $\gamma = 0$; c versus ω has prominent supernormal hump. Parameter values here apply to remaining Figures.

maximum $c \doteq 6.67$, about 60% greater than c_0 ($\doteq 4.18$), while speed falls off to less than 25% of c_0 at maximum frequency, $\omega_{max} \doteq 0.11$. The signaling problem we solve numerically is

$$\begin{aligned} v_t &= v_{xx} - v + H(v-v_T) - w + I_s(t)\delta(x) \\ w_t &= \varepsilon(v - \gamma w) \end{aligned} \tag{6}$$

with boundary conditions

$$v(-x,t) = v(x,t), \qquad v_t(L,t) + c_0\, v_x(L,t) = 0$$

and initial conditions

$$v(x,0) = 0 = w(x,0).$$

The radiation boundary condition at x = L ensures that each pulse runs off the end with velocity c_0. Here again, we find that the boundary influence is only local. As a numerical procedure we used a finite element methodology with piecewise linear basis functions for the space dependence (suggested by B. Kellogg, private communication) and an explicit or forward Euler difference method in time. To compute x - t and c - x trajectories, accuracy was essential. We found the finite element approach was better suited for the discontinuity $H(v-v_T)$ than say a centered second difference approximation for v_{xx} without special treatment of $H(v-v_T)$. Our mesh sizes were $\Delta x = 0.1$, $\Delta t = 0.0025$, and 0.3 was the value chosen for v^*.

The first numerical experiment (Fig. 7-A) with this model

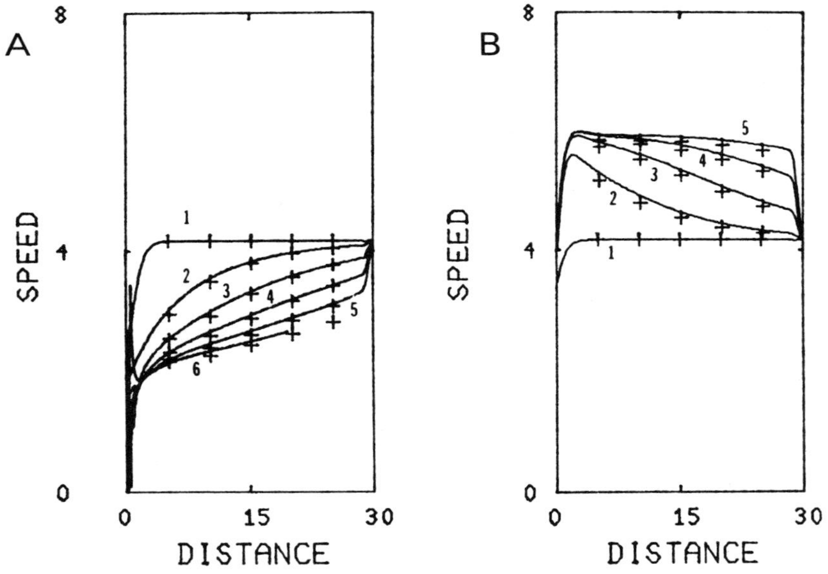

Fig. 7. Numerical solution of two FHN signaling problems (6); c - x plots of pulses for periodic stimulus with $\omega = 0.1$ (A), 0.08 (B) corresponding to ⌑, Δ of Fig. 6. Kinematic results indicated by crosses.

is for regular periodic I_s pulses at high frequency $\omega = 0.10$ (cf. ¤ in Fig. 6). This case of one pulse per each stimulus is analogous to that of Fig. 5-A,B for the HH model. The total integration time here, however, was not long enough to fully establish an apparent[2] periodic response. The transient phase is long, perhaps because ω is so close to ω_{max} for which the periodic wave train is presumably neutrally stable [28].

Our second example (Fig.7-B) is also for regular periodic stimulation but now at a frequency $\omega = 0.8$ (cf. Δ in Fig. 6) for which the corresponding wave train speed ($\doteq$ 5.8) exceeds c_0. This stimulating frequency lies on the monotone decreasing side of the supernormal hump of the $c - \omega$ curve. After P_1 with its solitary pulse speed, P_2 starts with speed close to 5.8 but then decelerates as it shortens the interpulse time by advancing toward P_1 faster than P_1 moves ahead. Successive pulses travel faster and experience less slowing. An apparent[2] periodic state is approached during which each pulse travels, over most of the region, with constant velocity as predicted from the dispersion relation. The larger is L, the longer time it takes to approximately establish the periodic state.

For the next case, $L = 75$ rather than 30 as above. The stimulus is now only two I_s pulses with interstimulus time corresponding to frequency $\omega = 0.0625$ (cf. + in Fig. 6) on the increasing branch of the supernormal hump. As above, P_1 develops as a solitary pulse (Fig. 8-A,B). P_2 first forms with speed greater than c_0, corresponding to + in Fig. 6. Thus, P_2 starts shortening the interpulse time since P_1 and thereby increases its speed as the left branch of Fig. 6 predicts. As the speed of P_2 passes over the hump (Fig. 8-B) it begins to slow while continuing to shorten the interpulse time interval (Fig. 8-A). The interstimulus time interval of 16 is decreased to less than 11 during pulse propagation. An

---

[2]We say "apparent" because for (6) with $\gamma = 0$ and $I_s \geq 0$, we find empirically that w near $x = 0$ is not strictly periodic but rather its time average over each stimulus period continues to grow slowly; for integration times here, this appears to be a local effect near $x = 0$.

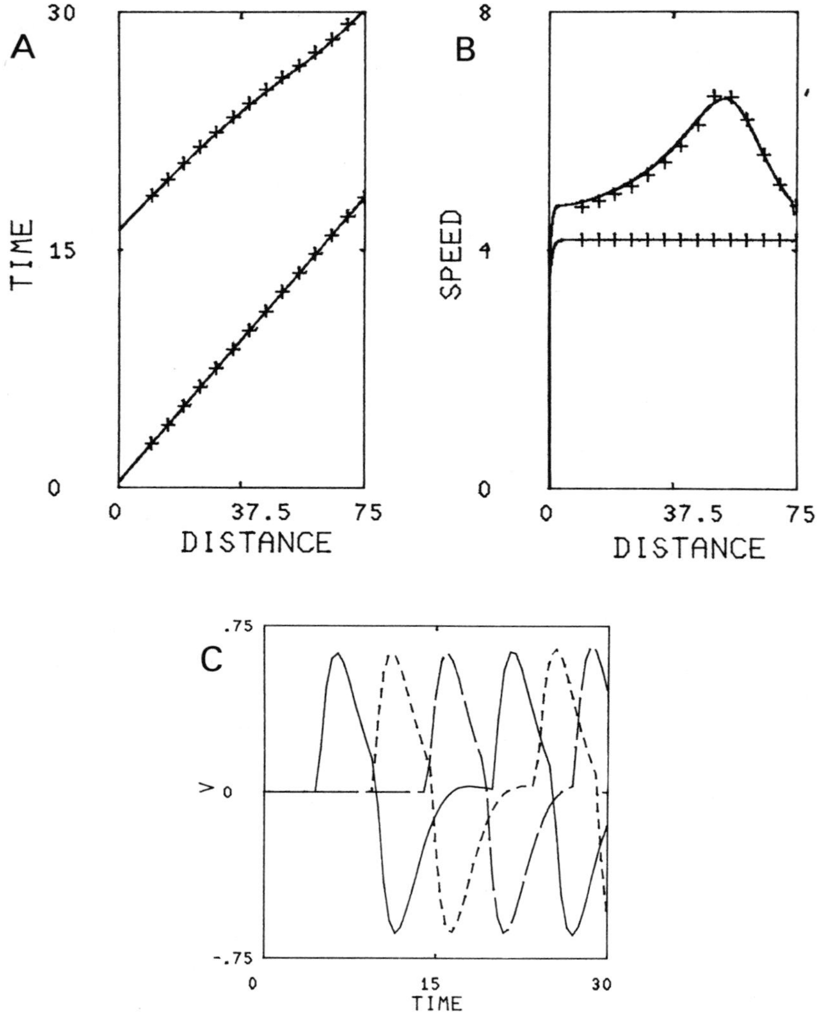

Fig. 8. Numerical solution of two-pulse FHN signaling problem (6); interstimulus time corresponds to + in Fig. 6. A and B illustrate x-t and c-x plots; crosses for kinematic approximation. C shows v versus t at x = 20,40,60.

intuitive understanding of the acceleration-deceleration phase may be obtained from Fig. 8-C in which v versus t is plotted at three locations x = 20,40,60 (solid, short dashes, long dashes respectively). In each case, the P_2 upstroke occurs in a phase of superexcitability, i.e. for $v > 0$, and $w < 0$ (not shown), and so the speed of P_2 exceeds c_0. At x = 20, the P_2 upstroke follows the v-overshoot peak of P_1. Thus P_2 is advancing into a region of increased superexcitability so it accelerates. As P_2 moves up the overshoot peak of P_1 it reaches a maximal supernormal speed at some point. Then, say at x = 60, even though P_2 still travels faster than c_0, it is advancing into a region of decreasing superexcitability and therefore decelerates.

This two-pulse numerical simulation is analogous to two-pulse experiments performed by neurophysiologists. Several cases of decreased conduction time of a pulse following a predecessor have been reported and attributed to supernormal conduction speed (e.g., see [32] and [22] for other references). Although the specific functional implications are yet to be fully described it is clear that decreases and/or increases in pulse speed occur during propagation (depending on the dispersion relation and the frequency range of operation) even along a uniform nerve. As more neurophysiologists become aware of these possibilities, more opportunities should arise for specific theoretical-experimental comparisons.

For the preceding calculation, P_2 travels faster than P_1. This cannot continue indefinitely since if P_2 gets too close to P_1 it must slow drastically according to Fig. 6. One may guess that if L were large enough, P_2 would asymptote to the speed c_0 in which case P_1 and P_2 would ultimately travel as a locked pair with fixed speed and spacing. To this question we will apply an approximate kinematics for individual pulses which we describe next.

IV. Approximate kinematics

To motivate our approximate treatment we refer back to Fig. 2 and recall, at least for large period wave trains, that pulse propagation speed is determined by the level of excitability at the upstroke location on the recovery trajectory of the predecessor. These different wave trains share an essentially common recovery trajectory, that of the solitary pulse, and differ primarily in time spent (t-duration) along this trajectory. Hence the single variable, time since preceding upstroke, describes not only speed of a pulse but also the approximate interpulse dynamics for these trains. We exploit the significance of this single variable to estimate the $x - t$ trajectory of an individual pulse upstroke. Suppose pulses are initiated at some location x_o and let $t^{n+1}(x)$ denote the arrival time of P_{n+1} at location $x \geq x_o$. Then, by definition, dt^{n+1}/dx is the reciprocal speed of P_{n+1} at x. As above, we say this is approximately determined by time since the preceding pulse and write

$$dt^{n+1}/dx = 1/c(t^{n+1}-t^n), \qquad x \geq x_o, \tag{7}$$

where for $c(\cdot)$ we use the dispersion relation (section II) to express speed as a function of temporal period or reciprocal frequency. Equation (7) is a recursive set of nonlinear ordinary differential equations for $t^{n+1}(x)$. As data, we must give the time at which P_{n+1} first appears at x_o: $t^{n+1}(x_o) = t_o^{n+1}$, and we also specify the trajectory of P_1: $t^1(x)$ for $x \geq x_o$. As presented above, the formulation does not provide a description for how pulses are initiated from initial and boundary data (e.g., the stimulus) or how pulses are influenced by spatial heterogenieties or boundaries into which they propagate. When a derivation for these kinematics is developed it may become clear how to treat some of these complications. Under special assumptions on relative time scales in the reaction dynamics (such as, $0 < \varepsilon << 1$ for the FHN case), singular perturbation methods have been used to describe pulse trajectories for some initial-boundary-value problems [20, 25]. We emphasize that no explicit time scale assumptions have been made for our pulse kinematics.

The signaling problems of the preceding section offer an opportunity to compare empirically this kinematic approximation with exact (i.e., numerical) solutions; additional examples may be found in [22]. In each case we assume P_1 is a solitary pulse so that $t^1(x) = t_o^1 + (x-x_o)/c_o$. The kinematic equations were solved numerically (forward Euler difference method), this solution evaluted at selected x locations and indicated by crosses on each of Figs. 5,7,8. For Fig. 5, we used $x_o = 1$ cm with $\Delta x = 0.05$ cm, $c_o = 12.3$ m/sec and for Figs. 7,8, $x_o = 5,10$ with $\Delta x = 0.2$ and $c_o = 4.176$. In each case we find very good quantitative agreement between the exact and approximate x - t trajectories as well as c - x plots. Moreover, although we expect the approximation to be best at lower frequencies, it works well over a substantial frequency range even toward higher frequencies (cf. Fig. 5-A,B and Fig. 7-A,B).

We next reconsider the experiment of Fig. 8 for which we had wondered if P_1 and P_2 would eventually form a two-pulse locked pattern if L were larger. Formal kinematics predict there is a stable locked pattern. For this case

$$\begin{aligned} dt^2/dx &= 1/c(t^2 - t^1) \\ dt^1/dx &= 1/c_o \end{aligned} \tag{8}$$

so that by subtraction

$$dT/dx = [c_o - c(T)]/[c_o\, c(T)] \tag{9}$$

where

$$T(x) = t^2(x) - t^1(x).$$

A locked pattern corresponds to $T = T^*$, a constant, x-independent solution to (9). This means $c(T^*) = c_o$, so that any crossing of the dispersion relation (c versus temporal period) at the level c_o is a possible locking interpulse time interval. For the scalar equation (9), an analysis of linear stability yields the simple condition that T^* provides a <u>stable</u> locked pattern if $dc(T^*)/dT > 0$ but unstable if $dc(T^*)/dT < 0$. In terms of Fig. 6 this means $dc(\omega^*)/d\omega < 0$ (where $\omega^* = 1/T^*$) for stability. Hence for Fig. 8, our approximate analysis implies that P_2 will lock with T^* corresponding to the c_o- crossing on the right branch of the supernormal hump (Fig. 6); the left branch c_o-crossing corresponds to an unstable interpulse time for two pulses. Observe that if c versus ω has a damped oscillatory approach to c_o as $\omega \to 0$ then an infinite number of two-pulse patterns may exist corresponding to the T_i^*, $c(T_i^*) = c_o$, with $T_{i+1}^* > T_i^*$. These different patterns, indexed by i, alternate between stable and unstable for increasing i. An unstable pattern is presumably one with P_2 following in a subexcitable recovery phase of P_1. Furthermore, one could construct a locked pattern with an arbitrary number of pulses whose interpulse time intervals are arbitrarily chosen from the T_i^*. Since the initial pulse would have speed c_o and since, for the kinematic description (7), a pulse cannot influence its predecessor, the pattern would have speed c_o. Any such pulse train is linearily stable if and only if each interpulse time in the train satisfies $dc(T_i^*)/dT > 0$. These approximate patterns are analogous to the finite pulse traveling wave solutions, homoclinic orbits, whose existence is proven by Evans, Fenichel, and Feroe (private communication) for a general nerve conduction equation when the solitary pulse exhibits a damped oscillatory return to rest (Fig. 2-C). Their rigorous analysis, which does not preclude a pulse from influencing its predecessors (through diffusion over long distances), shows asymptotically for large interpulse intervals that such stable pulse trains travel (slightly) faster than c_o.

By further kinematic analysis we may extend the two-pulse case of Fig. 8 to a longer stimulus train and a larger L value, say 200. For this we assume $t_o^{n+1} = t_o^n + 1/\omega_+$ where $\omega_+ = 0.0625$ for $n > 1$. Fig. 9 illustrates the kinematic approximation applied to nine successive pulses. An alternating transient pattern is observed. P_{2n} exhibits an acceleration-deceleration phase (Fig. 9-B) as its speed and interval $t^{2n} - t^{2n-1}$ pass over the supernormal hump (Fig. 6) as P_{2n} advances into its proper locking position with P_{2n-1}. On the other hand, P_{2n+1} travels slower than its initial speed. This follows because P_{2n}, as it speeds away, widens the gap from P_{2n} to P_{2n+1} and forces the speed of P_{2n+1} down the low frequency (left) branch of the supernormal hump. As the interval $t^{2n+1} - t^{2n}$ for P_{2n+1} increases, the speed decreases more and falls below c_o following the subnormal dip in Fig. 6. If the kinematics were extended for larger L one would see that, as $t^{2n+1} - t^{2n}$ continually increases, the speed of P_{2n+1} would eventually

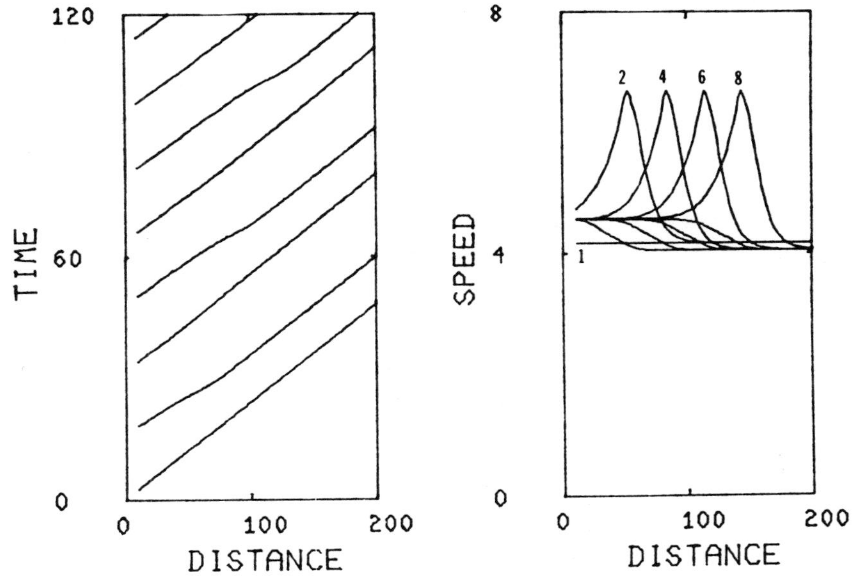

Fig. 9. Kinematic approximation for time periodic ($\omega = 0.0625$, case + in Fig. 6) initiation of pulses at $x_0 = 10$. Here, x - t (A) and c - x (B) plots illustrate tendency toward pairwise locking for this x and t range.

increase upward toward c_0 along the low frequency (left) branch of the subnormal dip. For a finite number of pulses, the regularly spaced pulses at x_0, initiated with speed $c(\omega_+)$, tend toward unevenly spaced locked pairs sufficiently far downstream with speed c_0, i.e. an imposition of pattern while initially there was only regularity. This is a consequence of the non-monotone nature of the dispersion relation. During propagation of a (finite) train over long distances, interpulse intervals with $dc/dT < 0$ (such as $1/\omega_+$) will not be maintained. Although with precise periodicity of t_0^n, for $n > 1$, successive pulses maintain uniform spacing longer. The mechanism for changes in spacing, as seen from (9), is differences in speed between successive pulses; hence, the longer P_{2n} travels with speed approximately $c(\omega_+)$, the longer will P_{2n+1} travel with the same speed. Yet, at a fixed finite x but after a sufficiently long time, the response will appear to be periodic in t with uniform interpulse times. Thus, the stimulus eventually forces the pulses to propagate over finite lengths at the uniform temporal spacing $1/\omega_+$. This regularity however would be broken if even a slight perturbation in stimulus timing were to occur. A break in the regular response pattern would follow; pairwise locking at unequal spacings would be seen transiently until gradually the regularity would be restored asymptotically for large t as in Fig. 9.

Non-monotonicity of the dispersion relation as a possible mechanism for pulse locking has been discussed by some previous investigators [3,18,30] but without rigorous or formal analysis. Here we have placed the phenomena into an explicit, empirically tested, kinematic framework. Some previous consideration of locking [3,18], was for propagation of pulses through a one dimensional ring geometry, e.g. a nerve fiber coupled end-to-end electronically or chemical waves in a topologically ring-shaped thin tube. For this, however, it seems more natural to focus on the dispersion relation of speed versus wavelength or wavenumber rather than speed versus temporal period or frequency as in [3,18]. We therefore consider an alternative kinematic approximation.

Kinematics for a ring geometry. Let $p^n(t)$ denote the position of P_n at time t on a ring of length L. For N pulses, oriented to propagate counterclockwise (positive direction here) around the ring, with $0 \le p^N(0) < \ldots < p^1(0) < L$ we write

$$dp^n/dt = c(p^{n-1} - p^n) \quad , \quad 2 \le n \le N \tag{10}$$
$$dp^1/dt = c(p^N + L - p^1).$$

Here, rather than introducing additional notation, we emphasize that $c(\cdot)$ is the dispersion relation expressing speed versus wavelength (spatial period) as in Fig 3-B. In this formulation, each pulse influences all other pulses on the ring as opposed to (7) for the linear geometry in which only successors are affected. By subtraction, one may obtain from (10) a system of equations for the N - 1 interpulse spatial intervals $P^n(t) = p^n(t) - p^{n+1}(t)$, $1 \le n \le N-1$:

$$dP^n/dt = c(P^{n-1}) - c(P^n), \qquad 2 \le n \le N-1$$
$$dP^1/dt = c(L - \sum_{n=1}^{N-1} P^n) - c(P^1). \tag{11}$$

We consider the special case of N=2 for which (11) reduces to

$$dP/dt = c(L-P) - c(P) \tag{12}$$

for $P(t) = p^1(t) - p^2(t)$. The possible t-independent equilibrium solutions for which P_1 and P_2 travel with the same speed **and fixed spacing P=P*, a constant, must satisfy c(L-P*)=c(P*).** The solution $P^* = L/2$ for equal spacing in the kinematic approximation is admissible for any L which exceeds twice the minimum wavelength, P_{min}, for the dispersion relation. This spacing is linearly stable as a solution to (12) if and only if $dc(P^*)/dP > 0$. Hence, for a strictly monotone curve as in the case $\varepsilon = 0.2$ of Fig. 3-B (upper branch), uniform spacing is stable for any $L > 2\ P_{min}$. For a curve which exhibits local maximum and/or minimum speeds, only certain ring lengths will allow stable uniform spacing. Thus, to allow for those ring lengths which do not, we consider the more general possibility of a steadily circulating pair of pulses which are unequally spaced, $L-P^* \neq P^*$. For simplicity, suppose the c versus P dispersion relation has a single maximum speed c_{max}, $c_{max} > c_o$, for $P = P_{c_{max}}$ so that $dc/dP < 0$ for $P_{c_{max}} < P \le \infty$ (e.g., consider case $\varepsilon = 0.5$ in Fig. 3-B and imagine the approach $c \to c_o$ as $P \to \infty$ to be monotone decreasing rather than damped oscillatory). To obtain the unequally-spaced locking patterns, it is conceptually easier to specify c, $c_o < c < c_{max}$, and determine the corresponding L(c) and $P^*(c)$ from the two intersections, $P_1(c) = P^*$ and $P_2(c) = L-P^*$ ($P_1 < P_2$) of the c-P curve with the horizontal line of constant speed c. From (12), the condition for stability is $dc(P_2)/dP + dc(P_1)/dP > 0$ or equivalently $dL/dc \bullet dc(P_2)/dP \bullet dc(P_1)/dP > 0$ where $L(c) = P_1(c) + P_2(c)$. Hence such a pattern is linearly stable if and only if $dL/dc < 0$. To easily check this condition, append to the c-P dispersion relation the curve $L(c)/2 = [P_1(c) + P_2(c)]/2$

versus c which specifies the mid-point of the left and right branches and which joins the c-P plot at its maximum $P_{c_{max}}$, c_{max}. Then stability follows if the c-P right branch is less steep than the left branch for the specified value of c. On the other hand, if the right branch falls off quicker in some range of c values, then the L(c)/2 curve bends leftward for this range (before ultimately bending rightward as $c \searrow c_o$) and the corresponding unequally spaced patterns are unstable. For such a ring length, however, one finds an additional unequally spaced pattern (slower speed) as well as an equally spaced one (faster speed); both of these patterns are stable for the same ring length. Thus, depending upon the c-P dispersion relation, there may be one or more stable locking patterns for a given ring length L. Here we have considered the simple case of N = 2 for a specified type of non-monotone behavior. For a dispersion relation like the $\varepsilon = 1.0$ case of Fig. 3-B and $N \geq 2$, the ring kinematics suggests a variety of stable and unstable patterns.

We remark that for the case of either a linear or ring geometry, our conclusions based on the formal kinematic description, are restricted only to the range of validity for that approximation.

V. Summary

Our discussion of pulse propagation in excitable systems has emphasized its dispersive nature. A solitary pulse propagating into a resting medium travels with fixed speed c_o. But in a train of more than one pulse, a non-leading pulse will not be advancing into a medium at rest so its speed in general will not equal c_o. To understand how the speed of a pulse may vary during propagation and perhaps asymptote to a speed other than c_o, we first considered a class of particular solutions to the reaction-diffusion equations, periodic traveling waves of uniformly spaced pulses. There is typically a one parameter family of such periodic wave trains with, for example, temporal or spatial period as the parameter. The speed of the train (and pulse amplitude) is determined by its period and this dependence constitutes a dispersion relation; for example,

$c = c(\omega)$ where ω the temporal frequency satisfies $0 < \omega < \omega_{max}$ where ω_{max} is the finite maximum frequency for steady periodic propagation. For an excitable system, the solitary pulse is seen as the limit of this family as $\omega \to 0$, i.e., as pulses become widely separated. Different long period wave trains have pulses with a similar upstroke-downstroke phase separated by a nearly identical recovery trajectory. Duration along this essentially common trajectory before the next upstroke varies from train to train; this duration determines the state of excitability at upstroke and hence propagation speed. This description suggests that one might apply the $c(\omega)$ relation to individual pulses in a, not necessarily periodic, train by interpreting ω as reciprocal time interval between successive upstrokes. Our empirical experience suggests that this may be reasonable over a substantial frequency range. For some illustrative examples of numerically solved signaling problems (full reaction-diffusion model with time-varying, spatially localized, stimulus presentation), instantaneous speed versus instantaneous frequency of a pulse at low to intermediate frequency ranges approximately satisfies the dispersion relation. Moreover, these observations suggest a kinematic description, based on the dispersion relation for periodic wave trains, to approximate individual pulse trajectories. Kinematic results agree very well quantitatively with pulse trajectories obtained from solutions of the full signaling problems.

In this paper we considered parameter ranges of models for which the computed dispersion relation was either strictly monotone, c decreasing with ω for stable wave trains, or had some non-monotone features (e.g., supernormal speeds for an intermediate range of frequencies and/or an oscillatory approach $c \to c_0$ as $\omega \to 0$); some mechanisms for these different characteristics were discussed. We also contrasted signal transmission in these different cases. For example, in the strictly monotone case, a pulse train initiated at position x_0 with a specific but not uniform, temporal pattern can lose pattern during propagation; pulse intervals widen and tend to become uniform. In contrast for a non-monotone case and stimulus frequency in some range of a local speed maximum say, a regular pattern imposed near x_0 may give way to non-uniformly, but specifically, spaced pulses, e.g., pairwise locking. The different effects are more or less significant depending on distance from the stimulus site, typical conduction times, range of speeds in the $c(\omega)$ relation, and for how long stimulus forcing is imposed.

Dispersion effects may be assessed from a description of speed versus temporal period as opposed to spatial period; which is more suitable may depend on the geometry of the medium and time scale of observation. We have considered propagation in only one space dimension but have discussed both linear and ring configurations to which different kinematic descriptions and stability notions seem natural. Our explicit numerical examples of dispersion relations and signaling problems (complete treatment and kinematic approximation) have been for nerve conduction models. Nevertheless, several chemical and biochemical systems exhibit excitability and for these we expect similar phenomena. Our results show how spatio-temporal pattern (perhaps interpretable as information) can be altered during propagation.

More experimental data on dispersive aspects of pulse propagation are needed to determine how significant are the effects for particular cases. Periodic stimulation and propagation over a range of frequencies should offer the most directly interpretable data. However, as the theoretical results suggest, estimates of $c(\omega)$ for the lower to intermediate frequency range may be obtained even from two-pulse experiments. In this latter case one should have measurements at several sites through the medium. For nerve conduction experiments one is often limited unfortunately to only two recording locations, one at the stimulus site and the other far downstream, which does not lead to direct description of the c versus ω relation. Results for the HH equation [22] may be more directly comparable to data from the experimentally convenient squid axon.

REFERENCES

1. Anan, Y. and N. Gō, Solitary wave and spatially locked solitary pattern in a chemical reaction system, J. Theoret. Biol. 80 (1979), 171-183.
2. Andronov, A. A., A. A. Vitt, and S. E. Khaikin, Theory of Oscillators, Pergamon Press, New York, 1966, 468-480.
3. Arshavskii, Y. I., M. B. Berkinblit, and V. L. Dunin-Barkovskii, Propagation of pulses in a ring of excitable tissues, Biophysics 10 (1965), 1160-1166.
4. Bell, J. and L. P. Cook, On the solutions of a nerve conduction equation, SIAM J. Appl. Math. 35 (1978), 678-688.
5. Carpenter, G. A., A geometric approach to singular perturbation problems with applications to nerve impulse equations, J. Differential Equations 23 (1977), 335-367.
6. Carpenter, G. A., Bursting phenomena in excitable membranes, SIAM J. Appl. Math. 36 (1979), 334-372.
7. Casten, R. H., H. Cohen, and P. Lagerstrom, Perturbation analysis of an approximation to Hodgkin-Huxley theory, Quart. Appl. Math. 32 (1975), 365-402.
8. Collins, M. A. and J. Ross, Chemical relaxation pulses and waves. Analysis of lowest order multiple time scale expansion, J. Chem. Phys. 68 (1978), 3774-3784.
9. Evans, J. W. and J. Feroe, Local stability theory of the nerve impulse, Math. Biosci. 37 (1977), 23-50.
10. Feroe, J., Temporal stability of solitary impulse solutions of a nerve equation, Biophys. J. 21 (1978), 103-110.
11. Field, R. J. and R. M. Noyes, Oscillations in chemical systems. IV. Limit cycle behavior in a model of a real chemical reaction, J. Chem. Phys. 60 (1974), 1877-1884.
12. FitzHugh, R., Impulses and physiological states in theoretical models of nerve membrane, Biophys. J. 1 (1961), 445-466.
13. FitzHugh, R., Mathematical models of excitation and propagation in nerve, in Biological Engineering (H. P. Schwan, ed.), McGraw-Hill, New York, 1969, 1-85.
14. Goldbeter, A., T. Erneux, and L. A. Segel, Excitability in the adenylate cyclase reaction in Dictyostelium discoideum, FEBS Letters 89 (1978), 237-241.

15. Greenberg, J. M., B. D. Hassard, and S. P. Hastings, Pattern formation and periodic structures in systems modeled by reaction-diffusion equations. Bull. Amer. Math. Soc. 84 (1978), 1296-1327.

16. Hastings, S. P., Some mathematical problems from neurobiology, Amer. Math. Monthly 82 (1975), 881-894.

17. Hodgkin, A. L. and A. F. Huxley, A quantitative description of membrane current and its application to conduction and excitation in nerve, J. Physiol. (Lond.) 117 (1952), 500-544.

18. Karfunkel, H. R. and C. Kahlert, Excitable chemical reaction systems. II. Several pulses on the ring fiber, J. Math. Biol. 4 (1977), 183-185.

19. ______________ and F. F. Seelig, Excitable chemical reaction systems. I. Definition of excitability and simulation of model systems, J. Math. Biol. 2 (1975), 123-132.

20. Keener, J. P., Waves in excitable media, preprint.

21. Kopell, N. and L. N. Howard, Plane wave solutions to reaction-diffusion equations, Stud. Appl. Math. 52 (1973), 291-328.

22. Miller, R. N. and J. Rinzel, The dependence of impulse propagation speed on firing frequency, dispersion, for the Hodgkin-Huxley model, preprint.

23. Nagumo, J. S., S. Arimoto, and S. Yoshizawa, An active pulse transmission line simulating nerve axon. Proc. IRE. 50 (1962), 2061-2070.

24. Ortoleva, P., and J. Ross, Theory of propagation of discontinuities in kinetic systems with multiple time scales: fronts, front multiplicity, and pulses. J. Chem. Phys. 63 (1975), 3398-3408.

25. Peskin, C. S., Partial Differential Equations in Biology, Courant Institute Lecture Notes, New York University, 1976, 55-68.

26. Reusser, E. J. and R. J. Field, The transition from phase waves to trigger waves in a model of the Zhabotinskii reaction, J. Am. Chem. Soc. 101 (1979), 1063-1071.

27. Rinzel, J. and J. B. Keller, Traveling wave solutions of a nerve conduction equation, Biophys. J. 13 (1973), 1313-1337.
28. Rinzel, J., Integration and propagation of neuroelectric signals, in Studies in Mathematical Biology (S. A. Levin, ed.) Math. Assoc. America, 1978, 1-66.
29. ________ and R. N. Miller, Numerical calculation of stable and unstable periodic solutions to the Hodgkin-Huxley equations, Math. Biosci., in press.
30. Scott, A. C. and S. D. Luzader, Coupled solitary waves in neurophysics, Physica Scripta (Sweden), in press.
31. Showalter, K., R. M. Noyes, and H. Turner, Detailed studies of trigger wave initiation and detection, J. Am. Chem. Soc., in press.
32. Swadlow, H. A. and S. G. Waxman, Variations in conduction velocity and excitability following single and multiple impulses of visual callosal axons in the rabbit, Exp. Neurol. 53 (1976), 128-250.
33. Tatterson, D. F. and J. L. Hudson, An experimental study of chemical wave propagation, Chem. Eng. Commun. 1 (1973), 3-11.
34. Troy, W. C., Mathematical modeling of excitable media in neurobiology and chemistry, in Theoretical Chemistry 4 (H. Eyring and D. Henderson, eds.), Academic Press, New York, 1978, 133-157.
35. Uppal, A., W. H. Ray, and A. B. Poore, On the dynamic behavior of continuous stirred tank reactors, Chem. Engng. Sci. 29 (1974), 967-985.
36. Winfree, A. T., Stably rotating patterns of reaction and diffusion, in Theoretical Chemistry 4 (H. Eyring and D. Henderson, eds.), Academic Press, New York, 1978, 1-51.
37. Winfree, A. T., The Geometry of Biological Time, Springer-Verlag, New York/Heidelberg, 1980.

Mathematical Research Branch, NIAMDD
National Institutes of Health
Bethesda, Maryland 20205

Current Problems in Combustion Research

Forman A. Williams

1. INTRODUCTION

Combustion research may be defined as the study of chemically reacting flows in which heat is liberated. Interest in the subject has been increasing during the past decade, first because of problems in air pollution that result from combustion and later because of the recognition of energy problems, for which combustion knowledge is pertinent in seeking solutions. This interest is exemplified not only by enhanced activity in design and experiment but also by intensified efforts in modeling and mathematical analysis.

"Modeling" has many different meanings, even within the restricted context of combustion; here it is defined as the construction of equations that are believed to describe various combustion processes--construction in forms sufficiently complete to enable solutions to be sought mathematically. The approaches adopted by applied mathematicians in seeking these solutions may be divided into two categories, numerical methods and analytical methods. In recent years, solutions by analytical methods based on perturbations have been advancing rapidly because applied mathematicians conversant with asymptotic analysis have "discovered" combustion. The present paper discusses approaches to some current problems in combustion, with emphasis placed largely on advances achievable by asymptotic analysis.

ISBN 0-12-669550-4

The equations of combustion will be exhibited first. Next, comments on numerical methods will be made. After that, asymptotic methods and some specific problems will be discussed. Instead of identifying outstanding problems in a separate section, the presentation distributes comments on current uncertainties, evaluations of existing work and statements of research needs throughout the paper. Reference citations are far from complete and are designed only to provide a few convenient entries into the literature.

2. THE EQUATIONS OF COMBUSTION

Exothermically reacting flows may be described by a set of partial differential equations [1,2]. Writing these equations and their boundary conditions in forms that provide well-posed problems constitutes one way to accomplish the modeling defined in the introduction. A nondimensional form that may be selected for the differential equations is the following:

2.1 Mass Conservation

If the nondimensional density ρ and velocity $\vec{v}$ are defined as the ratio of the dimensional density and velocity to the constant, representative values, ρ_0 and v, respectively, then mass conservation for a homogeneous medium may be written as

$$\omega \partial\rho/\partial t + \nabla \cdot (\rho\vec{v}) = 0 \; , \tag{1}$$

in which the ratio of flow time to transient time is $\omega = \ell/(v\tau)$, where ℓ and τ are representative scales of boundary length and transient time for the problem. The nondimensional time t and nondimensional gradient ∇ have been defined from the dimensional independent variables by use of τ and ℓ.

2.2 Momentum Conservation

To state an equation for conservation of momentum, let the nondimensional pressure p be defined as the ratio of the pressure to the representative value p_0, the nondimensional acceleration $\vec{f}$ from body forces be defined as the ratio of the body-force acceleration to the representative acceleration g, e.g. that of gravity, and the nondimensional shear-stress tensor $\overleftrightarrow{\tau}$ be defined as the ratio of the shear-stress tensor to $(\ell v \mu_0)$, where μ_0 is a representative value for the coefficient of viscosity. Then conservation of momentum may be written as

$$\omega \partial(\rho\vec{v})/\partial t + \nabla \cdot (\rho\vec{v}\vec{v}) = -M^{-2}\nabla p + F^{-1}\rho\vec{f} + R^{-1}\nabla \cdot \overleftrightarrow{\tau}, \qquad (2)$$

where $M = v/\sqrt{p_o/\rho_o}$ is a Mach number, $F = v^2/(g\ell)$ is a Froude number and $R = \rho_o v\ell/\mu_o$ is a Reynolds number for the flow. Physical understanding of the significance of these parameters is well developed [3].

2.3 Species Conservation and Chemical Kinetics

In chemically reacting flows, changes in concentrations of chemical species must be described. These concentrations may be defined by local, instantaneous values of the mass fractions Y_i of each species i (i = 1,...,N, there being N species in all). In addition to experiencing convection and transient accumulation, these species diffuse and react. If a nondimensional diffusion velocity $\vec{V}_i$ for species i (obeying $\sum_{i=1}^{N} Y_i\vec{V}_i=0$) is defined by dividing the dimensional diffusion velocity by v, then the equations for species conservation may be expressed in the form

$$\omega \partial(\rho Y_i)/\partial t + \nabla\cdot(\rho\vec{v}Y_i) = -\nabla\cdot(\rho Y_i\vec{V}_i) + \sum_{k=1}^{M} \nu_{ik} D_k f_k(p,T) e^{-\beta_k/T}$$

$$\prod_{j=1}^{N} X_j^{n_{jk}} [1 - K_k^{-1} g_k(p,T) e^{-\gamma_k/T} \prod_{j=1}^{N} X_j^{(m_{jk}-n_{jk})}], \quad i=1,...,N \qquad (3)$$

N-1 of which are linearly independent.

The summation term in (3) describes the effects of chemical reactions in a general, nondimensional form. In this term the nondimensional temperature T is the ratio of the local, instantaneous temperature to a constant temperature T_o, here called the flame temperature, whose value approximately equals the maximum temperature in the flow field; this nondimensionalization is important because most of the chemistry usually occurs near the maximum temperature. There are M different reactions, each identified by a subscript k. Important nondimensional parameters describing these reactions are Damköhler's first similarity group, D_k, the nondimensional activation energy, β_k, the nondimensional equilibrium constant, K_k and the nondimensional heat of reaction, γ_k. Here $D_k = \ell/(v\tau_{fk})$, where τ_{fk} is the characteristic time for occurrence of the k'th reaction in the forward direction. Reactions may also proceed in the reverse or backward direction, and $K_k = \tau_{bk}/\tau_{fk}$, where

τ_{bk} is the characteristic time for occurrence of the k'th reaction in the backward direction. The parameters D_k and K_k measure chemical rates. Important energetics of the chemistry are parameterized by β_k and γ_k, which measure the sensitivities of rates and equilibria to T. If E_k is the activation energy of reaction k and H_k is the heat released in this reaction, then $\beta_k = E_k/R_oT_o$ and $\gamma_k = H_k/R_oT_o$, where R_o is the universal gas constant.

Additional factors in the summation in (3) include ν_{ik}, a constant stoichiometric coefficient equal to the increase in the number of moles of species i in reaction k. The definitions have been constructed in such a way that the forward and backward rate functions, $f_k(p,T)$ and $g_k(p,T)$, are positive and of order unity. The most proper nondimensional measures of the concentrations whose products appear in the reaction rates are the mole fractions X_j of each species j ($X_j = (Y_j/W_j)/\sum_{i=1}^{N}(Y_i/W_i)$, where W_j is the molecular weight of species j). The constant exponents n_{jk} of these mole fractions represent the order of the forward reaction k with respect to species j; the constants m_{jk} are the corresponding orders of the backward reactions and obey $m_{jk} = n_{jk} + \nu_{jk}$ if the reaction is an elementary step. Detailed numerical studies often deal only with elementary steps [4], and for numerous elementary steps all rate parameters are known, with accuracies that typically vary from perfection for n_{jk} and m_{jk}, to better than a few percent for K_k, γ_k and β_k, to a factor of ten for D_k. Mathematical studies frequently approximate the detailed chemistry by a small number of steps, often one. In this case the effective rate parameters sometimes may be found approximately through analysis of the chemistry but usually must be measured from combustion experiments. Fewer of these effective rate parameters are known, but when they are available accuracies for D_k may exceed those in elementary steps. In combustion approximated by a one-step reaction, the orders n_{jk} need not be integers and are not simply related to the molecularities ν_{jk}, and the nondimensional measure β_k of the overall activation energy always is a large number, on the order of ten.

2.4 Energy Conservation

Conservation of energy, a key equation in combustion, may be expressed in several different forms, depending in part on whether attention is focused on thermal energy or on total (thermal plus chemical) energy. Working with total energy removes rates of chemical reactions from energy conservation and allows the equation to be written as

$$(\omega\rho\partial/\partial t + \rho\vec{v}\cdot\nabla)[\int c_p dT + \alpha \sum_{i=1}^{N} h_i Y_i + (1-\gamma^{-1})M^2 v^2/2]$$

$$= \omega(1-\gamma^{-1})\partial p/\partial t + P^{-1}R^{-1}\nabla\cdot(\lambda\nabla T) - \alpha \sum_{i=1}^{N} h_i \nabla\cdot(\rho Y_i \vec{V}_i)$$

$$+ (1-\gamma^{-1})M^2 R^{-1}\nabla\cdot(\vec{v}\cdot\overleftrightarrow{\tau}) + (1-\gamma^{-1})M^2 F^{-1}(\rho\vec{v}\cdot\vec{f} + \rho \sum_{i=1}^{N} Y_i \vec{V}_i\cdot\vec{f}_i)$$

$$+ B^{-1}\nabla\cdot\vec{q} \qquad (4)$$

Here the integral is the thermal enthalpy, the summation the chemical enthalpy and $v^2/2$ the kinetic energy; therefore changes in the total stagnation enthalpy are described. The ratio of the specific heat at constant pressure for the chemical mixture to a constant, reference specific heat c_{po} is denoted by c_p, and the ratio of the thermal conductivity of the mixture to a constant, reference value λ_o is λ. The ratio of constant-pressure to constant-volume specific heats at the reference condition is γ, provided that p_o, ρ_o and T_o refer to the same state. The nondimensional chemical enthalpy h_i is the ratio of the enthalpy of formation of per unit mass for species i at the temperature corresponding to the lower limit of the integral, to the overall heat released per unit mass of mixture in the representative combustion process considered. The parameter α is the ratio of this overall heat release to $c_{po}T_o$ and is of order unity in combustion.

Other previously undefined quantities in (4) are $\vec{f}_i$, the body-force acceleration on molecules of species i, divided by g, and $\vec{q}$, the ratio of the radiant energy flux to its representative value q_o. Additional new parameters that appear are the Boltzmann number, $B = \rho_o v c_{po} T_o / q_o$ and the Prandtl number, $P = \mu_o c_{po}/\lambda_o$. The importance of heat conduction in (4) may be characterized by a Peclet number, which equals PR. Terms involving M^2R^{-1} and M^2F^{-1} represent viscous dissipation and work due to body forces, respectively. The Dufour effect is excluded in (4) as insignificant.

2.5 Diffusion

Along with the conservation equations given above, expressions are needed for molecular transport, for example for the nondimensional diffusion velocity $\vec{V}_i$. The form best justified from the kinetic theory of gases is

$$R \sum_{j=1}^{N} S_{ij} X_i X_j (\vec{V}_j - \vec{V}_i)/d_{ij}(p,T) = \nabla X_i + (X_i - Y_i)\nabla p/p$$
$$+ \sum_{j=1}^{N} X_i X_j [\alpha_{ij} e_{ij}(p,T,X_\ell)/Y_i - \alpha_{ji} e_{ji}(p,T,X_\ell)/Y_j]\nabla T/T$$
$$+ M^2F^{-1}(\rho/p) \sum_{j=1}^{N} Y_i Y_j (\vec{f}_j - \vec{f}_i) \quad , \quad i=1,\ldots,N, \qquad (5)$$

of which N-1 are linearly independent. Here the functions $d_{ij}(p,T)$ of order unity and inversely proportional to p, are defined as the ratios of the coefficients of binary diffusion D_{ij} to representative, constant values D_{ijo}, and the Schmidt number for the species pair ij is $S_{ij} = \mu_o/(\rho_o D_{ijo})$. Terms on the left represent molecular diffusion, pressure-gradient diffusion, thermal diffusion and body-force diffusion, respectively. In the thermal-diffusion term, the parameters α_{ij} denote constant, representative thermal-diffusion ratios $D_{Tio}/(\rho_o D_{ijo})$, where D_{Tio} is a representative magnitude for the coefficient of thermal diffusion of species i in the mixture, and the functions $e_{ij}(p,T,X_\ell)$, of order unity, are $(D_{Ti}\rho_o D_{ijo})/(D_{Tio}\rho D_{ij})$, where D_{Ti} is the coefficient of thermal diffusion for species i. The parameters α_{ij} are typically of the order of one tenth [1].

3. DISCUSSION OF THE EQUATIONS

The unknowns in (1) through (5) are ρ, $\vec{v}$, Y_i, T and $\vec{V}_i$. Additional equations, such as the equation of state, the equation for the shear-stress tensor, etc., relate the other variables shown to these. Other quantities that appear, such as $\vec{f}_i$, are given in advance. Coupling to an equation for radiation transport may occur through $\vec{q}$; in an approximation in which absorption and stimulated emission are neglected, $-\nabla \cdot \vec{q}$

is a nondimensional rate of emission of radiation per unit volume, and $q_o = \varepsilon_o \sigma T_o^4$, where ε_o is a representative emissivity and σ the Stefan-Boltzmann constant. Modeling is completed by writing suitable boundary conditions for the problem.

The basis of the modeling at this level is firm, relying on known physics and on traditional developments in the kinetic theory of gases. However, the mathematical problem in general is too complicated to be solved, even numerically by the best electronic computers. Therefore in many problems further approximations must be introduced to obtain a set of manageable model equations. Of particular interest with respect to this further modeling are multiphase systems and turbulent systems.

3.1 Multiphase Combustion

The stated equations apply to gases that are dilute in the kinetic-theory sense. Often condensed fuels and products occur in combustion. The equations can be augmented by boundary conditions at interfaces and by differential equations applying within condensed phases to describe multiphase combustion. However, deterministic solutions to the resulting set cannot reasonably be sought if the condensed or gaseous phases are dispersed, as occurs, for example, in spray combustion or in the combustion of granular propellants or pulverized fuels. Statistical descriptions of dispersed phases become necessary [2], and less firmly based modeling is needed. Formulations often yield integrodifferential equations, e.g. with terms involving integration over dispersed-phase variables appearing in (1) through (4). For simple, laminar flows, the additional uncertainties in modeling often are not excessively severe; physical and geometrical simplifications then may lead to manageable problems with meaningful results. In complex or turbulent flows, uncertainties in required modeling are so great that analyses at best become equivalent to empirical correlations of experimental data over very limited ranges of conditions. The value of such investigations seems questionable.

3.2 Turbulent Combustion

Even in homogeneous systems, turbulence introduces severe problems of modeling in combustion. Since it is impractical to attempt to solve (1) through (4) deterministically for turbulence, additional hypotheses, termed "turbulence modeling", generally are introduced for the purpose of averaging over turbulent fluctuations in some sense [5,6].

The most common approaches to turbulence modeling are moment methods, in which equations like (1) through (4) are written for averages of the dependent variables, for some of their variances (e.g., turbulent kinetic energy or turbulent length scale) and occasionally for some higher moments. Formal development of moment methods encounters closure problems, in that an equation generated for one moment always contains higher moments. Approximations for these higher moments effectively comprise the turbulence modeling needed by moment methods. For nonreacting flows, these approximations are relatively well developed [7], and it is often quite meaningful to add turbulent viscosities, thermal conductivities and diffusion coefficients to the molecular transport coefficients in (2), (4) and (5) to describe evolutions of averaged variables.

However, in combustion the moment methods are not justified [5]. It is easy to show, from (3) for example, that any realistic description of chemistry necessitates retention of numerous higher moments. Moreover, for simplified problems in which alternatives to moment methods are available, clear demonstrations have been given recently that if diffusion approximations for turbulent transport are introduced then negative coefficients of turbulent diffusion are needed in portions of the flow field that cannot be defined precisely in advance [8, 9]. These difficulties establish that turbulence modeling commonly employed in combustion leads to formulations having highly questionable foundations. Therefore exercising vast computational routines for solving partial differential equations obtained from moment-method modeling, e.g. for combustion in the cylinder of a spark-ignition engine, appears to constitute a sterile activity that may correlate data through adjustment of parameters but is unlikely to advance fundamental understanding.

There are alternative approaches to turbulence modeling based on consideration of probability density functions for describing statistical distributions of dependent variables [5,6]. In the simplest versions of such approaches modeling consists mainly of selecting shapes for probability density functions, parameterized by a small number of variables. For example, in a premixed system that reacts rapidly once chemistry begins, the probability density function for the reactant concentration may be approximated as a delta function at the initial concentration, a delta function at zero concentration and a small, constant value in between [6,10]. Derivation of differential equations for evolution of the descriptors of the probability density functions employs suitable averages of (1) through (4) and necessitates very few modeling approximations of the moment-method type. For this reason, the approximation of probability density functions provides descriptions of turbulent combustion having less objectionable foundations than those of moment-method modeling; some understanding of combustion in simpler flows can be obtained by this approach.

A more ambitious use of probability density functions is to derive and solve differential equations for their evolution, without approximating their shapes [6]. Although closure problems arise here as well, foundations of the modeling are somewhat firmer. The mathematical problem of solving the derived equations can be challenging and generally needs numerical methods for electronic computers. Although this path is difficult, its pursuit eventually might provide the most satisfactory description of turbulent combustion.

There has recently been a flurry of activity directed toward developing novel approaches to turbulence modeling of reacting systems, through ideas generated by observations of "coherent structures" in turbulence [6,11,12]. The chemistry is viewed as occurring in flow structures, convected and deformed by turbulence but maintaining their identity over their lifetime; statistical approaches to describing the distributions of these structures are sought. Such approaches currently are incomplete, in that thorough and well-posed mathematical formalisms have not been defined clearly. The value of these novel approaches cannot be assessed properly unless definitions are made more precise.

Another class of approaches to modeling of turbulent combustion is based on formal expansions of (1) through (4) for limiting values of parameters [8,13]. Such expansions often are found to be perturbations of laminar flows and consequently may apply only for relatively limited ranges of turbulent conditions. However, their modeling approximations are the firmest of all, being based only on acceptance of the validity of (1) through (4) as a fundamental description of the fluctuating flow and on the assumption of existence of the expansion. Types of expansions that may be studied are numerous. Some are indicated in the following subsection.

3.3 Parameters in Combustion

Numerous nondimensional parameters appear in (1) through (5). Specifically, the quantities ω, M, F, R, D_k, β_k, K_k, γ_k, ν_{ik}, n_{ik}, m_{ik}, α, γ, P, B, S_{ij} and α_{ij} have been defined and eshibited. In multiphase flows, a variety of other parameters enter, such as r_j, a characteristic dimension of connected regions of the j'th phase, divided by ℓ, and m_j, the ratio of the mass of the j'th phase to the total mass at a reference condition. In turbulent flows, additional parameters that occur include $\varepsilon = (\overline{v'^2})^{1/2}/v$, the ratio of a representative root-mean-square fluctuation in velocity to the representative velocity v (which may be a laminar flame speed, for example) and $\delta = L/\ell$, where L is one of the characteristic length scales of turbulence (e.g., the integral scale) and ℓ might be taken as the thickness of a laminar flame. Further parameters, such as additional length ratios, may be introduced through boundary conditions. More than twenty parameters thus are seen to arise in combustion.

Progress is not made by a frontal attack on a system of equations with twenty parameters. It is essential to restrict attention to limits in which numerous parameters assume special values. Moreover, combustion situations encountered in practice correspond to limiting values of many parameters. Thus, asymptotic concepts may play a central role in formulation of tractable combustion problems. For example, small values of ε or δ^{-1} may be selected as a basis for studying turbulent flames. Depending on how many parameters are assigned limiting values, the problem becomes suitable for numerical or analytical methods.

3.4 Limiting Classes of Problems

There are too many limiting combinations of parameters for a thorough discussion of limiting cases to be given. However, some comments may be made.

Putting $\omega = 0$ defines steady problems. Of course, stationary turbulent flows require retention of ω terms except in equations for suitable averages. There are numerous important laminar problems with $\omega = 0$. Tractable problems with $\omega \neq 0$ have other simplifications, such as complete spatial homogeneity, one-dimensionality or two-dimensionality, effectively eliminating many other parameters.

The largest group of current combustion problems occurs in low-speed flows for which $M = 0$ is a reasonable limit; in such cases γ also is irrelevant. Exceptions are in supersonic combustion and detonations, where other approximations, such as $R = \infty$, are applicable and provide simplifications except in narrow regions such as shock waves.

In many problems, buoyancy and electrostatic forces are negligible, so that $F = \infty$ may be introduced to provide simplification. In others, such as fire problems with flows induced by natural convection, the Grashof number, a combination of F and R, is large, introducing boundary-layer phenomena [14]. In (2), it is seldom ever necessary to retain F and M in the same problem, and M^2F^{-1} in (4) and (5) is practically always negligible. Although there are a few problems where $R \to 0$ is a reasonable limit, in most situations $R \to \infty$ is more realistic. In two or more dimensions this does not enable terms in R^{-1} to be neglected; it introduces boundary-layer approximations instead. There are also problems, such as those for premixed laminar flames, having R of order unity.

Parameters describing relative magnitudes of molecular transport effects are P, S_{ij} and α_{ij}. In condensed phases, S_{ij} and the Lewis numbers $L_{ij} \equiv S_{ij}/P$ are large, while $P \to \infty$ for solids but typically is small for liquids; gases exhibit narrower variations, having P and S_{ij} of order unity, except for light molecules such as hydrogen, for which S_{ij} may decrease to values on the order of one tenth. Thus, in describing combustion of gases it is important to retain P and S_{ij}, treating R, RP_{ij} and RS_{ij} as being of the same order.

In (5), M^2F^{-1} and the term in $\nabla p/p$ generally are negligible, the latter partially because $X_i - Y_i$ seldom can be appreciable. Since α_{ij} is small, it may be expected that the term in $\nabla T/T$ also can be omitted. Although this is usually true, there are special problems for which thermal diffusion should be retained. For example, stability of premixed flames is modified by small departures of L_{ij} from unity, and α_{ij} in (5) thereby may exhibit quantitatively large effects on flame stability [15]. In certain diffusion flames, such as those surrounding burning aluminum droplets, condensed products are formed in sizes initially much smaller than a molecular mean-free-path, and these small particles may escape from the hot reaction zone by diffusing as a heavy molecule under the influence of ∇T. In premixed flames, hydrogen atoms may be attracted preferentially to the hottest zone by thermal diffusion. In problems such as the latter two, thin zones may exist, due to α_{ij}, which would not exist if $\alpha_{ij} = 0$. Since very little study has been devoted to such questions, it would be of interest to investigate the influences of α_{ij} on combustion.

In most problems, α_{ij} is negligible and only ∇X_i remains on the right-hand side in (5). Often as a further approximation it is assumed that $S_{ij}/d_{ij}(p,T)$ is the same for each pair (i,j). This leads to Fick's law in the form

$$RS_{ij}Y_i\vec{V}_i = -d_{ij}(p,T)\nabla Y_i \quad , \quad i = 1, \ldots , N , \tag{6}$$

separately relating the diffusion velocity of each species to its own concentration gradient. Use of (6) in (3) and (4) eliminates $\vec{V}_i$ entirely as a dependent variable. This approach is useful unless influences of differing coefficients of diffusion for different species are to be investigated. There are problems in which these differences can be important. For example, they influence stability of premixed flames and explain why cellular flames occur in hydrogen-oxygen systems not only under fuel-lean conditions [16].

In studying influences of differing diffusion coefficients for different species, often (6) is used with $S_{ij}/d_{ij}(p,T)$ assigned different values for each i, corresponding to introduction of a multicomponent diffusion coefficient for each species in the mixture. This prevalent practice is mathematically incorrect; proper introduction of multicomponent diffusion coefficients requires more complicated equations, except in extreme limits, such as for systems having a large excess of a single species. Relatively few studies have employed (5) with only ∇X_i on the right-hand side. More investigations of this type would be desirable, for example for purposes of improving descriptions of cellular flames in premixed systems. Furthermore, steady, one-dimensional equations for describing premixed-flame structure by numerical methods could be solved just as easily if the correct formulation (5) were employed instead of an improper formulation involving multicomponent diffusion coefficients.

The radiation parameter B in (4) is large if the emissivity ε_o is sufficiently small. Often $B = \infty$ is reasonably accurate in combustion. Effects of radiant energy losses in combustion may be studied by perturbation methods in which B^{-1} is a small parameter. Very few studies of this type have been made. Their pursuit could help to clarify effects of radiant loss on flame extinction, for example. There are situations (largely turbulent or particle-containing flows such as are found in fires or in furnace combustion) in which energy transfer occurs primarily by radiation; treating B^{-1} as small in these cases might be questionable.

The remaining parameters listed in the previous section, aside from those describing multiphase or turbulent characteristics, are all related to the chemistry. The quantities ν_{ik}, n_{ik} and m_{ik} all are of order unity, although some may be zero. Irreversible reactions are described by the limit $K_k \to \infty$. Investigations have been made of phenomena such as equilibrium broadening of diffusion flames by considering expansions for large but finite values of K_k within the approximation of a one-step reaction [17]. Although reversibility often is important in chemical reactions, the amount of fuel present at equilibrium in combustion systems almost always is negligible. Therefore chemistry more complex than a one-step model is needed if realistic descriptions of near-equilibrium effects such as diffusion-flame broadening are to be obtained. It should be possible to develop simplified chemical models that describe effects of equilibrium dissociation through expansion for suitable $K_k \to \infty$.

The Damköhler group D_k is large in vigorously reacting systems. Rate broadening of diffusion flames has been described for one-step reactions through expansions for large values of D_k [17]. Also, ideas of large D_k form the bases of many treatments of turbulent flames [6]. However, when possible it is desirable to retain D_k as a parameter, rather than focusing attention on limiting values for it. In this way, a wider range of combustion phenomena can be described and understood.

The remaining two parameters relating to chemistry are the nondimensional heat release α and the nondimensional activation energy β_k. These may be termed Frank-Kamenetskii parameters, after the author who defined combustion as the science of large heat release and large activation energy [18]. By definition, α is between 0 and 1; in combustion, realistically $1-\alpha$ is a small parameter. Little has been done to attempt to exploit this fact for achieving simplifications in descriptions of combustion processes. On the contrary, by neglecting density variations, analyses often implicitly presume that α is small. In numerical studies, correct values of α usually are included because computations are not simplified significantly by restricting α to limiting values. However, through analytical methods, studies of $1-\alpha$ as a small parameter of expansion might help to improve understanding of heat-release aspects of combustion phenomena.

By far the most useful parameter of expansion in combustion problems has proven to be β_k. The fact that at least some of the β_k's are large can make numerical integrations difficult. However, through analytical methods, expansions for large β_k have added clarity to many aspects of combustion phenomena [17]. These include ignition, extinction, flame propagation and flame stability. Asymptotic methods, based on large β_k, have provided predictions with accuracies sufficient for comparison with experiment. In recent years, a large body of literature has been generated on activation-energy asymptotics [19]. A few specific combustion problems in which it is useful to take advantage of large β_k will be discussed in the following section.

4. EXAMPLES OF SPECIFIC PROBLEMS

The problems to be addressed below all concern one-step, irreversible reactions of high nondimensional activation energy. The reaction subscript k is omitted from the symbols for brevity. Although attention is focused on the use of asymptotic analysis for large β, relevance of other methods also is mentioned occasionally.

4.1 Thermal Explosions

A homogeneous mixture, capable of reacting exothermically, is placed in a vessel at time zero, and it is desired to calculate the temperature history. First neglect spatial variations, take $N = 2$, $h_1 = 1$, $h_2 = 0$, $c_p = 1$ and $\gamma = 1$; then (4) gives $T + \alpha Y_1 = 1$. Use of this in (3) with the further selections $\nu_1 = -1$, $X_1 = Y_1$, $n_1 = n$, $f = \alpha^{n-1}$ produces the problem

$$dT/dt = D(1-T)^n e^{-\beta/T} \quad , \quad T(0) = 1-\alpha \quad . \tag{7}$$

Although numerical integration is straightforward, use of an asymptotic expansion with $(1-\alpha)(\alpha\beta)^{-1}$ and $(1-\alpha)\beta^{-1}$ small gives, in a first approximation, for T near $1-\alpha$,

$$T = 1 - \alpha - [(1-\alpha)^2/\beta] \ln[1 - \beta D\alpha^n (1-\alpha)^{-2} e^{-\beta/(1-\alpha)} t], \tag{8}$$

leading to a first approximation for the explosion time,

$$t = (1-\alpha)^2/[\beta D\alpha^n e^{-\beta/(1-\alpha)}]. \tag{9}$$

From (7), $T \to 1$ as $t \to \infty$ for $n > 0$. Hence (8) must fail before T reaches unity. Nevertheless (9) is useful because T increases more rapidly with t as T increases, until $1-T$ becomes of order β^{-1}. Corrections to (8) and (9) can be generated through use of multiple time scales.

To describe thermal explosions more realistically, elaborations of (7) may be studied. Heat losses may be considered, in formulations ranging from two simultaneous ordinary differential equations to simultaneous partial differential equations. If losses are too great, rapid explosion does not occur, and there is bifurcation to a solution exhibiting slow reaction. With $\gamma \neq 1$, $\partial p/\partial t$ in (4) couples the explosion to momentum conservation (2) and leads to pressure waves [20]. These complications remain topics of current research.

4.2 Steady, Planar Premixed Flames

It is desired to calculate the structure and propagation speed for a one-dimensional reaction wave with $\omega = 0$, and $M = 0$. First consider $N = 2$, $h_1 = 1$, $h_2 = 0$, $c_p = 1$, $R = 1$, $P = 1$, $S_{ij} = 1$, $d_{ij} = 1$, $\alpha_{ij} = 0$, $\lambda = 1$, $B = \infty$, $K = \infty$, $\nu_1 = -1$, $X_1 = Y_1 \equiv Y$, $n_1 = n$ and $f = \alpha^{n-1}$. With $Y_1(-\infty) = 1$, $Y_1(\infty) = 0$, $T(-\infty) = 1-\alpha$ and $T(\infty)$ bounded, (4) and (6) lead to $T + \alpha Y_1 = 1$, and (3) becomes

$$d^2Y/dx^2 - dY/dx = \Lambda F(Y) \quad , \quad Y(-\infty) = 1, \ Y(\infty) = 0 \quad , \qquad (10)$$

where

$$F(Y) = Y^n e^{-\alpha\beta Y/(1-\alpha Y)} \quad , \qquad (11)$$

and where Λ is a constant proportional to $De^{-\beta}$, which is inversely proportional to the square of the speed of propagation. Evidently (10) is a two-point boundary-value problem for an ordinary differential equation, and solutions are anticipated to exist for a particular value of Λ, the burning-rate eigenvalue.

In fact, as formulated, solutions to this problem do not exist, but they do in an asymptotic sense of $\alpha\beta \to \infty$ [17]. There is a convective-diffusive zone, where $F(Y)$ is negligible, extending over an x interval of order unity. Downstream from that there is a reactive-diffusive zone, where dY/dx is negligible in first order, and where Y is of order $\alpha\beta$; this zone extends over an x interval of order $(\alpha\beta)^{-1}$. The expansion $\Lambda = \Lambda_0(\alpha\beta)^{n-1} + \Lambda_1(\alpha\beta)^{n-2} + \ldots$ is obtained through matched asymptotic expansions.

Extensions of this problem in many directions remain subjects of current research. The ambiguity in the formulation is being clarified further by investigating a time-dependent problem in which the mixture reacts ahead of the wave [21]. Flame speeds for two-reactant systems with one-step reactions have been given [22], and extensions to chemistry of more complex types have been considered by both numerical and asymptotic methods. Investigations of effects of heat loss and of consequent extinction of the flame have been pursued [23]. Steps have been taken toward elimination of virtually all of the restrictions listed at the outset, but in many cases more work remains to be done.

4.3 Cellular and Pulsating Flames

Premixed flames having cellular shapes often are observed. An approach to describing them may be defined roughly by considering, among other things, $R = P = \rho = 1$, $L_{ij} \equiv S_{ij}/P = L_i \neq 1$ and by deriving from (3), (4) and (5), for flames propagating in the x direction, equations like

$$\left.\begin{aligned} &\partial T/\partial x = \nabla^2 T + \alpha D Y_1 Y_2 \, e^{-\beta/T} \,, \\ &\partial Y_i/\partial x = L_i^{-1}\nabla^2 Y_i + \nu_i D Y_1 Y_2 \, e^{-\beta/T} \,, \quad i = 1,2,\ldots,N \\ &T(-\infty,y,z) = T_o \,, \quad Y_i(-\infty,y,z) = Y_{io} \,, \quad i = 1,2,\ldots,N \\ &T_x(\infty,y,z) = Y_{ix}(\infty,y,z) = 0 \,, \quad i = 1,2,\ldots,N \\ &T \text{ and } Y_i \text{ bounded as } y,z \to \infty \,, \quad i = 1,2,\ldots,N \,. \end{aligned}\right\} \quad (12)$$

This boundary-value problem for partial differential equations, which is ill posed in the same sense that (10) is ill posed, would be appropriate for describing steadily propagating cellular flames. Since there always are planar solutions in an asymptotic sense, a search for nonplanar solutions is an investigation in bifurcation theory.

The stability of nonplanar solutions to (12) currently is uncertain. Successful analyses have treated the stability of planar solutions. The most recent approach is through an asymptotic expansion for large $\alpha\beta$, again identifying reactive-diffusive and convective-diffusive zones, and matching [24]. For $N = 1$, dispersion relations exhibit cellular instabilities if $L_1 < 1 - 2/(\alpha\beta)$. Cellular structure occurs if the reactant diffuses more readily than heat, i.e., the instability is thermo-diffusive. The use of $\rho = 1$ rules out hydrodynamic instabilities. Effects of relative diffusion of reactants have been clarified by analysis for $N = 2$ [25]. If L_1 is sufficiently large, then an instability occurs with a nonzero imaginary part of the growth rate, and transversely moving waves are predicted. For $L_1 \to \infty$, there are pulsating planar solutions whose characteristics have been explained by bifurcation analyses [26]. This limit applies to combustion waves in solids; both pulsating and spinning waves have been observed and described [27,28].

Various phenomena concerning cellular flames remain to be investigated. One is merely the solution of (12), which has not been obtained. Another is the influence of strong variations of ρ on cellular behavior; in realistic flames, ρ in the burnt gas is an order of magnitude less than that in the fresh mixture. Possibly $1-\alpha$ is a useful small parameter of expansion for studying these effects. Influences of using (5), instead of the diffusion approximations in (12), on occurrence of cellular phenomena also would be of interest to study. Effects of more complex chemistry on cellular structure deserve more attention; preliminary indications suggest that certain kinetics tend to eliminate cellular instabilities [29]. There is need for much more work on cellular and pulsating phenomena.

4.4 Ignition

Ignition of reactive materials typically is described by initial-boundary-value problems for partial differential equations. A representative problem, corresponding to $\omega = RP = \rho = 1$, $L_{ij} = \infty$ (a solid material), is

$$\left.\begin{array}{ll} \partial T/\partial t = \partial^2 T/\partial x^2 + \alpha D Y^n e^{-\beta/T} & , \\ \partial Y/\partial t = -\nu D Y^n e^{-\beta/T} \quad (t > 0,\ x > 0) & , \\ Y(x,0) = 1 \ ,\ T(x,0) = 1-\alpha \quad (x > 0) & , \\ Y(\infty,t) = 1 \ ,\ T(\infty,t) = 1-\alpha \quad (t > 0) & , \\ T_x(0,t) = 1 \ \text{ or } \ T(0,t) = 1 \quad (t > 0) & . \end{array}\right\} \tag{13}$$

These problems typically have an inert zone, transient-diffusive in character, away from the boundary at which the ignition stimulus is applied, and a narrower reactive-diffusive zone adjacent to the boundary. For large β they also have an initial stage, often inert, followed by an ignition stage. Asymptotic formulas for ignition times can be obtained through use of a criterion of thermal runaway in the ignition stage; this has been done for (13), [30]. Although many problems of this type have been solved, there are others remaining to be investigated. To follow the history beyond the ignition stage is difficult.

4.5 Extinction

Extinction problems also are amenable to asymptotic analysis for large β. Premixed-flame extinction by heat loss has been mentioned previously. Similar methods are applicable to extinction in nonpremixed systems, such as burning fuel droplets in oxidizing atmospheres or counterflow diffusion flames [31]. For the latter, results of asymptotic analyses have been employed for extracting parameters of overall rates of combustion reactions from experimental data on extinction [32]. Thus, useful, quantitative interplays between theory and experiment ensue. Needs for additional work include studies of influences of radiant losses and of more complex chemistry on diffusion-flame extinction.

4.6 Wrinkled Laminar Flames in Turbulent Flows

Possibilities of using expansion methods for investigating premixed turbulent flames have been mentioned in Sections 3.2 and 3.3. Under a variety of restrictions [8], a representative problem that emerges is

$$\left.\begin{array}{l} \partial Y/\partial t + \partial Y/\partial x + \vec{v}\cdot\nabla Y - \nabla^2 Y = -\Lambda F(Y) \quad , \\ Y(-\infty,y,z,t) = 1 \quad , \quad Y(\infty,y,z,t) = 0 \quad , \\ Y \text{ homogeneous in } y \text{ and } z \text{ and stationary in } t. \end{array}\right\} \qquad (14)$$

Here F is given by (11), the fluctuating velocity vector $\vec{v}$ is a zero-mean, divergence-free random variable with the same statistical properties stated in (14) for Y, and Λ is a constant to be determined, inversely proportional to the square of the turbulent flame speed. The equation in (14) includes the physical phenomena of accumulation, convection, diffusion and chemical production of heat or reacting species. The problem is one in stochastic partial differential equations.

Investigations of (14) have employed first the singular expansion for large $\alpha\beta$, followed by a regular expansion in gradients, for large δ. Conclusions have been drawn concerning structures and propagation speeds of turbulent flames consisting of wrinkled laminar flames [8,13]. However, there are many questions outstanding, such as structures of the turbulent flames for conditions under which cellular instabilities of the corresponding laminar flames occur, and influences of variations in ρ on the turbulent flames. Alternative expansions, such as expansions for small intensities ε of turbulence, have not been fully explored. Thus, there are opportunities for further advances in theories of turbulent combustion by use of asymptotic methods.

5. CONCLUSIONS

The coverage of combustion problems given here is incomplete. Yet it can be seen that numerous areas within the field require further study by mathematical methods. There are so many parameters in the equations of combustion that despite many recent advances, much remains to be done.

REFERENCES

1. Hirschfelder, J.O., C.F. Curtiss, and R.B. Bird, Molecular Theory of Gases and Liquids, John Wiley and Sons, New York, 1954.
2. Williams, F.A., Combustion Theory, Addison-Wesley Publishing Co., Reading, Mass., 1965.
3. Landau, L.D., and E.M. Lifshitz, Fluid Mechanics, Addison-Wesley Publishing Co., Reading, Mass., 1959.
4. Westbrook, C.K., and F.L. Dryer, A comprehensive mechanism for methanol oxidation, Combustion Sciences and Technology 20 (1979), 125-140.
5. Libby, P.A., and F.A. Williams, Turbulent flows involving chemical reactions, Annual Reviews of Fluid Mechanics 8 (1976), 351-376.
6. Libby, P.A., and F.A. Williams, editors, Turbulent Reacting Flows, Springer-Verlag, Berlin (to appear-1980).
7. Bradshaw, P., editor, Turbulence, Springer-Verlag, Berlin, 1976.
8. Clavin, P., and F.A. Williams, Theory of premixed flame propagation in large-scale turbulence, Journal of Fluid Mechanics 90 (1979), 589-604.
9. Libby, P.A., and K.N.C. Bray, Implications of the laminar flamelet model in premixed turbulent combustion, Combustion and Flame (to appear-1980).
10. Bray, K.N.C., and J.B. Moss, A unified statistical model of the premixed turbulent flame, Acta Astronautica 4 (1977), 291-320.
11. Roshko, A., Structure of turbulent shear flows: A new look, AIAA Journal 14 (1976), 1349-1357.
12. Spalding, D.B., Mathematical models of turbulent flames: A review, Combustion Science and Technology 13 (1976), 3-25.

13. Clavin, P., and F.A. Williams, Effects of Lewis number on propagation of wrinkled flames in turbulent flows, Proceedings of the Seventh International Colloquium on Gasdynamics of Explosions and Reactive Systems, J.R. Bowen, editor (to appear-1980).

14. Williams, F.A., Mechanisms of fire spread, Sixteenth Symposium (International) on Combustion, The Combustion Institute, Pittsburgh, 1977, 1281-1294.

15. Garcia, P., and P. Clavin, Dufour and Soret effects and stability properties of a premixed non-adiabatic flame, Proceedings of the Seventh International Colloquium on Gasdynamics of Explosions and Reactive Systems, J.R. Bowen, editor (to appear-1980).

16. Mitani, T., and F.A. Williams, Studies of cellular flames in hydrogen-oxygen-nitrogen mixtures, Combustion and Flame (to appear-1980).

17. Williams, F.A., Theory of combustion in laminar flows, Annual Reviews of Fluid Mechanics 3 (1971), 171-188.

18. Frank-Kamenetskii, Diffusion and Heat Transfer in Chemical Kinetics, Plenum Press, New York, 1969.

19. Buckmaster, J., and G.S.S. Ludford, Theory of Laminar Flames, Cambridge University Press (to appear-1980).

20. Kassoy, D.R., The supercritical spatially homogeneous thermal explosion-initiation to completion, Quarterly Journal of Mechanics and Applied Mathematics 31 (1978), 99-112.

21. Zeldovich, Y.B., Flame propagation in a mixture reacting at the initial temperature, preprint, USSR, Chernogolovka, 1978.

22. Mitani, T., Propagation velocities of two-reactant flames, Combustion Science and Technology (to appear-1980).

23. Joulin, G., and P. Clavin, Analyse asymptotique des conditions d'extinction des flammes laminares, Acta Astronautica 3 (1976), 233-240.

24. Sivashinsky, G.I., Diffusional-thermal theory of cellular flames, Combustion Science and Technology 15 (1977), 137-145.

25. Joulin, G., and T. Mitani, Linear stability analysis of two-reactant flames, Combustion and Flame (to appear-1980).

26. Matkowsky, B.J., and G.I. Sivashinsky, Propagation of a pulsating reaction front in solid fuel combustion, SIAM Journal on Applied Mathematics 35 (1978), 465-478.

27. Merzhanov, A.G., A.K. Filonenko, and I.P. Borovinskaya, New phenomena in combustion of condensed systems, Proceedings of the Academy of Sciences, USSR, Physical Chemistry Section 208 (1973), 122-125.

28. Sivashinsky, G.I., On spinning propagation of combustion waves, SIAM Journal on Applied Mathematics (to appear-1980).

29. Liñán, A., personal communication, 1979.

30. Liñán, A., and F.A. Williams, Ignition of a reactive solid exposed to a step in surface temperature, SIAM Journal on Applied Mathematics 36 (1979), 587-603.

31. Liñán, A., The asymptotic structure of counterflow diffusion flames for large activation energies, Acta Astronautica 1 (1974), 1007-1039.

32. Krishnamurthy, L., F.A. Williams, and K. Seshadri, Asymptotic theory of diffusion-flame extinction in the stagnation-point boundary layer, Combustion and Flame 26 (1976), 363-377.

This work was supported in part by the United States Air Force Office of Scientific Research under Grant Number AFOSR77-3362.

Department of Applied Mechanics
and Engineering Sciences
University of California, San Diego
La Jolla, CA 92093
and
Department of Mechanical and
Aerospace Engineering
Princeton University
Princeton, NJ 08544

Measurement and Estimation of Rate Constants

David M. Golden

1. INTRODUCTION

There are probably as many chemical processes as one can write balanced equations. Even excluding these that are intuitively unlikely on steric grounds, there will always remain many more elementary processes than measured rate constants. This simple truth becomes more apparent if rate constants as a function of temperature and pressure are desired.

For the purposes of this symposium, I shall assume that we all agree that rate constants for elementary processes as a function of temperature and pressure (and perhaps nature of the other species) are the desired goal for use as input to a computer code designed to simulate a complex chemical process. Thus, the goal should be to develop an empirically based framework for estimation of most rate constants and an arsenal of measurement techniques which allow the determination of sensitive values with sufficient precision for the tasks at hand.

In this discussion, we will address the status of this goal for homogeneous gas-phase processes.

2. THERMOCHEMICAL KINETICS

Background

The development of an empirically based framework for extrapolation and estimation is intimately tied to the available methods for measuring rate constants. These are discussed separately for organizational convenience. This framework has

ISBN 0-12-669550-4

been given[1] the name "Thermochemical Kinetics" by S. W. Benson. I have discussed "Thermochemical Kinetics" in the following manner with reference to both tropospheric measurements [2] and combustion problems [3] recently, but I will repeat it here for completeness.

Thermochemistry

It is impossible to begin a discussion of the theoretical basis for critical evaluation and extrapolation of thermal rate data without first discussing methods for estimating thermochemical quantities, such as $\Delta H^{o}_{f,T}$, ΔS^{o}_{T}, and $C^{o}_{p,T}$ for molecules.

Group Additivity

When a sufficient data base exists, we have found [4] the method of group additivity to best fit the need for accuracy and ease of operation. The basic concept and assumptions involved in the group additivity method are as follows:

For the disproportionation reaction

$$RNN'R + SNN'S \rightleftharpoons RNN'S + SNN'R$$

any additivity approximation assumes that $\Delta\Phi = \Delta\Phi_{\sigma}$, where Φ is any molecular property, and $\Delta\Phi_{\sigma}$ is the contribution to that property due to symmetry changes and optical isomerism. For the molecular properties of interest here, $\Delta H_T \rightarrow 0$, $\Delta C_{p,T} \rightarrow 0$, and $\Delta S_T \rightarrow S_{\sigma} = R \ln K_{\sigma}$, where $K_{\sigma} = \sigma(RRNN'R)\,\sigma(SNN'S)/\sigma(RNN'S)\,\sigma(SNN'R)$, $\sigma(X)$ being the symmetry number including both internal and external symmetry. An additional term for entropy of mixing, due to the existence of optical isomers, must also be included.

If the molecular framework NN′ is two atoms or greater, these relationships imply the additivity of group properties, which include all nearest-neighbor interactions, since a group is defined as an atom together with its ligands (e.g., in the group C-$(H)_3$(C), the central C atom is bonded to three H atoms and one C atom). Thus the equation

$$CH_3OH + CH_3CH_2OCH_3 \rightleftharpoons CH_3CH_2OH + CH_3OCH_3$$

implies the additivity of the properties of the groups C-$(H)_3$(C), C-$(H)_3$(O), O-(C)(H), C-$(H)_2$(C)(O), and O-$(C)_2$, if the appropriate $\Delta\phi = \Delta\phi_{\sigma}$.

We have developed group additivity methods that permit the estimation, for many organic chemicals in the gas phase, of heats of formation to ± 1 kcal/mole, and of entropies and heat capacities to ± 1 cal/(mole-K), from which free energies of formation can be derived to better than ± 2 kcal/mole.

It should be noted that entropy and heat capacity are molecular properties that can be accurately estimated under much less stringent conditions than energy (or enthalpy). Thus the method of bond additivity seems to work quite well (± 1 cal/(mole-K)) for estimating the former properties, but not at all well (± 4 kcal/mole) for the latter.

Structural Considerations and Model Compounds

If sufficient thermochemical data are lacking for the estimation of group properties, entropy and heat capacity can often be adequately estimated from structural parameters of the molecule. (Enthalpy estimates are more difficult, requiring a better knowledge of potential functions than is usually available.) The methods of statistical thermodynamics may be used to calculate C_p^o and S^o directly for those molecules where a complete vibrational assignment can be made or estimated.

Also, "reasonable" structural and vibrational frequency "corrections" to the corresponding established thermodynamic properties of "reference" compounds may be made. A suitable choice of reference compound, i.e., one similar in mass, size and structure to the unknown, assures that the external rotational and translational entropies and heat capacities of the reference and unknown compounds will be the same and that many of the vibrational frequencies will be similar. The basic assumption is that S^o and C_p^o differences can be closely estimated by considering only low-frequency motions thought to be significantly changed in the unknown. Fortunately, entropies and heat capacities are not excessively sensitive to the exact choice of these vibrational frequencies, and estimates of moderate accuracy may be made with relative ease.

Kinetics

The extension of thermochemical estimation techniques to the evaluation of kinetic data rests largely on the validity of transition state theory.

The transition state theory expression for a thermal rate constant is,

$$k = (\kappa T/h)\exp[-\Delta G_T^{o\ddagger}/RT]$$

(the units are sec^{-1} for the first order and atm^{-1} for the second order) and,

$$\Delta G_T^{o\ddagger} = \Delta H_{300}^{\ddagger} - T\Delta S_{300}^{o\ddagger} + \langle\Delta C_p^{\ddagger}\rangle[(T-300) - T\ln(T/300)]$$

(In the ideal gas approximation we can drop the standard state notation on $\Delta H^{\ddagger}$ and $\Delta C_p^{\ddagger}$.) If the empirical temperature dependence is represented by

$$k = AT^B\exp(-C/T)$$

$$A = \left[\kappa/h(300)^{\langle\Delta C_p^{\ddagger}/R\rangle}\right]\exp\left[(\Delta S_{300}^{\ddagger} - \langle\Delta C_p^{\ddagger}\rangle)/R\right]$$

$$B = (\langle\Delta C_p^{\ddagger}\rangle + R)/R$$

$$C = \left(\Delta H_{300}^{\ddagger} - \langle\Delta C_p^{\ddagger}\rangle(300)\right)/R$$

κ = Boltzmann's constant

h = Planck's constant

$\Delta S_{300}^{\ddagger}$ = entropy of activation at 300 K, standard state of 1 atm.

$\Delta H_{300}^{\ddagger}$ = enthalpy of activation at 300 K

$\langle\Delta C_p^{\ddagger}\rangle$ = average value of the heat capacity at constant pressure of activation over the temperature range 300--T °K.

If we wish to express second-order rate constants in concentration units instead of pressure units, we must multiply by RT in the appropriate units. This has the effect of writing:

$$k = A'T^{B'}\exp(-C/T)$$

where $A' = AR$ and $B' = B + 1 = (\langle\Delta C_p^{\ddagger}\rangle + 2R)/R$

Thus, simple "Arrhenius behavior" which will be sufficient for lower tropospheric temperatures is characterized for first-order reactions by $\Delta C^{\ddagger} = -R$; ($\Delta C_p^{\ddagger} = \Delta C_v^{\ddagger} = \Delta C^{\ddagger}$), and for second-order reactions using concentration units by $\Delta C_p^{\ddagger} = -2R$ (or $\Delta C_v^{\ddagger} = -R$).

In the case of simple Arrhenius behavior:

$k = A \exp(-B/T)$

$\log A = \log(e\kappa \langle T \rangle / h) + \Delta S^{\ddagger}/R; \quad B = (\Delta H^{\ddagger} + R\langle T \rangle)/R$

Thus, the quanties $\Delta H^{\ddagger}$, $\Delta S^{\ddagger}$, and $\Delta C_p^{\ddagger}$ are of interest. We apply similar methods to those already discussed with respect to thermochemistry to view rate data in a rational framework. These techniques are discussed in some detail by Benson[1], but certain points are worthy of re-emphasis here.

We begin by classifying reactions as unimolecular or bimolecular. (The only termolecular processes of interest to us will be energy-transfer controlled bimolecular processes.)

Unimolecular Processes

Simple Fission:	$AB \rightarrow A + B$
Complex Fission:	Molecule → Molecule + Molecule (or radical)
Isomerization:	Intramolecular atom rearrangement

Bimolecular Processes

Direct Metathesis:	$A + BX \rightarrow AX + B$
Addition:	A + Molecule → Stable Adduct (reverse of complex fission)
Association:	$A + B \rightarrow A - B$ (reverse of simple fission)

The first thing to notice is that of all these reactions, only direct metathesis reactions are not subject to becoming energy transfer limited at high temperatures and low pressures (i.e., in the "fall-off" region!). This means that not only does the so-called high pressure rate constant need to be estimated or known, but the extent of fall-off, as well. Methods are available for making fall-off corrections[5].

In hydrocarbon reactions in the troposphere, we may expect that most direct metathesis reactions will involve the exchange of a hydrogen atom between larger groups. A simple, semi-empirical prescription exists for estimating the value of $\Delta S^{\ddagger}$ for these types of reactions. First, one realizes that these values are limited between the "loosest" possible model (A-factor equals gas kinetic collision frequency) and the "tightest" possible model in which $R\cdots H\cdots R'$ is represented by the molecule $R\text{-}R'$. Experience using data in the range

$300K < T < 700K$ has taught us that generally the $\Delta S^{\ddagger}$ value corresponds to a transition state only slightly looser than the tightest possible value.

Since the other two classes of bimolecular processes are the reverse of unimolecular reactions, we may consider them in that direction. (The equilibrium constant is either known or estimable.) Once again, using experimental results as our guide, we note that model transition states which correspond to the values of $\Delta S^{\ddagger}$ are generally "tight". That is, we may visualize them as minor modifications of the reactant molecule, usually involving some increase in rotational entropy due to slight enlargement of certain bonds. The dominant entropic feature is usually the stiffening of internal rotations as a result of multiple bond formation or ring formation [1].

Bond scission reactions present a particular problem, since it is particularly difficult to locate a transition state. Recent work [6] both experimental and theoretical, indicates that these reactions can be modeled with a transition state which becomes tighter as the temperature rises. Lower limits of A-factors can be estimated fairly accurately for H-atom metathesis reactions by making use of model transition states [1]. We illustrate the method for the atmospherically important reaction of OH with HNO_3 which is supposed to proceed via: $HO + HONO_2 \rightarrow H_2O + NO_3$. Reported values [7] are $k \sim 10^8\ M^{-1}\ s^{-1}$ independent of temperature. Our analysis suggests that the reported rate constant does not represent the reaction as written above.

$$HO + HONO_2 \rightarrow \left[H\!\diagup^{O}\cdot H\cdot {}^{O}\!\diagdown NO_2\right]^{\ddagger} \rightarrow H_2O + NO_3$$

	$\Delta S^{0\ddagger}_{300}$ [a]	$\Delta C^{\ddagger}_{p,300}$ [a]	$\Delta C_{p,400}$ [a]
Reference Reaction [TS = $CH_3ONO \approx HOONO_2$]	−35.4	− 1.6	1.0
Spin	1.4	——	——
Symmetry	0	——	——
External Rotation (2 x 2 x 1.5)[b]	1.7	——	——
O—O(1000 cm^{-1}) → O·H(r.c)	− 0.1	− 0.4	− 1.0
H·O(2000 cm^{-1})	0	0	0.2
O—O—N(400 cm^{-1}) → O·H·O(600 cm^{-1})	0.5	0.2	0.1
O·H·O(600 cm^{-1})	1.0	1.0	1.6
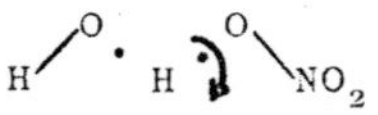 [c]	>2.2	1.0	1.0
	> −28.7	0.2	2.9

[a] cal $mole^{-1}$ deg^{-1} (standard state, 1 atm)

[b] Estimated increase of product of inertia.

[c] 10° off linear.

$$\log[A_{300}/M^{-1}\ s^{-1}] \geq 13.2 + [-28.7 + 8.35]/4.58 \geq 8.8$$

$$\log[A_{400}/M^{-1}\ s^{-1}] \geq 13.3 + [-28.7 + 1.6 \ln 4/3 + 8.92]/4.58 \geq 9.1$$

$$E_{300} = \Delta H^{\ddagger}_{300} + 2R(.300) = \Delta H^{\ddagger}_{300} + 1.2 \text{ kcal mole}^{-1}$$

$$E_{400} = \Delta H^{\ddagger}_{300} + 1.6(.1) + 2R(.4) = \Delta H^{\ddagger}_{300} + 1.8$$

$$E_{400} - E_{300} = 0.6 \text{ kcal mole}^{-1}$$

3. MEASUREMENT TECHNIQUES

Bimolecular

Bimolecular reactions of interest are usually radical-molecule or radical-radical interactions, although occasionally a molecule-molecule reaction will be an elementary process. (Gas-phase ion-molecule reactions are of interest sometimes, and they can be included in the same general framework with the realization that attractive forces are longer ranged.) Since bimolecular rate constants can be measured only if concentrations are known, the standard "trick" is to arrange for "pseudo first-order" conditions by running the reaction in a large excess of one of the reactants. This allows the relative measurement of the other reactant to determine a first-order rate constant, k^{I}, which, when divided by the concentration of the overwhelming component, will yield k^{II}, the second-order rate constant of interest. Most experimental techniques take advantage of the pseudo-first-order trick over the ranges of T and P accessible.

Several methods exist for concentration (or relative concentration) measurements. Most prominent are mass spectroscopy and all the wavelength ranges of optical spectroscopy, both absorption and fluorescence.

The experimental environment can also vary widely: Flow tubes that monitor time as distance after an interaction region, real-time measurements that are rapid enough to determine actual concentration changes, shock tubes that create high-temperature conditions rapidly, and some recently exploited infrared laser-heating techniques (vide infra).

Unimolecular

Unimolecular reactions are measured using many of the same general techniques as for bimolecular reactions. In general, a substantial energy barrier must be surmounted in order for these processes to proceed, so that moderate-to-high temperatures are required. In addition, since unimolecular processes are dependent on the total pressure and the specific nature of other constituents of the gaseous mixture, all these properties need to be under experimental control.

Very Low-Pressure Pyrolysis (VLPP)

A particular technique that can be used is the Very Low-Pressure Pyrolysis (VLPP) technique developed at SRI International[8] . VLPP is ideally suited for the measurement of the rate of initial bond-breaking reactions in the pyrolysis of organic molecules where secondary reactions often interfere with the characterization of the initial step. The technique has been described in detail previously [8,9].

The procedure consists of allowing the reactant to flow through the inlet system at a controlled rate and switching valves in the inlet lines so that the reactant alternately flows through a Knudsen cell reactor or through the bypass directly into the mass spectrometer. Since the flow rate is held constant, the difference between the mass spectrometer signals for the reactant in the bypass position and the reactor position corresponds to the amount of reactant that is decomposed as it flows through the reactor.

The data are interpreted with the aid of various steady-state expressions derived as shown below. At low flow rates, the treatment for a simple irreversible unimolecular decomposition is appropriate, and the extraction of rate parameters is straightforward. At higher flow rates (higher reaction pressures) and smaller escape apertures, rapid bimolecular reactions can compete with unimolecular decomposition and with escape from the reactor. These interactions must be included in the analysis. The reason for using VLPP at pressure high enough for secondary reactions to occur is shown by the description given below of our recent study for the pyrolysis of 1-ethyl naphthalene [10]. Briefly, observation of competition between unimolecular bond scission and radical recombination amounts to measurement of an equilibrium constant:

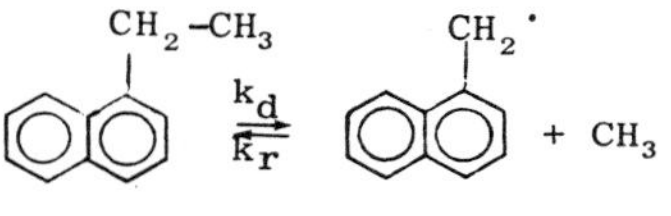

$$\begin{array}{cc}\text{CH}_2\text{--CH}_3 & \text{CH}_2{}^{\bullet} \\ \bigcirc\!\!\bigcirc \;\; \underset{k_r}{\overset{k_d}{\rightleftharpoons}} \;\; \bigcirc\!\!\bigcirc \;\; + \; \text{CH}_3 \end{array} \qquad (1)$$

Reliable measurement of an equilibrium constant can provide very good third-law values for ΔH°, since these values are not subject to the systematic errors that can markedly affect the slope of Arrhenius plots. Thus, by providing for measurement of an equilibrium constant that is otherwise not readily measured at higher pressures or in static systems, the VLPP system allows two largely independent measurements of bond strength and, therefore, a valuable internal consistency check.

For irreversible, unimolecular decomposition of a substance A, only three rate processes are considered:

$$\begin{aligned} \text{inlet} &\xrightarrow{R_A} \text{A} \\ \text{A} &\xrightarrow{k_d} \text{products} \\ \text{A} &\xrightarrow{k_{eA}} \text{mass spectrometer} \end{aligned} \qquad (2)$$

where R_A is the rate at which the reactant is allowed to flow into the reactor and k_{eA} is the first-order rate constant describing escape from the reactor.

Steady-state analysis provides the expression:

$$\frac{A_{escaped}}{A_{reacted}} = \frac{(A)_{ss}}{(A)_{o,ss} - (A)_{ss}} = \frac{k_{eA}}{k_d} \qquad (3)$$

where $(A)_{o,ss}$ is the steady-state concentration in the reactor when there is no decomposition.

When recombination of the radical fragments produced by unimolecular decomposition competes with escape from the reactor, the following reactions must be considered:

$$\begin{aligned} &\text{inlet} \xrightarrow{R_A} \text{A} \\ &\text{A} \underset{k_r}{\overset{k_d}{\rightleftharpoons}} \text{B}^{\bullet} + \text{C}^{\bullet} \\ &\left.\begin{aligned} \text{A} &\xrightarrow{k_{eA}} \\ \text{B}^{\bullet} &\xrightarrow{k_{eB}} \\ \text{C}^{\bullet} &\xrightarrow{k_{eC}} \end{aligned}\right\} \text{escape to mass spectrometer} \end{aligned} \qquad (4)$$

The controlled flow rate into the reactor is given by R_A, and the first-order escape rate constants for the various fragments are related by square roots of their masses so that the escape constant for the 1-naphthylmethyl radical ($B^{\bullet}$) is related to that for 1-ethylnaphthalene by

$$k_{eB} = k_{eA}\left(\frac{156}{141}\right)^{1/2} \simeq k_{eA}(1.05) \qquad (5)$$

Steady-state analysis of this sequence for 1-ethyl naphthalene provides an expression similar to equation (3):

$$\frac{A_{escaped}}{A_{reacted}} = \frac{(A)_{ss}}{(A)_{o,ss} - (A)_{ss}} = \frac{k_{eA}}{k_d} + \frac{k_r F(A)}{k_d (3.39)} \qquad (6)$$

Equation (6) indicates that decomposition and competitive recombination will provide a straight line of slope k_r/k_d and intercept k_{eA}/k_d. In the limiting case where recombination is unimportant, equation (6) reduces to equation (3).

Photolytic Radical Production

We have recently [11] combined the phenomenon of infrared multiphoton dissociation of organic molecules with the VLPP technique to produce a method for determining the rate constant for radical-molecule reactions at temperatures determined by reactor wall temperatures and completely independent of any need to heat the radical precursor. (Current powerful dye lasers will allow this technique to be useful for UV-vis photochemical radical production as well.) The current application has been to reactions of CF_3 radicals from CF_3I, but extension to aliphatic and aromatic systems requires only time and funds.

In this experiment, as previously, the effusive molecular beam was mechanically chopped in the second (differentially pumped) chamber before it reached the ionizer of the quadrupole mass filter (Finnigan 400). The signal was demodulated by a lock-in amplifier (PAR 128A) whose output was now stored in a signal averager (PAR 4202), which also served to trigger the laser. The two-aperture Knudsen cell was fitted with KCl windows and had an optical pathlength of 20.5 cm and a volume of approximately 105 cm^3. The cell was coated with Teflon by rinsing with a finely dispersed Teflon slurry in a water/aromatic solvent mixture (Fenton Fluorocarbon, Inc.) and curing at 360 C.

The Lumonics TEA-laser (Model K-103) was operated at the R(16) line of the 9.6μ transition at a pulse repetition frequency of .25 Hz, slow enough to permit > 99% of the reaction products to escape from the cell before the next laser shot. The multimode output of the laser consisted of a pulse of approximately 5.0 J, directed through the photolysis cell after being weakly focused by a concave mirror (fl = 10 m), which gave a beam cross section of 2.67 cm^2 at the KCl-entrance window of the cell.

The typical experiment consisted of averaging the time-dependent mass spectroscopic signal intensity of the products for a number of laser shots (10-100) as a function of the flow rate of the reactant gases at constant CF_3I flow rate and constant energy per pulse. Although experiments could be performed on a single-shot basis, the signal/noise ratio was improved by averaging the results of a number of laser shots.

The total yield of product formed in the reaction of interest was then determined by integration of the accumulated time-dependent signal of the signal averager on a strip-chart recorder fitted with an electronic integrator (Linear Instruments, Inc.).

Using the apparatus described above, reaction products could be observed as well as the transient depletion of CF_3I. An advantage of the low-pressure technique is that the effects of secondary reactions are minimized, although they must be considered in the data analysis.

The following chemical reaction mechanism is appropriate for studying the reaction $CF_3 + Br_2 \rightarrow CF_3Br + Br$:

$$CF_3I + nh\nu \rightarrow CF_3 + I$$

$$CF_3 + Br_2 \xrightarrow{1} CF_3Br + Br$$

$$CF_3 \xrightarrow{k_w} \text{(wall loss \underline{not} yielding } CF_3Br)$$

No heterogeneous first-order reaction of CF_3 to produce CF_3Br is included, since the data interpretation does not suggest it. As in the usual data treatment for VLPP studies, each molecular and radical specie escapes from the reactor with a characteristic first-order escape rate constant; for CF_3I, Br_2, CF_3, and CF_3Br, the escape rate constants are k_2, k_3, k_4, and k_5.

Analysis of the reaction mechanism and solution of the appropriate differential equations give an expression for the time-dependent mass spectrometer signal due to CF_3Br. The total yield of CF_3Br, (Y), is related to the rate constants as follows:

$$Y^{-1} = \left(\alpha\beta V\left[CF_3I\right]_0\right)^{-1}\left[1 + \frac{k_4 + k_w}{k_1\left[Br_2\right]}\right]$$

where pseudo-first-order conditions are assumed to hold. In this expression, α is a mass spectrometric sensitivity factor, β is the fraction of the initial CF_3I that is dissociated by the laser pulse, and V is the volume of the cell. The initial $[CF_3I]_0 = F_{CF_3I}/(V \cdot k_2)$, where F_{CF_3I} is the flow rate of CF_3I into the reactor, and V is the reactor volume; similarly, $[Br_2] = F_{Br_2}/(V \cdot k_3)$. For each F_{Br_2}, two escape apertures can be used, giving two different values for the escape rate constants, corresponding to the "big" and "small" apertures. Plots of Y^{-1} versus $F_{Br_2}^{-1}$ give two straight lines with intercepts c_s and c_b (small and big apertures) given by i = s or b):

$$c_i = \frac{k_{2i}}{\alpha\beta VF_{CF_3I}}$$

The slopes of the straight lines are given by

$$m_i = \frac{c_i V(k_{4i} + k_w)k_{3i}}{k_1}$$

Laser-Powered Homogeneous Pyrolysis (LPHP)

In VLPP experiments aimed at obtaining unimolecular rate information, we rely on the walls of the VLPP reactor to be the source of heat through gas-wall collisions, while at the same time being non-catalytic for the destruction of the substrate. Sometimes this latter condition is not met. Therefore, we may, in addition to VLPP, use the technique of laser-powered homogeneous pyrolysis [12]. This technique is a valuable complement to the VLPP procedure, since it essentially provides a "wall-less" reactor. The total pressure in the reactor is on the order of 100 torr, consisting mostly of bath gas and SF_6. An IR laser is used to heat the strongly absorbing sensitizer (SF_6), using a wavelength at which the substrate does not absorb. The SF_6 transfers its thermal energy by collision to the substrate molecules, and decomposition takes place. As described by Shaub and Bauer [12], the technique worked well for compounds of relatively high vapor pressure, using a static reactor system and a CW laser; we are currently adapting the technique for application to poly-nitro aromatics and other low-vapor-pressure substrates, using a pulsed CO_2 laser and a flow system with GC detection. The reasons for these modifications are: (1) When operation is with a pulsed laser, reaction times are short because of rapid cooling by contact with the off-axis cell contents, and secondary reactions are either unimportant or can be minimized by suitable choice of a scavenger, and (2) when low-vapor-pressure substrates are being studied, quantitative recovery and measurement of products and unreacted starting material is simpler with a flow system.

We have preliminary evidence from this technique which indicates that nitrotoluenes decompose by NO_2-aromatic bond scission. This seems to be true of those with ortho-methyl substitution as well, in contrast to earlier reports.

4. DISCUSSION

There are various limitations on the usefulness of thermochemical kinetics. These range from quantitative to qualitative uncertainties. We need to test some of the preceding ideas with experiments conceived for just that purpose. Since the ideas are based on the transition state theory formalism, it is important to address the question of limits of validity of transition state theory. In general, these testing reactions should be measured under conditions where isolated reactions can be observed, as the extraction of individual rate constants from complex reacting systems is fraught with difficulty.

There are many examples of reactions for which rate constants have been studied, but product studies are lacking.

Thus, in reactions of OH with olefins current smog and combustion models must arbitrarily decide on branching ratios. This is equally true in aromatic systems.

In all of the above discussion of estimation of rate data, the importance of thermochemical values for all species has been emphasized. It is particularly important to have a good set of values for the entropy and heat of formation of organic free radicals.

Very few spectroscopic assignments exist for modest-to-large size organic free radicals. Entropies (and heat capacities) have generally been estimated by methods discussed earlier. Uncertainties arise from changes in hindered rotation barriers and changes in skeletal bending frequencies.

Many bimolecular reactions are not direct, and involve a bound complex as an intermediate. These give rise to what seem to be unusual parameters for bimolecular processes [13].

A simple treatment for the pressure dependence of unimolecular processes, which would lend itself to easy use in large models, is seemingly close at hand, but care must be exercised.

In summary, we have a framework for the codification and extrapolation of rate data, but much testing and modification will be necessary.

REFERENCES

1. Benson, S. W., Thermochemical Kinetics, 2nd Ed., John Wiley and Sons, Inc., New York, 1976.
2. Golden, D. M., in Chemical Kinetics Data Needs for Modeling the Lower Troposphere, Proceedings of a Workshop at Reston, VA, May 1978. NBS Special Publication 557, US Government Printing Office, Washington, DC, August 1979.
3. (a) Golden, D. M., "Estimation of Rate Constants of Elementary Processes-A Review of the State of the Art," Fourteenth Symposium (International) on Combustion, The Combustion Institute, Pittsburgh, PA, 1973.

(b) Golden, D. M., in Summary Report on the Workshop on High Temperature Chemical Kinetics, NBS Special Publication 531, US Government Printing Office, Washington, DC, December 1977.

(c) Golden, D. M., "Pyrolysis and Oxidation of Aromatic Compounds" in Progress in Astronautics and Aeronautics, Vol. 62, 233 (1978), Chapt. III, Craig T. Bowman and Jorgen Birkeland, Ed., Proceedings of Project SQUID Workshop on Alternative Hydrocarbon Fuels: Combustion and Chemical Kinetics, September 1977, Columbia, MD.

4. Benson, S. W., F. R. Cruickshank, D. M. Golden, G. R. Haugen, H. E. O'Neal, A. S. Rodgers, R. Shaw, R. Walsh, Chem. Rev., 69, 279 (1969).

5. (a) Golden, D. M., Richard K. Solly, and Sidney W. Benson, J. Phys. Chem., 75, 1333 (1971).
(b) Troe, J., J. Phys. Chem., 83, 114 (1979).

6. (a) Golden, D. M., A. C. Baldwin, and K. E. Lewis, Int. J. Chem. Kinetics, 11, 529 (1979).
(b) Quack, M., and J. Troe, Ber. Bunsenges, Phys. Chem., 81, 329 (1977).

7. (a) Smith, I.W.M., and R. Zellner, Int. J. Chem. Kinetics, Symp. No. 1, 341 (1975).
(b) Margitan, J. J., F. Kaufman, and J. G. Anderson, Int. J. Chem. Kinetics, Symp. No. 1, 281 (1975).

8. Golden, D. M., S. W. Benson, and G. N. Spokes, Angew. Chem. (International Edition), 12, 534 (1973).

9. Golden, D. M., M. Rossi, and K. D. King, J. Amer. Chem. Soc., 101, 1223 (1979).

10. Trevor, P., D. M. Golden, and D. F. McMillen, to be published.

11. Rossi, M., J. R. Barker, and D. M. Golden, J. Chem. Phys. 71, 3722 (1979); Chem. Phys. Lett., 65, 523 (1979).

12. Shaub, W. M., and S. H. Bauer, Int. J. Chem. Kinetics, 7, 509 (1975); Shaub, W. M., Ph.D. Thesis, Cornell Univ., 1975.

13. Golden, D. M., J. Phys. Chem., 83, 108 (1979).

The author was partially supported by the National Aeronautics and Space Administration under Contract NAS-100 (JPL 954815), the Department of Energy under Contract No. DE-AC03-79ER10483, the Air Force Office of Scientific Research under Contract No. F49620-78-C-0107, and the Army Research Office under Contract No. DAAG 29-78-C-0026.

Department of Chemical Kinetics
SRI International
Menlo Park, CA 94025

Nonisobaric Flame Propagation

George F. Carrier, Francis E. Fendell and Philip S. Feldman

1. INTRODUCTION

There are many contexts in which it is important to understand the propagation of deflagration waves in configurations in which the pressure undergoes significant changes [1,4,6,7,9,13]. In the internal combustion engine, for example, the deflagration phenomenon occurs in an approximately cylindrical volume of time-dependent length, and the pressure changes are central to its successful operation. Furthermore, whatever the container geometry may be, the kinematics of the motion are artifacts of earlier stages of the process and are largely unknown and, in any detail, uncontrolled. We study here, first, the propagation of a one-dimensional flame in a cylinder whose length is L(t), using the crudest description of the relevant processes which will leave us with physically realistic descriptions, and then we argue for a use of the result of that study in the investigation of more complicated configurations. Of particular interest is the hope that the performance in the latter case may require only an approximate knowledge of only the most macroscopic aspects of the kinematic fields.

2. THE ONE-DIMENSIONAL PROBLEM.

Let a combustible gas mixture occupy a cylinder whose cross-section is perpendicular to the x coordinate axis, and let the axial extent of the domain be $0 < x < L(t)$. We

ISBN 0-12-669550-4

presume the flame speed to be known as a function of the local thermodynamic state just upstream of the flame (flame speed is the rate [cm/sec] at which the flame moves relative to the particles it is engulfing); we consider phenomena in which the flame speed and $\dot{L}(t)$ are both slow compared to the acoustic speed of the gas so that, to a reasonable degree of approximation, the pressure, p, is a function of time and not of position; we take the flame to be indefinitely thin in our description [5,15] and adopt the heat release per unit mass, H, as a result of the combustion; and, tentatively, we study a flame which is plane and which proceeds from $x = 0$ to $x = L(t)$ during the time of interest. If there should be any interest in doing so, one could subsequently embed within this framework a description of the flame structure, constructed from a quasi-steady phenomenology and using steady flame results which are already familiar.

It is useful to associate a "space" coordinate ψ which is attached to the particles. A useful and conventional definition, which we adopt, is given by

$$\rho = \psi_x, \quad \rho u = -\psi_t, \quad \psi(0,t) = 0. \tag{2.1}$$

Note that, at time t, $\int_0^x \psi_x \, dx \equiv \psi(x,t)$ is just the mass of the gas to the left of x at that time and is therefore the "tag" for that particular slice of gas. Furthermore, the position at time t of a slice of gas whose mass coordinate is ψ is given by

$$x(\psi,t) = \int_0^x \frac{d\psi}{\rho(\psi,t)} \quad . \tag{2.2}$$

Figure 1 shows the manner in which the region is subdivided by the moving flame and it also schematically delineates the variation with time of the gas-filled domain (i.e., $0 < x < L(t)$).

We imagine for the moment that p(t) will increase monotomically with time in $0 < t < t_f$. In that case the foregoing x-t picture can be drawn in ψ,p coordinates as is shown in Fig. 2.

The flame locus will be described by $\psi = \Psi(p)$ or $p = p_1(\psi)$ as convenience dictates.

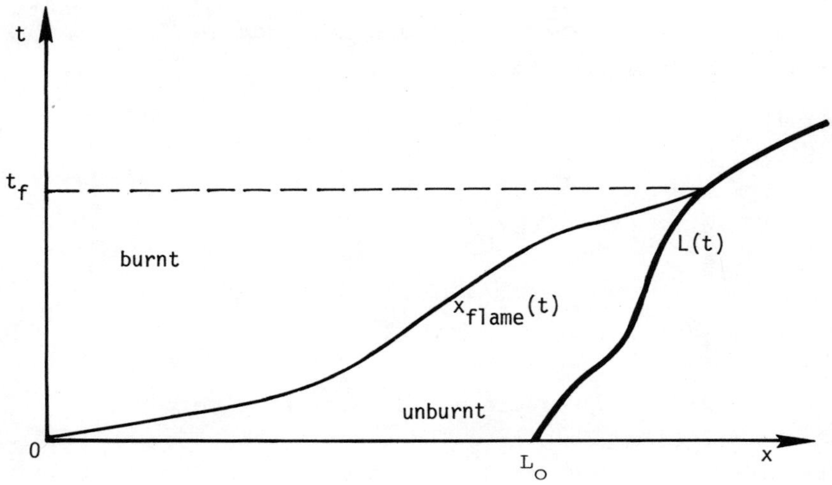

FIGURE 1. A schematic, in the physical, (x,t) plane, of the prescribed piston position L(t) and the to-be-determined flame position $x_{flame}(t)$.

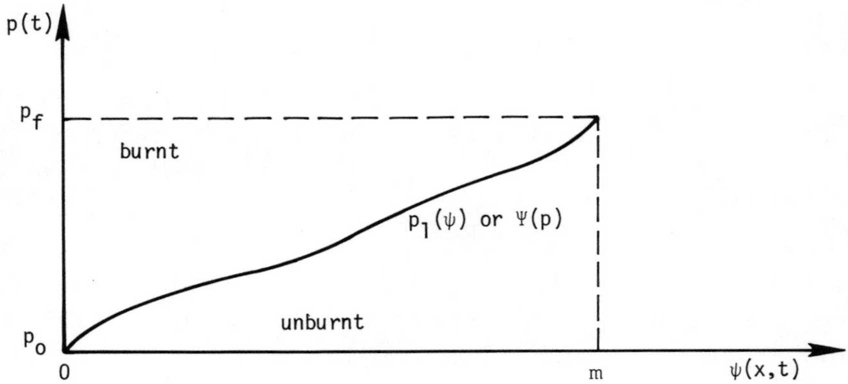

FIGURE 2. The schematic of Fig. 1 sketched in $[\psi(x,t),p(t)]$ coordinates.

In $\psi > \Psi(p)$, where the gases have not yet burned, the density and enthalpy are given by:

$$\frac{\rho_u}{\rho_o} = (p/p_o)^{1/\gamma}, \tag{2.3}$$

$$\frac{h_u}{h_o} = (p/p_o)^{(\gamma-1)/\gamma} \equiv (p/p_o)^{\alpha} , \tag{2.4}$$

where

$$h_o = \frac{\gamma}{\gamma-1}\frac{p_o}{\rho_o} = \frac{a_o^2}{\gamma-1} ; \tag{2.5}$$

p_o and ρ_o are the pressure and density at time zero.

As the flame completes its sweep across a thin slice of particles, their thermodynamic state changes from that just before the encounter to one in which H units of energy per unit mass have been added. That is: since this process is one at constant pressure, at $\psi = \Psi(p)$,

$$h_{after} = h_o(p/p_o)^{\alpha} + H \equiv h^*(p), \tag{2.6}$$

and

$$(\rho/\rho_o)_{after} = (\rho_u/\rho_o) \frac{(h_u/h_o)}{(h/h_o)_{after}}$$

$$= (p/p_o)^{1/\gamma} \frac{(p/p_o)^{\alpha}}{(p/p_o)^{\alpha} + \frac{H}{h_o}}$$

$$= \frac{p/p_o}{(p/p_o)^{\alpha} + \frac{H}{h_o}} \equiv \frac{\rho^*(p)}{\rho_o} . \tag{2.7}$$

Equations (2.6) and (2.7) can also be presented as follows: at $p = p_1(\psi)^+$,

$$h[\psi, p_1(\psi)] = h_o[p_1(\psi)/p_o]^{\alpha} + H \equiv h^{**}(\psi), \tag{2.8}$$

and

$$\frac{\rho[\psi, p_1(\psi)]}{\rho_o} = \frac{p_1(\psi)/p_o}{\left[p_1(\psi)/p_o\right]^{\alpha} + \frac{H}{h_o}} \equiv \frac{\rho^{**}(\psi)}{\rho_o} . \tag{2.9}$$

In $\psi < \Psi(p)$, i.e., in $p > p_1(\psi)$, for constant γ, [8]

$$\rho/\rho^{**}(\psi) = [p/p_1(\psi)]^{1/\gamma},$$

i.e.,

$$\rho/\rho_o = \frac{p_1(\psi)/p_o}{[p_1(\psi)/p_o]^{\alpha} + \frac{H}{h_o}} [p/p_1(\psi)]^{1/\gamma}; \tag{2.10}$$

and

$$h/h^{**}(\psi) = [p/p_1(\psi)]^{\alpha} ,$$

i.e.,

$$\frac{h}{h_o} = \left\{ \left[\frac{p_1(\psi)}{p_o}\right]^{\alpha} + \frac{H}{h_o} \right\} [p/p_1(\psi)]^{\alpha} . \tag{2.11}$$

Since $\psi_x = \rho$, it follows (as we have already noted) that, for fixed time t, $dx = d\psi/\rho$ and, in particular, using Eqs. (2.3) and (2.9),

$$L(t) = \int_0^{\Psi(p)} [p_1(\psi)/p]^{1/\gamma} \frac{d\psi}{\rho_o\left(\frac{p_1(\psi)/p_o}{[p_1(\psi)/p_o]^{\alpha} + \frac{H}{h_o}}\right)}$$

$$+ \int_{\Psi(p)}^{m} (p_o/p)^{1/\gamma} \frac{d\psi}{\rho_o} , \tag{2.12}$$

where m (as indicated in Fig. 2) is $\rho_o L(0)$. Multiplication of (2.12) by $(p/p_o)^{1/\gamma}$ and differentiation with regard to p yield (note that $p_1\{\Psi(p)\} \equiv p$),

$$\left[L(p/p_o)^{1/\gamma}\right]_p = \frac{1}{\rho_o} \Psi'(p) \left[1 + \frac{H}{h_o}\left(\frac{p_o}{p}\right)^{\alpha}\right] - \frac{\Psi'(p)}{\rho_o} . \tag{2.13}$$

For a given mixture, the flame speed, w, is an explicit function of p and T and we postulate that it is known, whether it be on empirical or on theoretical grounds.

Accordingly, we can write

$$w = \dot{x}_{flame} - \dot{x}_{particle\ just\ ahead\ of\ the\ flame}.$$

But, in $\Psi < \psi < m$ (i.e., in the unburned gas),

$$x_{particle} = L - \frac{m-\psi}{\rho_u} ,$$

so that,

$$\dot{x}_{particle} = \dot{L} - (m-\psi) \frac{d}{dt}\left(\frac{1}{\rho_u}\right) ,$$

and $x_{flame} = L - \frac{m-\Psi(p)}{\rho_u}$,

so that

$$\dot{x}_{flame} = \dot{L} - (m-\Psi)\,\frac{d}{dt}\left(\frac{1}{\rho_u}\right) + \frac{\dot{\Psi}}{\rho_u} \quad .$$

Thus,

$$w = \dot{\Psi}/\rho_u = \Psi'(p)\ \ \dot{p}(t)/\rho_u \quad , \tag{2.14}$$

and one can readily verify this by noting that the mass engulfed per unit time (i.e., $\dot{\Psi}$) divided by ρ_u is indeed, w, the rate of encroachment in units of velocity.

Noting that the left side of Eq. (2.13) can be written as

$$\frac{1}{\gamma p}\,(p/p_o)^{1/\gamma}\,L(t) + (p/p_o)^{1/\gamma}\,\dot{L}(t)/\dot{p}(t) \quad , \tag{2.15}$$

and using Eq. (2.14) to eliminate $\Psi'(p)$ from Equation (2.13), we obtain

$$\frac{\dot{p}(t)}{\gamma p(t)} + \frac{\dot{L}(t)}{L(t)} = \frac{H}{h_o}\,\frac{w[p(t)]}{L(t)}\,\left[\frac{p_o}{P(t)}\right]^{\alpha} . \tag{2.16}$$

We denote w as a function of only p because T is explicitly known in terms of p in the unburned gas.

For a given L(t) and a given w[p(t)], it is clear that p(t) can be calculated readily and, if one wants it, $\Psi(p)$ can be found by using the results for p(t) to express $\dot{p}$ as a function of p and then integrating Eq. (2.14) for Ψ.

For purposes of automotive-combustion research, there have been constructed rapid-compression-expansion devices with square pistons for which the one-dimensional development given here is particularly apropos [16].

3. SO WHAT

To us, the most interesting use for the foregoing result arises from an observation which was first brought to our attention by Frank Marble† in connection with an effort [2,3] (which still continues) to understand turbulent diffusion

†who also asserts that the idea goes back to Y.B. Zeldovich.

flames. In that effort one postulates that the "turbulent flame" is a thin, convoluted, laminar flame, undergoing enormous distortion and area change with time, but consuming fuel locally in a manner which is no different than that of a plane flame stretching at the same rate. We see no reason that this view should not apply equally well to flames in premixed gases and to that end we suggest that the foregoing be used in the following way.[†]

Let the cross-section of our simple cylinder be S so that its volume is $S \cdot L(t)$. Let the flame area be A(t) so that the volume rate of engulfment of gas by the flame is $A(t) \cdot w[p(t)]$. This is precisely equivalent to the situation in which a plane flame of area S moves at a flame speed W, where

$$W = \frac{A(t)}{S} w[p(t)] \quad .$$

Accordingly, the replacement of w by W in Eq. (2.16) will then allow the calculation of the pressure history for any given A(t). It is clear that we have sidestepped the difficult problem, that of finding the kinematic fields, but the foregoing ruse does permit the use of any estimate (again empirical or theoretical) of this one aspect of the velocity field, obtained without any encumbering coupling to the thermal field, to approximate via the modified Eq. (2.16) the pressure history which accompanies the burning in the highly turbulent interior of various combustion chambers.

To the extent that one is optimistic about the possibility of controlling this macroscopic aspect, i.e., A(t), of mixing, he can calculate from Eq. (2.16) that area A(t) for which a given p(t), chosen to be optimal for some compromise between efficiency, peak power, emission limitations and the like, would obtain.

[†]Other unsteady one-dimensional formulations of turbulent premixed burning in a finite adiabatic enclosure are given by Sirignano, with piston motion [14], and by Sorenson, without piston motion [17].

To illustrate the effect of various choices of A(t) on p(t), we have solved Eq. (2.16) for, (see [10])

$$L(t) = \frac{2r}{(CR)-1} + r\left[1 - \cos\theta(t) + R\left\{1-\left[1-\left(\frac{\sin\theta(t)}{R}\right)^2\right]^{\frac{1}{2}}\right\}\right],$$

and

$$\theta = 2\pi nt + \theta_o. \tag{2.17}$$

where (CR) is compression ratio, n is engine revolutions per unit time, r is crank radius, R is the ratio of connecting rod length to crank radius, θ_o is crank angle at "ignition" (more precisely, onset of flame propagation), taken as time t=0. Clearly at normal reciprocating-piston-type-engine speeds the laminar flame speed is far too slow for practicality, and this fact influences parameter assignment below. Equation (2.17) approximates piston clearance plus displacement, with displacement vanishing at top dead center (θ=0). We have adopted a laminar model, with [11,12]

$$\text{(a)}\quad W[p,t] = w(p) = w_o(p/p_o)^{\bar{\beta}}, \quad w_o, \bar{\beta} \text{ const.}, \tag{2.18}$$

where a power-law dependence fits hydrocarbon-air-mixture data adequately, and a "turbulent" model with A(t)/S assigned, quite arbitrarily, as follows:

$$\text{(b1)}\quad W[p,t] = \frac{A(t)}{S}\, w(p) = K\,\frac{\Psi(t)[m-\Psi(t)]}{m^2}\, w(p), \; K \text{ const.}, \tag{2.19a}$$

or,

$$\text{(b2)}\quad W[p,t] = \frac{A(t)}{S}\, w(p) = K_1\,\frac{X(t)[L(t)-X(t)]}{[L(t)]^2}\, w(p),$$

$$K_1 \text{ const.}; \tag{2.19b}$$

here X(t) is an abbreviation for x_{flame}, and K_1 and K are taken to be assigned. Clearly the burning terminates at time t = t_f where $\Psi(t_f)$ = m, or equivalently $x_{flame}(t_f)$ = L, and thenceforth

$$\frac{d}{dt}\left(Lp^{1/\gamma}\right) = 0 \Rightarrow Lp^{1/\gamma} = \text{const.}$$

4. A MODIFIED PROBLEM

There are variations on the calculation of Section 2 which may be of interest in some contexts. For example, if the gas composed of the combustion products has a significantly different ratio of specific heats (say, σ) than the value γ which prevails for the unburnt mixture, the calculation is modified in accord with the following equations, each of which is numbered n' so that it corresponds to n in the analysis of Section 2. The equations are:

$$\rho/\rho^{**}(\psi) = [p/p_1(\psi)]^{1/\sigma},$$

i.e.,

$$\rho/\rho_o = \frac{p_1(\psi)/p_o}{[p_1(\psi)/p_o]^{\alpha} + \frac{H}{h_o}} \left[\frac{p}{p_1(\psi)}\right]^{1/\sigma}, \tag{2.10'}$$

and

$$\frac{h}{h^{**}(\psi)} = [p/p_1(\psi)]^{\beta} \quad \text{where } \beta \equiv 1 - \frac{1}{\sigma},$$

i.e.,

$$\frac{h}{h_o} = \left\{\left[\frac{p_1(\psi)}{p_o}\right]^{\alpha} + \frac{H}{h_o}\right\} [p/p_1(\psi)]^{\beta}. \tag{2.11'}$$

Equation (2.12) becomes

$$L(t) = \int_o^{\Psi(p)} [p_1(\psi)/p]^{1/\sigma} \frac{d\psi}{\rho_o\left(\frac{p_1(\psi)/p_o}{[p_1(\psi)/p_o]^{\alpha} + \frac{H}{h_o}}\right)} + \int_{\Psi(p)}^{m} (p_o/p)^{1/\gamma} \frac{d\psi}{\rho_o}. \tag{2.12'}$$

and, again, after multiplication by $(p/p_o)^{1/\sigma}$, differentiation with respect to t and substitution of

$$\dot{\Psi} = \rho_u w = \rho_o (p/p_o)^{1/\gamma} w(p), \tag{2.14}$$

we get (with $p/p_o \equiv P$)

$$\frac{\dot{P}}{P}\left[\frac{1}{\sigma}+\left(\frac{1}{\gamma}-\frac{1}{\sigma}\right)\frac{m-\Psi}{\rho_o L(t)}(P)^{-1/\gamma}\right] + \frac{\dot{L}}{L} = \frac{H}{h_o}\frac{w(P)}{L}(P)^{-\alpha}\ . \qquad (2.13')$$

Thus (2.13') and (2.14) are a pair of simultaneous first-order equations from which one can extract P(t) and Ψ(t). Clearly, other special cases can be treated in very similar ways. Results of numerical integration are given in Figs. 3-9.

5. CONCLUDING REMARK

Other automotive-combustion-related phenomena that might be usefully included within formulations of the type undertaken here include blowby (mass loss out of the cylinder past the piston rings), stratification of the charge (non-uniform initial stoichiometry of the premixture), and heat loss (especially through the wall adjacent to the burned gas).

REFERENCES

1. Bradley, D., and A. Mitcheson, Mathematical solutions for explosions in spherical vessels, Combustion & Flame 26 (1976), 201-217.
2. Bush, W.B., and F.E. Fendell, Diffusion-flame structure for a two-step chain reaction, J. Fluid Mech. 64 (1974), 701-724.
3. Carrier, G.F., F.E. Fendell, and F.E. Marble, The effect of strain rate on diffusion flames, SIAM J. Appl. Math. 28 (1975), 463-500.
4. Fiock, E.F., and C.F. Marvin, The measurement of flame speeds, Chem. Revue 21 (1937), 367-387.
5. Flamm, L., and H. Mache, Die Verbrennung eines explosiven Gasgemisches in geschlossenem Gefäss, Sitzungsberichte der kaiserlichen Akademie der Wissenschaften, Wien 126 (1917), 9-44.
6. Guenoche, H., Flame propagation in tubes and in closed vessels, Unsteady Flame Propagation (G. Markstein, ed.) Pergamon Press, New York, 1964, 107-181.

7. Jost, W., Explosion and Combustion Processes in Gases, transl. by H.O. Croft, McGraw-Hill, New York, 1946.
8. Lavoie, G.A., J.B. Heywood, and J.C. Keck. Experimental and theoretical study of nitric oxide information in internal combustion engines, Combustion Sci. & Tech. 1 (1970), 313-326.
9. Lewis, B., and G. von Elbe, Combustion, Flames and Explosion of Gases, Academic Press, New York, 1951.
10. Lichty, L.C., Internal Combustion Engines, 5th Ed. McGraw-Hill, New York, 1939.
11. Metghalchi, M., and J.C. Keck, Laminar burning velocity of isooctane-air, methane-air and methanol-air mixtures at high temperature and pressure, Mechanical Engineering Dept. Report, Massachusetts Institute of Technology, 1978.
12. Obert, E.F. Internal Combustion Engines and Air Pollution, Intext Educational, New York, 1968.
13. Rallis, C.J. and G.E.B. Tremeer, Equations for the determination of burning velocity in a spherical constant volume vessel, Combustion & Flame 7 (1963), 51-61.
14. Sirignano, W.A., One-dimensional analysis of combustion in a spark-ignition engine, Combustion Sci. & Tech. 7 (1973), 99-108.
15. Sivashinsky, G.I., Hydrodynamic theory of flame propagation in an enclosed volume, Acta Astronautica 6 (1979), 631-645.
16. Smith, O.I., C.K. Westbrook, and R.F. Sawyer, Lean limit combustion in an expanding chamber, Seventeenth Symposium (International) on Combustion, Combustion Institute, Pittsburgh, 1978, 1305-1313.
17. Sorenson, S.C. Modeling turbulent transient combustion, Paper No. 780639, Society of Automotive Engineers, Warrendale, Pennsylvania, June 1978.

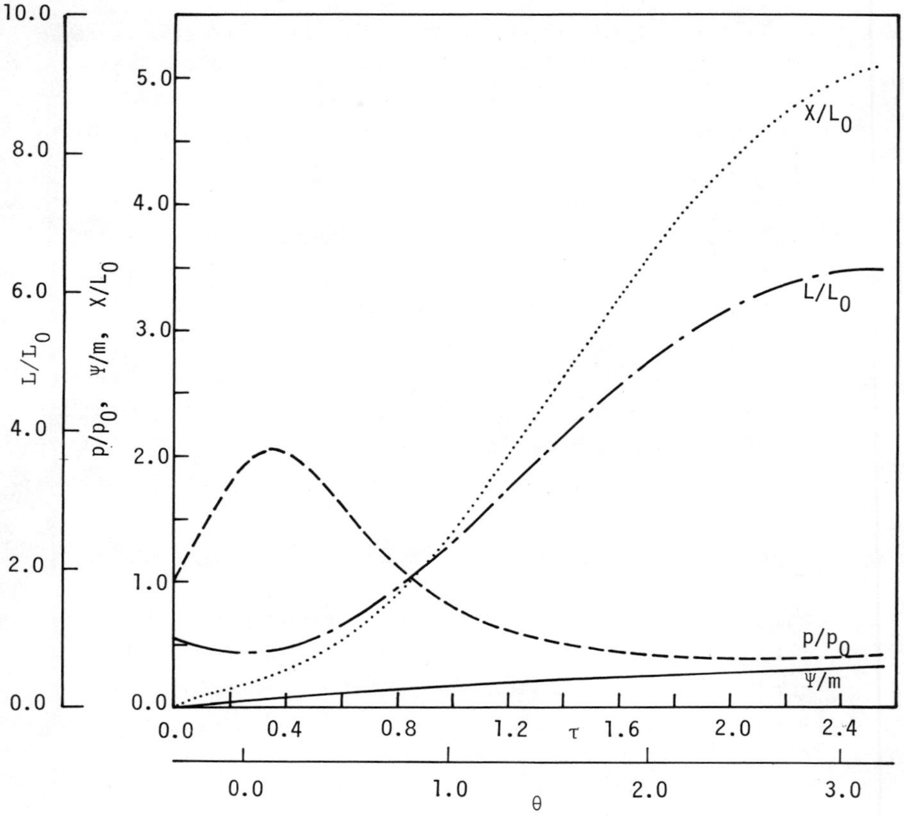

FIGURE 3. L/L_o, p/p_o, Ψ/m, and $X/L_o \equiv \frac{x_{flame}}{L_o}$ vs. "time" $\tau = [Hw_o/h_oL_o]t$ and vs. crank angle θ. In this and subsequent figures the "nominal" values referred to are: CR = 7.0, R = 4.0, r = 5.08 cm, $w_o = 2\cdot4\cdot10^3$ cm/min, n = 2000 rpm, $\theta_o = -.3491$ rad, $\gamma = \sigma = 1.4$, $H/h_o = 7.0$, $\overline{\beta} = -0.3$, and $K = K_1 = 40$. W, for this figure, is given by (2.18).

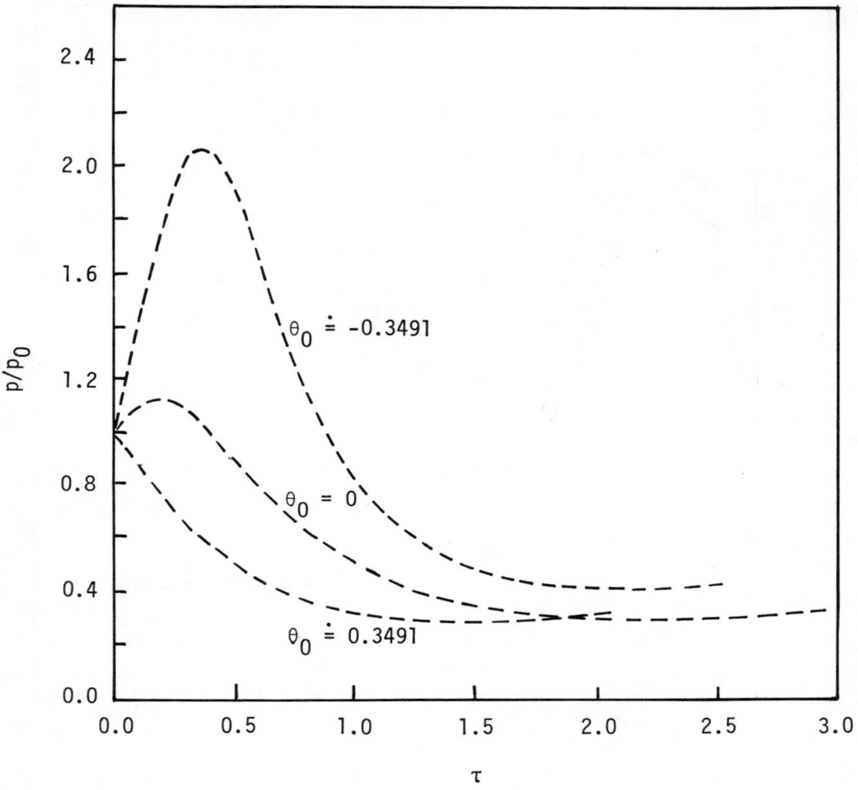

FIGURE 4. Results for the nominal case (see Fig. 3) with W given by the laminar expression (2.18); the initial value of crank angle, θ_o, is varied. It is recalled that θ is crank angle and τ is nondimensionalized time (see Fig. 3).

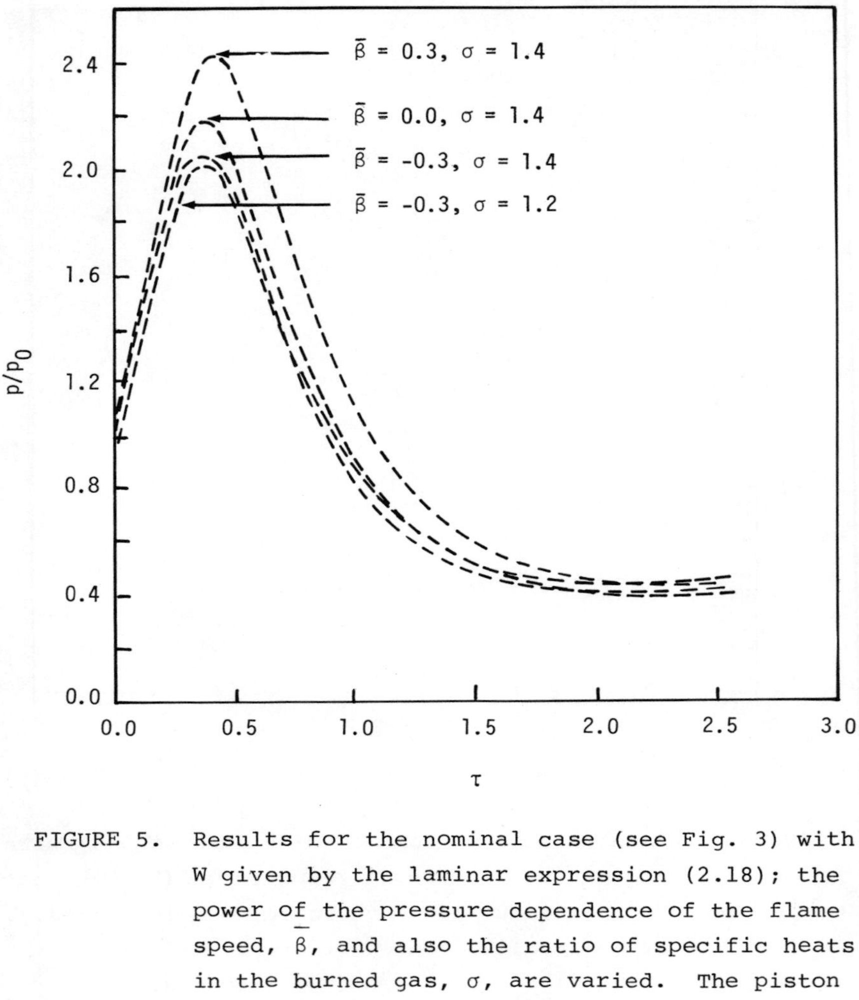

FIGURE 5. Results for the nominal case (see Fig. 3) with W given by the laminar expression (2.18); the power of the pressure dependence of the flame speed, $\bar{\beta}$, and also the ratio of specific heats in the burned gas, σ, are varied. The piston speed is too rapid for the laminar flame speed to afford much pressure rise during the portion of the cycle available for useful work. The distinction of σ and γ is not of consequence.

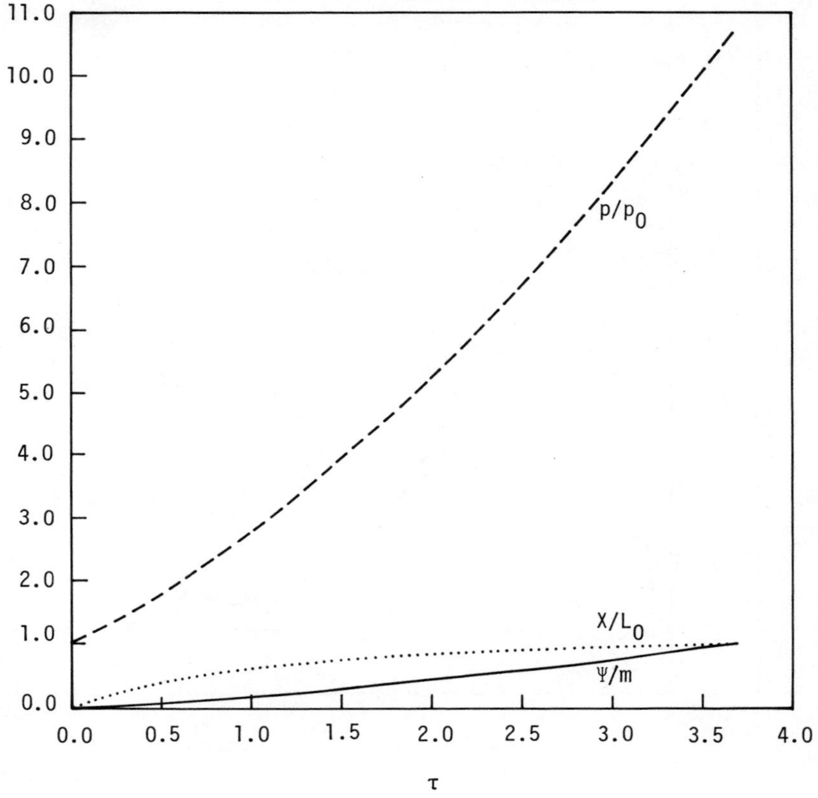

FIGURE 6. Results for the nominal case (see Fig. 3), except $n = 0$ [constant-volume container: $\theta(\tau) = \theta_o$, $L(\tau) = L_o$]; the function W is given by the laminar expression (2.18).

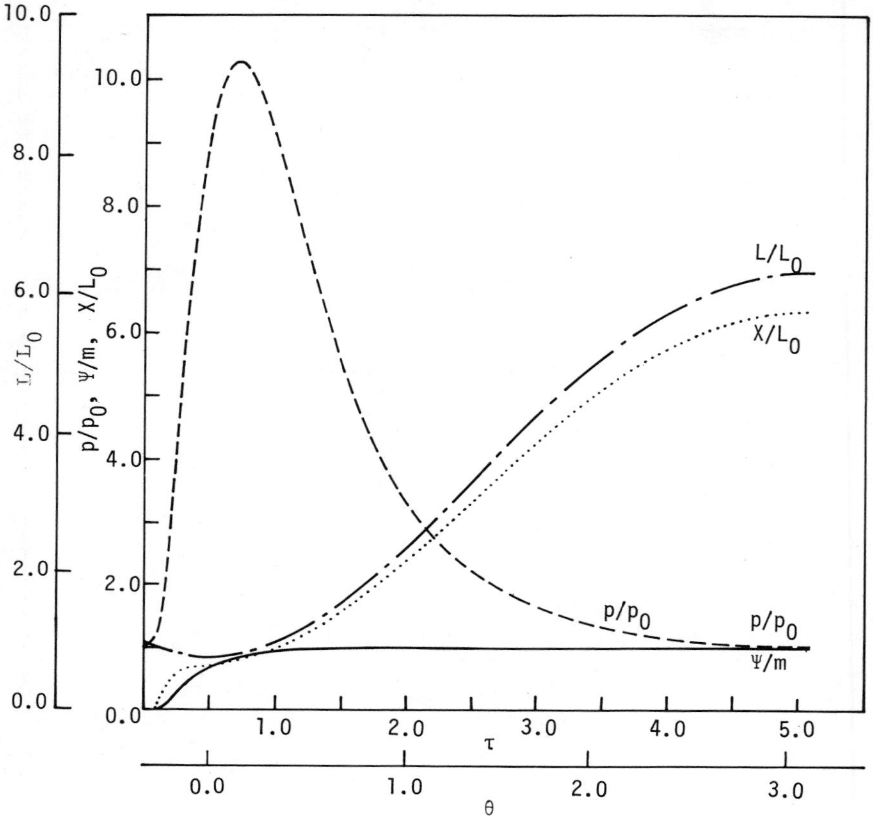

FIGURE 7. Results for the nominal case (see Fig. 3), except $n = 1000$ rpm and $\sigma = 1.2$ and $\varepsilon = 0.001$, with W given by the augmented-flame-area expression (2.19a). It is recalled that ε is defined by $X(0) = \varepsilon L_o$. The augmented-flame-area expression yields significant pressure rise during the portion of the cycle available for useful work.

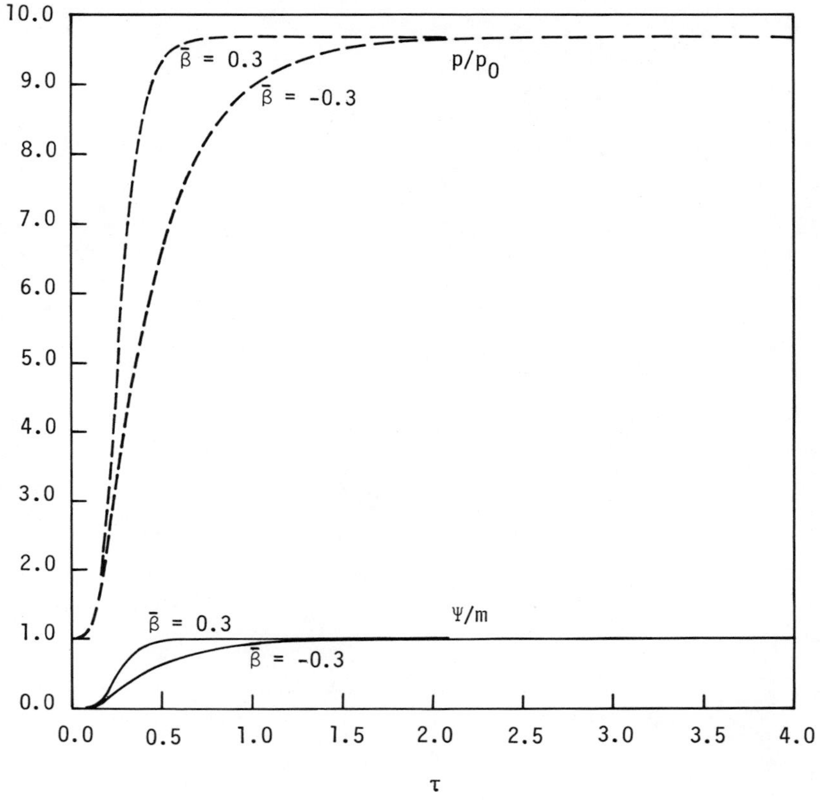

FIGURE 8. Results for the nominal case (see Fig. 3), except $n = 0$ [constant-volume container: $\theta(\tau) = \theta_o$, $L(\tau) = L_o$] and $\sigma = 1.2$ and $\varepsilon = 0.001$, with W given by the augmented-flame-area expression (2.19a); the power of the pressure dependence of the flame speed, $\bar{\beta}$, is varied, and while this parameter alters modestly the rate of pressure rise, it does not alter the ultimate pressure attained.

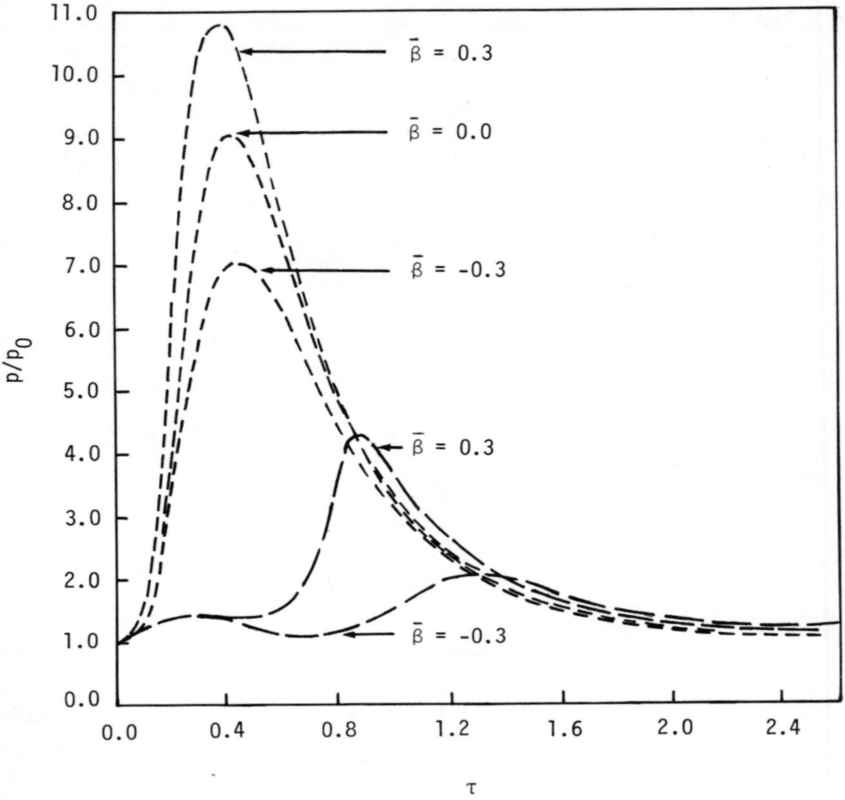

FIGURE 9. Results for the nominal case (see Fig. 3), except $\sigma = 1.2$ and $\varepsilon = 0.001$, with W given by either the augmented-flame-area expression (2.19a), denoted by short-dash curves, or the augmented-flame-area expression (2.19b), denoted by long-dash curves; the power of the pressure dependence of the flame speed, $\bar{\beta}$, is varied. The expression adopted is of important consequence, with results for (2.19a) typifying good performance and for (2.19b), almost a misfire.

Portions of this work were supported by the Army Research Office under contract DAAG29-77-C-0032, by the Dept. of Energy under contract E(04-3)-1261, by TRW Automotive Worldwide under authorization 1710-7494, and by NSF under contract MCS78-07598.

George F. Carrier
Division of Applied Sciences
Harvard University
Cambridge, Mass. 02138

and

Francis E. Fendell
and
Philip S. Feldman
Engineering Sciences Laboratory
TRW Defense and Space Systems Group
Redondo Beach, Calif. 90278

Diffusion and Reaction in Carbon Burning

Neal R. Amundson and Eduardo Mon

ABSTRACT

The combustion of a carbon particle in a fixed environment is described by a set of partial differential equations of the diffusion-reaction type with an associated set of non-linear boundary conditions. Certain stoichiometric and time-constant arguments allow the problem to be reformulated as an initial value problem at the boundary of a non-linear ordinary differential equation. The steady state structure of the numerical solution is presented and is shown to include multiple steady states. In particular, during the course of burn-off some states bifurcate while others become extinct. The temperature history of the particle is then subject to transitions and shifts between steady states, leading to behaviour that is best described as pathological in nature.

1. PROLEGOMENA

We consider an impervious, spherical particle surrounded by a finite, stagnant film of gas known as a boundary layer. A combination of physicochemical rate processes occur in the boundary layer as oxygen diffuses from the ambient and reacts with the carbon at the surface to form carbon monoxide. The latter then must diffuse outwards from the surface and on the way it meets the incoming oxygen, reacting homogeneously to form carbon dioxide. In addition, the carbon dioxide diffuses to the ambient fluid, but also to the surface where it produces additional carbon monoxide. The first two reactions are exothermic, the latter endothermic. The form of their

ISBN 0-12-669550-4

rate expressions and the nomenclature that we follow are shown in table I. There are thus a total of four gaseous

Table I. Notational Conventions

Index	Species	Reaction	Rate Expression
$i = 1$	O_2	$2C + O_2 \rightarrow 2CO$	$R_1 = k_1 e^{-E_1/RT} c_1$
$i = 2$	CO_2	$C + CO_2 \rightarrow 2CO$	$R_2 = k_2 e^{-E_2/RT} c_2$
$i = 3$	CO	$CO + \frac{1}{2}O_2 \rightarrow CO_2$	$R_3 = k_3 e^{-E_3/RT} c_3 c_1^{\frac{1}{2}}$

species: O_2, CO, CO_2 and N_2. The latter acts as an inert. The diffusion of these species occurs simultaneously with the transport of thermal energy, which arises from the heats of reaction, by conduction and radiation from the particle surface. Before presenting additional details, it will be useful to formulate the general problem under consideration. Following a Fickian law, the equations that describe this system are

$$\frac{\partial c_1}{\partial t} = \frac{D}{r^2}\frac{\partial}{\partial r}\left(r^2\frac{\partial c_i}{\partial r}\right) + \nu_i R_3; \quad a(t) < r < b(t) \tag{1}$$

$$cC_p\frac{\partial T}{\partial t} = \frac{k}{r^2}\frac{\partial}{\partial r}\left(r^2\frac{\partial T}{\partial r}\right) + (-\Delta H_3)R_3; \quad a(t) < r < b(t) \tag{2}$$

With the following set of boundary conditions for $r = a(t)$, $t > 0$

$$-D\frac{\partial c_i}{\partial r} = \sum_{j=1}^{2} \beta_{ij} R_j \tag{3}$$

$$-k\frac{\partial T}{\partial r} + \frac{a(t)\rho_c}{3M_c} C_s \frac{dT_s}{dt} = (-\Delta H_1)R_1 + (-\Delta H_2)R_2 - q_R \tag{4}$$

$$\frac{\rho_c}{M_c}\frac{da(t)}{dt} = -(2R_1 + R_2) \tag{5}$$

Equation (4) lumps the intraparticle heat conduction problem through the engineering assumption of a uniform particle temperature, due to the high thermal conductivity of the solid relative to the gas while equation (5) gives the rate of change of the particle radius as obtained from a carbon balance. At the external edge of the boundary layer, a set of constant conditions is specified

$$c_i = c_{ib}, \quad T = T_b; \quad r = b(t), \quad t > 0 \tag{6}$$

and

$$c_i = c_{io}(r), \quad T = T_o(r); \quad a < r < b, \quad t = 0 \tag{7}$$

$$a = a_o; \quad t = 0 \tag{8}$$

The radiation heat flux, q_R, takes the form

$$q_R = \sigma\varepsilon(T_s^4 - T_m^4)$$

where T_m is a model parameter that accounts for the possibility of having the surroundings at a temperature other than T_b, such as would be the case when the particle is a part of a large ensemble. Since the domain of interest is time-dependent, the problem is of the moving boundary type. However, it is reasonable to express the location of the outer edge of the boundary layer as being directly proportional to the particle radius

$$b(t) = [\delta+1]a(t) \tag{9}$$

We are then able to immobilize the boundary with the coordinate transformation

$$\rho = 1 + \delta - \frac{r}{a(t)} \tag{10}$$

Further simplification may be attained with the introduction of a dimensionless time

$$\theta = \int_o^t \frac{\alpha}{a^2(s)} ds \tag{11}$$

where $\alpha = D = k/\rho C_p$. The form of the spatial operator will be changed with

$$u_i = \frac{c_i}{c}(1+\delta-\rho) \tag{12}$$

$$v = \frac{T}{T_e}(1+\delta-\rho) \tag{13}$$

In the fixed domain $\rho\varepsilon[0,\delta]$ and dimensionless form the equations become

$$\frac{\partial u_i}{\partial \theta} = \frac{\partial^2 u_i}{\partial \rho^2} + \nu_i \Phi(\rho,\theta); \quad 0 < \rho < \delta \tag{14}$$

$$\frac{\partial v}{\partial \theta} = \frac{\partial^2 v}{\partial \rho^2} + \beta\Phi(\rho,\theta); \quad 0 < \rho < \delta \tag{15}$$

$$\frac{\partial u_i}{\partial \rho} = y(\theta)\sum_{j=1}^{2} n_j \beta_{ij} e^{-\varepsilon_j/v} u_j - u_i; \quad \rho = \delta, \ \theta > 0 \tag{16}$$

$$P\frac{dv}{d\theta} + \frac{\partial v}{\partial \rho} = y(\theta)\sum_{j=1}^{2} Q_j e^{-\varepsilon_j/v} u_j + sy(\theta)[v^4 - v_m^4] - v; \quad \rho = \delta, \ \theta = 0 \tag{17}$$

$$\frac{dy(\theta)}{d\theta} = -y^2(\theta)\sum_{j=1}^{2} m_j e^{-\varepsilon_j/v} u_j; \quad \rho = \delta,\ \theta > 0 \tag{18}$$

$$u_i = u_{ib},\ v = v_b; \quad \rho = 0,\ \theta > 0 \tag{19}$$

$$u_i = u_{io}(\rho),\ v = v_o(\rho); \quad 0 < \rho < \delta,\ \theta = 0 \tag{20}$$

$$y = 1; \quad \theta = 0 \tag{21}$$

where

$$\Phi(\rho,\theta) = \frac{Ay^2(\theta)}{(1+\delta-\rho)^{1/2}} \exp[-\varepsilon_3 \frac{(1+\delta-\rho)}{v}] u_3 u_1^{1/2} \tag{22}$$

and

$$\beta = \frac{(-\Delta H_3)}{cC_p T_e}; \quad \varepsilon_j = \frac{E_j}{RT_e}; \quad Q_j = (-\Delta H)_j k_j \frac{a_o c}{kT_e};$$

$$A = a_o^2 \frac{c^{3/2}}{\alpha} k_3; \quad n_j = a_o \frac{k_j}{\alpha}; \quad s = \frac{\sigma\varepsilon a_o}{k} T_e^3;$$

$$P = \alpha\frac{\rho_c C_s}{3M_c k}; \quad m_j = (3-j) M_c k_j \frac{ca_o}{\alpha\rho_c}$$

The time dependence of the dimensionless particle radius may be obtained formally from (18),

$$y(\theta) = [1 + \int_0^{\theta} (\sum_{j=1}^{2} m_j e^{-\varepsilon_j/v} u_j) d\tau]^{-1} \tag{23}$$

Thus the general problem consists of a set of simultaneous, non-linear partial differential equations with an associated set of non-linear time dependent boundary conditions. Of particular interest to engineers is the quasi-steady state solution of the system, that is, the solution of the equations obtained by setting the time partial derivatives to zero while freezing the value of y. The latter is necessary because y is a monotonically decreasing function of θ for all θ. The equations then simplify to

$$\frac{d^2u_i}{d\rho^2} + \nu_i \Phi(\rho) = 0; \quad 0 < \rho < \delta \tag{24}$$

$$\frac{d^2v}{d\rho^2} + \beta\Phi(\rho) = 0; \quad 0 < \rho < \delta \tag{25}$$

$$\frac{du_i}{d\rho} = y \sum_{j=1}^{2} n_j \beta_{ij} e^{-\varepsilon_j/v} u_j - u_i; \quad \rho = \delta \tag{26}$$

$$\frac{dv}{d\rho} = y \sum_{j=1}^{2} Q_j e^{-\varepsilon_j/v} u_j + sy(v^4 - v_M^4) - v; \quad \rho = \delta \tag{27}$$

$$u_i = u_{ib}, \quad v = v_b; \quad \rho = 0 \tag{28}$$

Even under these circumstances, the non-linearities preclude the possibility of an analytical solution. Such complexities led early investigators to introduce further assumptions based on physically limiting cases. Thus, most models may be classified within one of two categories. In the single film theory, the carbon monoxide oxidation or flame is negligible,

$$\Phi(\rho) = 0, \quad 0 < \rho < \delta \tag{29}$$

while in the double film theory there is complete oxidation in a flame front,

$$\Phi(\rho) = 0, \quad 0 < \rho < \rho_F$$

$$\Phi(\rho) \to \infty, \quad \rho = \rho_F$$

$$\Phi(\rho) = 0, \quad \rho_F < \rho < \delta$$

Both cases are amenable to analytical solution and were initially considered in 1931 by Burke and Schumann [1], [2], although a variety of models with minor modifications have followed. The general case, corresponding to the exact solution of equations (24)-(28), with no a priori specifications on the form of $\Phi(\rho)$ will be referred to as the continuous flame theory. The numerical solution of those equations, with $s = 0$ (radiation equilibrium), was obtained by Caram and Amundson [3].

2. A CONTINUOUS FLAME THEORY

Before considering the solution of (24)-(28), it will be useful to introduce an additional refinement. Recognizing that this is a multicomponent system, we introduce a correction to account for the difference between the diffusivities of the components in the mixture,

$$D_i = \gamma_i D \tag{30}$$

where

$$\gamma_1 = \gamma_3 = \gamma, \quad \gamma_2 = 1$$

Equations (24) and (26) then become

$$\gamma_i \frac{d^2 u_i}{d\rho^2} + \nu_i \Phi(\rho) = 0. \quad 0 < \rho < \delta \tag{31}$$

$$\gamma_i \frac{du_i}{d\rho} = y \sum_{j=1}^{2} n_j \beta_{ij} e^{-\varepsilon_j / v} u_j - u_i, \quad \rho = \delta \tag{32}$$

Obviously, to solve a set of four coupled non-linear ordinary differential equations with non-linear boundary conditions is not a simple numerical task. However, it is possible to reduce this complexity by obtaining certain relations between the concentration and temperature profiles. We rewrite (31) as

$$\frac{\gamma_i}{\nu_i}\frac{d}{d\rho}\left[\frac{du_i}{d\rho}\right] = -\Phi(\rho) \tag{33}$$

separating variables, and integrating from the surface to an arbitrary point in the boundary layer yields

$$\frac{\gamma_i}{\nu_i}\frac{du_i}{d\rho} - \frac{\gamma_i}{\nu_i}\left[\frac{du_i}{d\rho}\right]_{\rho=\delta} = -\int_0^{\rho}\Phi(\rho')d\rho' \tag{34}$$

Since the right-hand side is independent of i, it follows that

$$\frac{\gamma_i}{\nu_i}\frac{du_i}{d\rho} - \frac{\gamma_i}{\nu_i}\left[\frac{du_i}{d\rho}\right]_{\rho=\delta} = \frac{\gamma_k}{\nu_k}\frac{du_k}{d\rho} - \frac{\gamma_k}{\nu_k}\left[\frac{du_k}{d\rho}\right]_{\rho=\delta},\quad i \neq k \tag{35}$$

and

$$\frac{\gamma_i}{\nu_i}\frac{du_i}{d\rho} - \frac{\gamma_i}{\nu_i}\left[\frac{du_i}{d\rho}\right]_{\rho=\delta} = \frac{1}{\beta}\frac{dv}{d\rho} - \frac{1}{\beta}\left[\frac{dv}{d\rho}\right]_{\rho=\delta} \tag{36}$$

This yields three linearly independent relations

$$\gamma\frac{du_3}{d\rho} + \frac{du_2}{d\rho} = \left[\gamma\frac{du_3}{d\rho} + \frac{du_2}{d\rho}\right]_{\rho=\delta} \tag{37}$$

$$2\gamma\frac{du_1}{d\rho} + \frac{du_2}{d\rho} = \left[2\gamma\frac{du_1}{d\rho} + \frac{du_2}{d\rho}\right]_{\rho=\delta} \tag{38}$$

$$\frac{du_2}{d\rho} - \frac{1}{\beta}\frac{dv}{d\rho} = \left[\frac{du_2}{d\rho} - \frac{1}{\beta}\frac{dv}{d\rho}\right]_{\rho=\delta} \tag{39}$$

which when integrated from 0 to ρ become

$$\gamma(u_3-u_{3b}) + (u_2-u_{2b}) = \left[\gamma\frac{du_3}{d\rho} + \frac{du_2}{d\rho}\right]_{\rho=\delta}\rho \tag{40}$$

$$2\gamma(u_1-u_{1b}) + (u_2-u_{2b}) = \left[2\gamma\frac{du_1}{d\rho} + \frac{du_2}{d\rho}\right]_{\rho=\delta}\rho \tag{41}$$

$$(u_2-u_{2b}) - \frac{1}{\beta}(v-v_b) = \left[\frac{du_2}{d\rho} - \frac{1}{\beta}\frac{dv}{d\rho}\right]_{\rho=\delta}\rho \tag{42}$$

Notice that it is then only necessary to solve a single conservation equation since the remaining unknowns follow from (40)-(42). However the values $u_i(\delta) = u_{is}$, $v(\delta) = v_s$ are still needed. The boundary conditions (32) may be written explicitly for each species as

$$\gamma\left[\frac{du_1}{d\rho}\right]_{\rho=\delta} = -u_{1s}(n_1f_1+1) \tag{43}$$

$$\left[\frac{du_2}{d\rho}\right]_{\rho=\delta} = -u_{2s}(n_2f_2+1) \tag{44}$$

$$\gamma\left[\frac{du_3}{d\rho}\right]_{\rho=\delta} = 2f_1n_1u_{1s} + 2f_2n_2u_{2s} - u_{3s} \tag{45}$$

where $f_i = ye^{-\varepsilon_i/v_s}$. Using (43)-(45) in (40)-(42) and setting $\rho = \delta$ results in

$$2f_1n_1\delta u_{1s} + (f_2\delta n_2-\delta-1)u_{2s} - (\delta+\gamma)u_{3s} = u_{2b} + \gamma u_{3b} \tag{46}$$

$$(2f_1n_1\delta+2\gamma-2\delta)u_{1s} + (f_2n_2\delta+\delta+1)u_{2s} = 2\gamma u_{1b} + u_{2b} \tag{47}$$

$$\frac{Q_1f_1\delta}{\beta}u_{1s} + \left(f_2n_2\delta+\frac{Q_1f_2\delta}{\beta}+\delta+1\right)u_{2s} = \frac{v_s}{\beta} - \frac{v_b}{\beta} - \frac{sy\delta}{\beta}(v_s^{\,4}-v_m^{\,4}) + \frac{v_s\delta}{\beta} \tag{48}$$

a set of algebraic equations linear in the u_{is}, non-linear in v_s. We thus write

$$u_{is} = u_{is}(v_s,v_b) \tag{49}$$

then from (40)-(42),

$$u_i = u_i(\rho,v,v_s) \tag{50}$$

The problem is reduced to the solution of a single differential equation

$$\frac{d^2v}{d\rho^2} + \frac{\beta Ay^2}{(1+\delta-\rho)^{\frac{1}{2}}}\exp\left[-\varepsilon_3\left(\frac{1+\delta-\rho}{v}\right)\right]u_3(\rho,v,v_s)u_1^{\frac{1}{2}}(\rho,v,v_s) = 0 \qquad 0 < \rho < \delta \tag{51}$$

$$\frac{dv}{d\rho} = y\sum_{j=1}^{2} Q_je^{-\varepsilon_j/v}u_j(v) + sy(v^4-v_m^{\,4}) - v, \quad \rho = \delta \tag{52}$$

$$v = v_b; \quad \rho = 0 \tag{53}$$

While the system is now solvable in principle, certain numerical difficulties arise due to stiffness. The success of a numerical procedure, as a result, is highly dependent on the accuracy of the estimated v profile and particularly on the value of the surface temperature, v_s.

A useful form of presenting the solution will be to fix the u_{ib} and show the results as a locus of solutions in the $v_s - v_b$ plane. The problem then consists of providing a reasonable guess of v_s for a given v_b. In this respect, a necessary condition that v_s be a reasonable temperature is that $u_{is} \geq 0$, in order to satisfy the physical requirement that concentrations can not be negative. Hence the curves,

$$u_{is}(v_s, v_b) = 0 \quad i = 1,2,3 \tag{54}$$

which result from (49), define each a curve along which the surface concentration is zero, on one side positive, and on the other side negative. Any feasible solution of (51)-(53) must lie within the enclosure or <u>feasibility region</u> formed by these curves.

Equations (54) are linear in v_b, except when $v_m = v_b$, but in either case the resulting zero concentration curves are readily obtained by fixing v_s and solving for v_b. Fig. 1 shows a schematic representation of the resulting feasibility region. For the case of zero carbon dioxide concentration in the ambient ($u_{2b} = 0$), it is possible to establish rigorously certain relationships between the zero concentration curves and the restricted single and double film theories discussed previously. Namely, the locus of the single-film theory coincides with the $u_{2s} = 0$ curve whereas the locus of the double-film theory coincides with the $u_{1s} = 0$ curve. The $u_{3s} = 0$ curve results from a special case of the double-film theory where the flame front becomes located at the surface of the particle. Note that two pairs of the curves approach each other asymptotically while the third pair intersect in the finite plane.

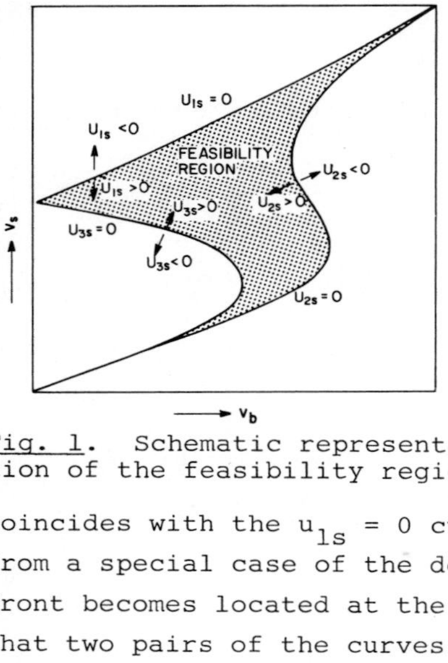

<u>Fig. 1</u>. Schematic representation of the feasibility region

3. STRUCTURE OF THE STEADY STATE

Three different loci that result from the numerical solution of (51)-(53) are shown in Fig. 2, along with their respective feasibility regions. The solution below $v_s = 1.0$ has been omitted because this portion generally corresponds to the unignited state of the particle and is therefore of minor physical interest. In all cases presented here we have set $u_{2b} = u_{3b} = 0$. Taking into account non-zero values of these quantities is by no means more difficult, but neither are new features of behaviour introduced. Values of the parameters used in the calculations are given in table II.

Table II. Parameter Values Used in the Calculations

$A = 5.395 \times 10^9 a_o^2$	$n_1 = 1.5 \times 10^9 a_o$	$P = 1.16 \times 10^3$
$\beta = 9.46 \times 10^7$	$n_2 = 2.008 \times 10^{12} a_o$	$Q_1 = 2.002 \times 10^8 a_o$
$\varepsilon_1 = 17.966$	$m_1 = 5.445 \times 10^5 a_o$	$Q_2 = -2.034 \times 10^{10} a_o$
$\varepsilon_2 = 29.790$	$m_2 = 3.644 \times 10^8 a_o$	$s = 7.867 \times 10^2 a_o$
$\varepsilon_3 = 15.098$		

Locus A of Fig. 2 corresponds to the case of independent diffusion and radiation equilibrium ($\gamma = 1$, $s = 0$) and displays multiple steady states--either one or three values of v_s are possible for a fixed value of v_b. In addition, there is agreement with the single and double film theories at low and high temperatures, respectively, but a wide transition range at intermediate temperatures. Accounting for the differences in diffusivity between the species causes some quantitative difference (locus B), although the effect of radiation is far more drastic (locus C). In this particular case, allowing for radiant interaction eliminates the multiplicity.

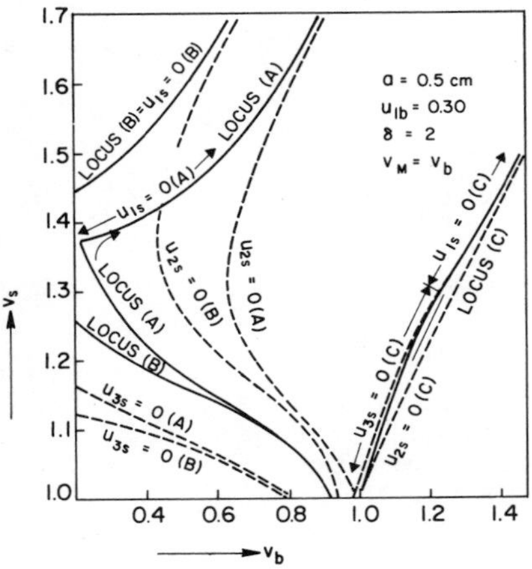

Fig. 2. Some feasibility refions and loci of solutions. (A) $s=0, \gamma=1$; (B) $s=0, \gamma=1.31$; (C) $s\neq0, \gamma=1.31$

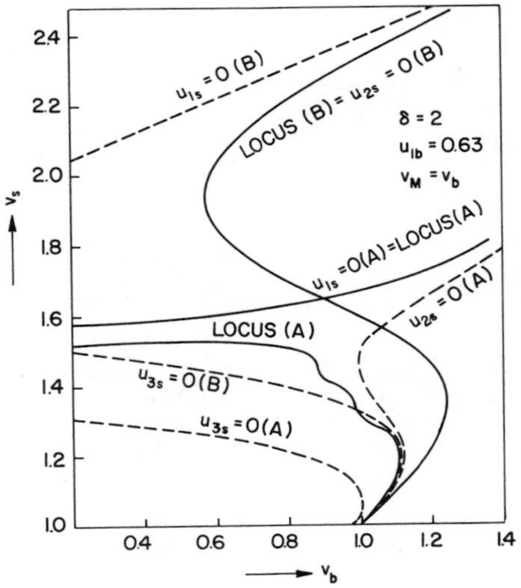

Fig. 3. Feasibility regions and loci of solutions for two different sized particles. (a) a=500 μm (B) a=50 μm.

Another set of $v_s - v_b$ loci obtained for a higher value

of the ambient oxygen content, and different particle sizes are shown in Fig. 3. Although again s $\neq$ 0, the feasibility region is now considerably wider, a result brought about primarily by decreasing the particle size. A significant result that emerges from comparing loci 3A and 3B is that the locus of the smaller particle coincides everywhere with the zero CO_2 curve, indicating that a critical size exists below which a particle burns according to the one-film theory, with no oxidation in the boundary layer.

If we decrease the thickness of the boundary layer for the larger particle of Fig. 3, the multiplicity increases to five in a narrow range of v_b (Fig. 4), although it is possible to obtain such multiplicity in a wider range of v_b, as shown in Fig. 5(A). A unique steady state results for the

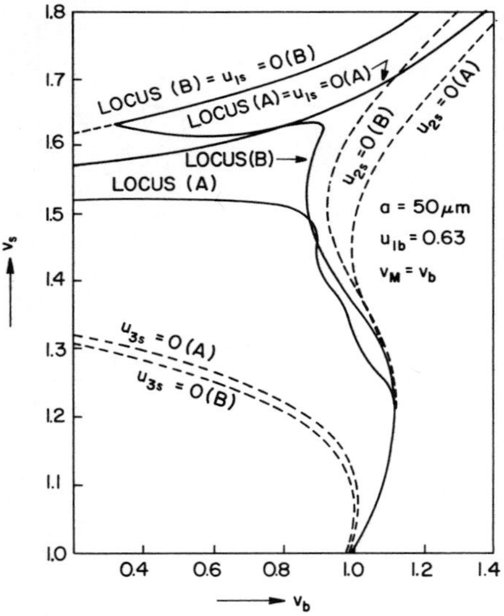

Fig. 4. Effect of the boundary layer thickness on the feasibility region and locus of solutions. (A) δ=2; (B) δ=1

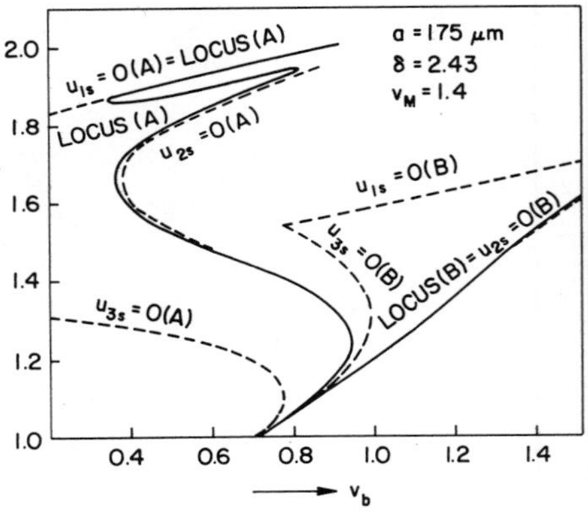

Fig. 5. Effect of the ambient oxygen concentration on the feasibility region and locus of solutions. (A) u_{1b}=0.63, (B) u_{1b}=0.15

same particle by lowering u_{1b} (Fig. 5(B)).

The dynamic behaviour of the particle is best presented in a different parameter plane, v_s - z, where z, the fractional decrease in particle radius, is 1 - y. Therefore, it will be worthwhile to consider the structure of the steady state solution in such a plane. We choose first a particle with the $v_s - v_b$ locus of Fig. 6 (single film theory). The shape of the locus in the new plane will depend on which value of v_b is chosen to be fixed and the range of possibilities is best illustrated by selecting the three different values of v_b indicated by the dotted lines in Fig. 6. They correspond to 1.0, 1.2 and 1.3. The points numbered in the v_s - v_b plane will thus lie on the ordinate of the v_s - z plot. Notice that at this initial value of the particle size there are three different steady states which correspond to the 1.0 and 1.2 values of the ambient temperature. At v_b = 1.3, however, there is a unique steady state. The result is shown in

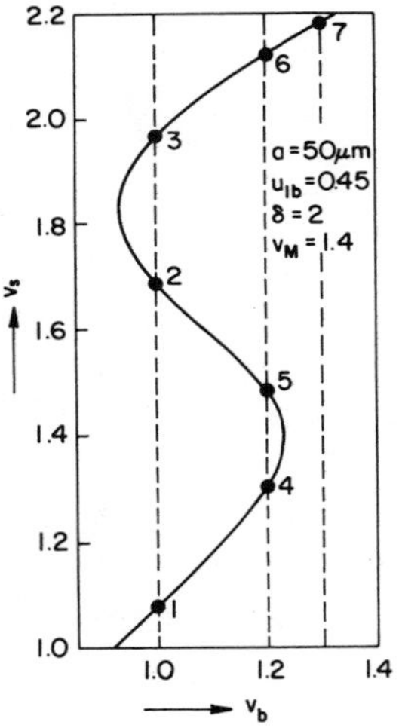

Fig. 6. Locus of steady state solutions in the v_s-v_b plane for a particle burning according to the one-film theory.

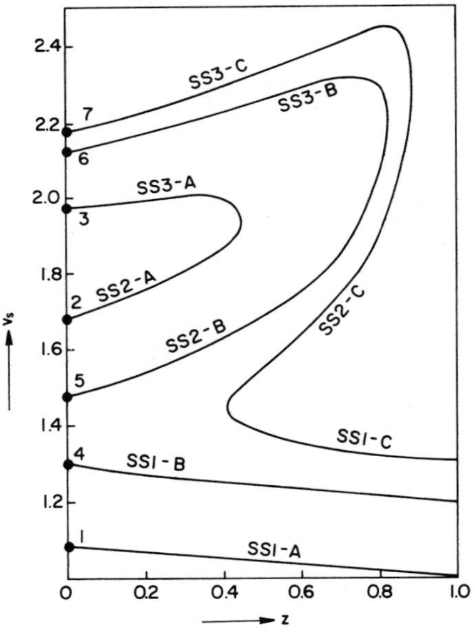

Fig. 7. Locus of steady state solutions in the v_s-z plane for the particle of Fig. 6 at three different values of v_b. (A) v_b=1.0; (B) v_b=1.2; (C) v_b=1.3.

Fig. 7 for the three chosen v_b values. The upper two steady states of loci A and B converge and disappear at a given size of the particle. Locus C which has a unique steady state initially, posesses a bifurcation at z = 0.41 and three steady states in the range $0.41 \le z \le 0.88$. The figure then indicates that the variation in the number of steady states during burn-off may be 3-1 or 1-3-1.

A similar approach may be followed for the more general case of five steady states. We have chosen to study the particle with the $v_s - v_b$ locus of Fig. 4(B) which is reproduced for our present purposes in Fig. 8. Fixing v_b at 0.9 yields the v_s - z of Fig. 9. Here the points marked 3-6 correspond to those similarly marked in Fig. 8. The results have been

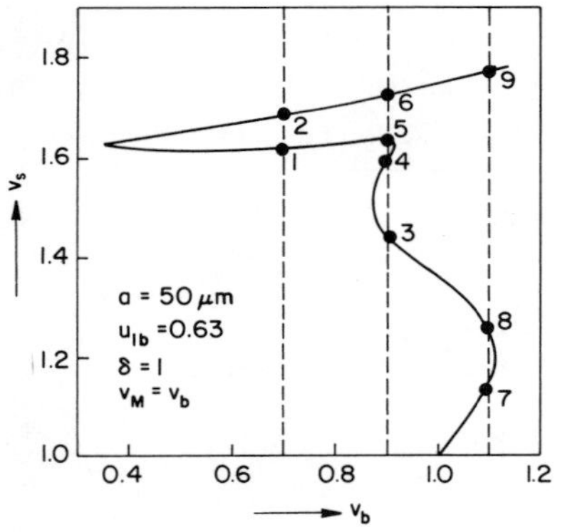

Fig. 8. Locus of steady state solutions in the v_s-v_b plane for a particle with maximum steady state multiplicity of five.

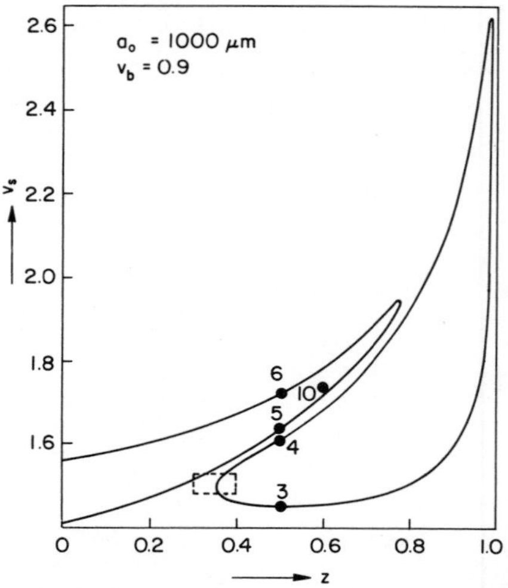

Fig. 9. Locus of steady state solutions in the v_s-z plane for a particle initially twice as large as the particle of Fig. 8 at v_b=0.9.

extended to larger particle sizes so that the initial radius is 1000 μm and the 500 μm particle of Fig. 8 corresponds to z = 0.50 in Fig. 9. It should be understood that we are omitting the additional unignited steady state below v_s = 1.0. The resulting variation pattern in the number of steady states is 3-5-3-1, with a bifurcation point at z = 0.35.

Selecting v_b = 0.7 for the same particle yields the steady state structure of Fig. 10. The loops formed by the steady states do not overlap as in Fig. 9, but instead there is a gap between the loops so that the resulting pattern is 3-1-3-1. The higher v_b value, 1.1, yields the locus of Fig. 11, with two bifurcation points and a variation in the number of steady states of 1-3-5-3-1.

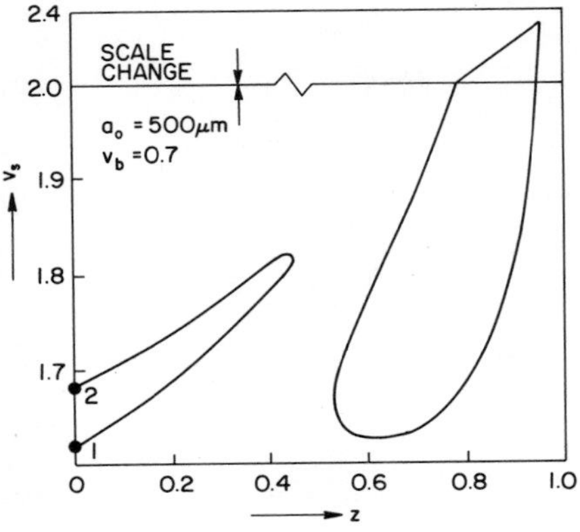

Fig. 10. Locus of steady state solutions in the v_s-z plane for the particle of Fig. 8 at v_b=0.7.

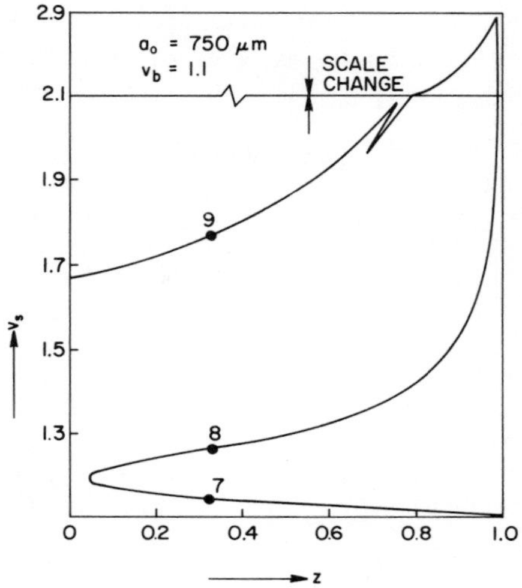

Fig. 11. Locus of steady state solution in the v_s-z plane for a 750 μm initial-radius particle burning under the conditions of Fig. 8 at v_b=1.1.

We will not address the question of stability in detail, although the results of a linear analysis [4] indicate that in the case of the one-film theory, when only three steady states are possible, the intermediate is unstable, the upper and lower stable. When five steady states are possible, the indication is that if they are numbered progressively, counting from either above or below, then even numbered states are unstable while odd numbered states are stable.

4. DYNAMICS

In the following treatment we assume, given the relatively large difference in thermal time constants between the solid particle and the gas-phase boundary layer, that the gas is in a quasi-steady state whether the system of particle and boundary layer is in a quasi-steady state or not. Mathematically, this amounts to setting all time partial derivatives to zero in equations (14)-(15), but retaining the time derivatives of the particle temperature and radius in equations (17) and (18). With the diffusivity correction, the system becomes

$$\gamma_i \frac{\partial^2 u_i}{\partial \rho^2} + \nu_i \Phi(\rho,\theta) = 0 \ ; \quad 0 < \rho < \delta \tag{55}$$

$$\frac{\partial^2 v}{\partial \rho^2} + \beta \Phi(\rho,\theta) = 0 \ ; \quad 0 < \rho < \delta \tag{56}$$

$$\gamma_i \frac{\partial u_i}{\partial \rho} = y(\theta) \sum_{j=1}^{2} n_j \beta_{ij} e^{-\varepsilon_j/v} u_j - u_i \ ; \quad \rho = \delta, \ \theta > 0 \tag{57}$$

$$P\frac{dv_s}{d\theta} + \frac{\partial v}{\partial \rho} = y(\theta) \sum_{j=1}^{2} Q_j e^{-\varepsilon_j/v} u_j + sy(\theta)[v^4 - v_M^4] - v \qquad \rho = \delta, \ \theta > 0 \tag{58}$$

$$\frac{dy(\theta)}{d\theta} = -y^2(\theta) \sum_{j=1}^{2} m_j e^{-\varepsilon_j/v} u_j \ ; \qquad \rho = \delta, \ \theta > 0 \tag{59}$$

$$u_i = u_{ib}, \quad v = v_b \ ; \quad \rho = 0, \quad \theta > 0 \tag{60}$$

and the initial conditions

$$y(0) = 1 \ , \quad v_s(0) = v_{so} \tag{61}$$

It is useful to consider first the dynamics of the single film theory. In particular, when $\Phi = 0$, $u_{2b} = 0$, $n_2 = 0$, we obtain

$$\frac{\partial^2 u_1}{\partial \rho^2} = 0 \ , \quad \frac{\partial^2 v}{\partial \rho^2} = 0 \ ; \quad 0 < \rho < \delta \tag{62}$$

$$\gamma\left[\frac{\partial u_1}{\partial \rho}\right]_{\rho=\delta} = -u_{1s}(n_1 f_1 + 1) \ ; \quad \rho = \delta \tag{63}$$

$$\frac{dv_s}{d\theta} = -\frac{1}{P}\left[\frac{\partial v}{\partial \rho}\right]_{\rho=\delta} + \frac{y}{P} Q_1 e^{-\varepsilon_1/v_s} u_{1s} + \frac{sy}{P}[v_s^4 - v_m^4] - \frac{v_s}{P} \tag{64}$$

$$\frac{dy}{d\theta} = -y^2 m_1 e^{-\varepsilon_1/v_s} u_{1s} \tag{65}$$

The oxygen conservation equation may be integrated from $\rho = 0$ to $\rho = \delta$ to give

$$u_{1s} = \frac{u_{1b}}{1 + \frac{\delta}{\gamma}(n_1 f_1 + 1)} \tag{66}$$

Similarly the energy equation yields

$$\left[\frac{\partial v}{\partial \rho}\right]_{\rho=\delta} = \frac{v_s - v_b}{\delta} \tag{67}$$

These expressions may then be used to eliminate u_{1s} and

$[\frac{\partial v}{\partial \rho}]_{\rho=\delta}$ in (64), (65). The result is

$$\frac{dv_s}{d\theta} = \frac{v_b}{\delta P} - \frac{v_s}{P}(1+\frac{1}{\delta}) + y\frac{Q_1 e^{-\varepsilon_1/v_s} u_{1b}}{P[1+\frac{\delta}{\gamma}(n_1 f_1+1)]} + \frac{sy}{P}(v_s{}^4 - v_m{}^4) \tag{68}$$

$$\frac{dy}{d\theta} = -y^2 \frac{m_1 e^{-\varepsilon_1/v_s} u_{1b}}{[1+\frac{\delta}{\gamma}(n_1 f_1+1)]} \tag{69}$$

subject to (61). This is a non-linear initial value problem, whose solution must be obtained numerically.

A set of transients with varying initial temperatures are shown in Fig. 12 for the particle with the quasi-steady state structure of Fig. 7(B). When the particle is initially heated to the temperature of the upper state, it burns essentially next to it until this state disappears (Fig. 12(A)). The temperature then drops sharply as there is a shift to the lower state.

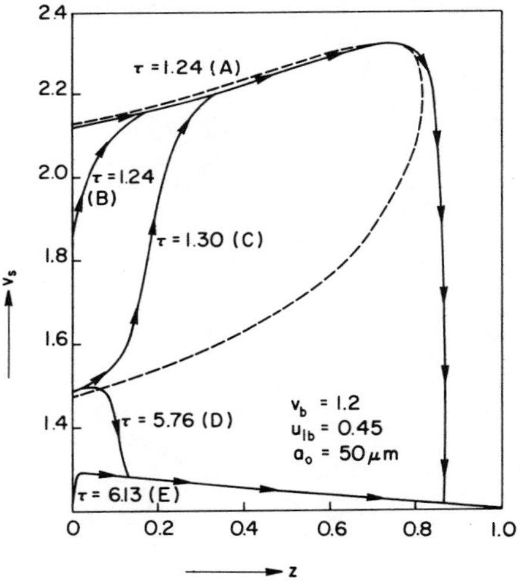

Fig. 12. Transients at various initial particle temperatures for the steady state structure of Fig. 7B.

With an initial particle temperature of 1.485, near the intermediate steady state, combustion proceeds initially parallel to this state but gradually moves away towards the upper state (Fig. 12(C)). The temperature history then coincides with that of Fig. 12(A), burning along the upper state but shifting to the lower as the upper disappears.

Decreasing the starting temperature by only one degree causes the particle to burn initially as in Fig. 12(C), but as the temperature drifts away from the intermediate state, the shift is to the lower state (Fig. 12(D)). As expected, any initial temperature above 1.485 will cause the particle to approach the upper state, whereas any intial temperature below 1.485 will cause it to approach the lower, of which

Fig. 12(B) and 12(E) are two examples. These results reiterate the conclusion that the intermediate state is unstable.

Fig. 13 shows several transients when the same particle burns at a lower ambient temperature, $v_b = 1.0$. While the behaviour is similar to the previously examined case, now there is some overshoot of the loop formed by the upper states. That is, there is some lag between the disappearance of the upper states and the shift to the lower. The transient labeled B approaches the upper state slowly and as a result never quite reaches it. Considerable burning of the particle then occurs away from any of the steady states.

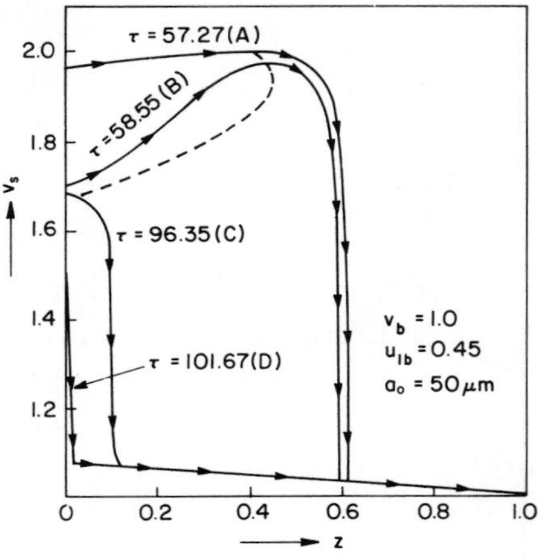

Fig. 13. Transients at various initial particle temperatures for the steady state structure of Fig. 7A.

The high ambient temperature case is studied in Fig. 14. In this figure all transients have been started from room temperature. In the size range $0 \le z \le 0.33$, all transients approach the unique, upper steady state with the usual shift to the lower state once the upper becomes extinct. However, in the neighborhood of the bifurcation point, the transients are deflected, as indicated by Fig. 14(C), may cross the unstable state and shift to the lower state. For $z > 0.33$, all particles would burn along the lower state. An unexpected result of this steady state structure is that the larger of two particles (a_o= 500 μm) may have a

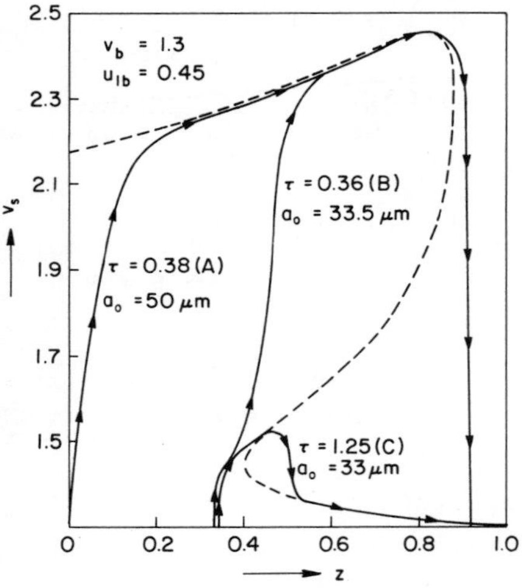

Fig. 14. Transients at various initial particle sizes for the steady state structure of Fig. 7C.

significantly shorter lifetime than the smaller particle ($a_o = 33$ μm).

Next consider the general case. Since the conservation equations for $0 < \rho < \delta$ retain the form of the steady state continuous flame theory, the previously derived relationships (40)-(42), have similar form

$$\gamma(u_3-u_{3b}) + (u_2-u_{2b}) = [\gamma\frac{\partial u_3}{\partial \rho} + \frac{\partial u_2}{\partial \rho}]_{\rho=\delta}\ \rho \tag{70}$$

$$2\gamma(u_1-u_{1b}) + (u_2-u_{2b}) = [2\gamma\frac{\partial u_1}{\partial \rho} + \frac{\partial u_2}{\partial \rho}]_{\rho=\delta}\ \rho \tag{71}$$

$$(u_2-u_{2b}) - \frac{1}{\beta}(v-v_b) = [\frac{\partial u_2}{\partial \rho} - \frac{1}{\beta}\frac{\partial v}{\partial \rho}]_{\rho=\delta}\ \rho \tag{72}$$

We may evaluate $[\frac{\partial u_i}{\partial \rho}]_{\rho=\delta}$ from (57), set $\rho = \delta$ in (70)-(72) and solve for the surface concentrations, u_{is}.

$$u_{is} = u_{is}(v_s, [\frac{\partial v}{\partial \rho}]_{\rho=\delta}) \tag{73}$$

where $[\frac{\partial v}{\partial \rho}]_{\rho=\delta}$ is an unknown since it cannot be expressed as in the steady state case in terms of u_{is}, v_s due to the time derivative that now appears in the boundary condition (58). The energy equation becomes

$$\frac{\partial^2 v}{\partial \rho^2} + \Omega(\rho,y,v,v_s,[\frac{\partial v}{\partial \rho}]_{\rho=0}) = 0\ ;\quad 0 < \rho < \delta \tag{74}$$

$$v(0) = v_b\ ,\ v(\delta) = v_s \tag{75}$$

which must be solved to evaluate $[\frac{\partial v}{\partial \rho}]_{\rho=\delta}$ for each specified, instantaneous value of v_s. This problem is thus coupled to

$$\frac{dv_s}{d\theta} = \chi(v_s,y,[\frac{\partial v}{\partial \rho}]_{\rho=\delta}) \tag{76}$$

$$\frac{dy}{d\theta} = \Gamma(v_s,y,[\frac{\partial v}{\partial \rho}]_{\rho=\delta}) \tag{77}$$

$$v_s(0) = v_{so},\quad y(0) = 1 \tag{78}$$

In a step-by-step forward integration of (76)-(78), each step requires N repetitive solutions of (74)-(75), N being the characteristic number of functional evaluations associated with the numerical procedure.

A significant result that arises in connection with the form of these equations is the non-uniqueness of the solution of (74). The numerical results indicate that either one or three different solutions are possible depending on the parameters. Thus, for a fixed value of the solid temperature, v_s, there can be more than one possible set of composition and temperature profiles that satisfy the quasi-steady state conditon of the boundary layer.

This multiplicity is shown in Fig. 15, where we compare the three transient solutions of (74) to the five steady states at an arbitrary point (labeled 10) of the v_s - z plane of Fig. 9. Notice that each transient bears a qualitative resemblance to one of the three upper steady states (states 3-5). The number of possible transient solutions under fixed ambient conditions is determined by the value of the particle size and temperature. Thus Fig. 16 illustrates the variation of $\frac{dv_s}{dt}$ caused by v_s for two different particle sizes that correspond to z = 0.60 and z = 0.30 in Fig. 9. For the 400 μm particle, multiplicity occurs in the range $1.61 \leq v_s \leq 1.75$. However, for the 700 μm particle, the range of multiplicity has decreased and by z = 0.25, there is a unique solution. It is interesting that although for the 700 μm particle, three transient solutions are possible for $1.49 \leq v_s \leq 1.58$, there are only two ignited steady state solutions.

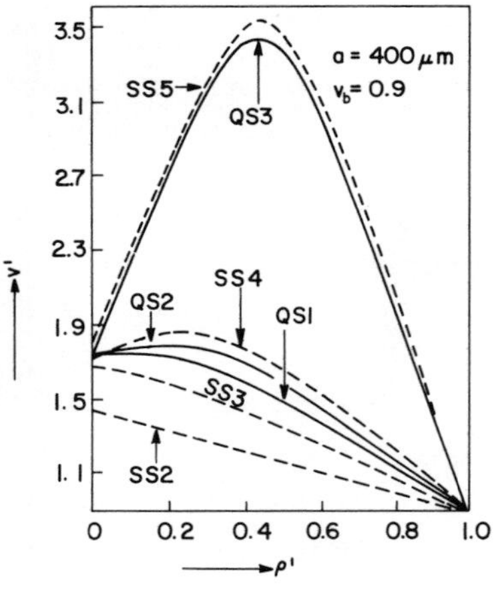

Fig. 15. Comparison of the transient profiles predicted by equation (74) with the steady state profiles at point 10 of Fig. 9. [v'=v/(2-ρ); ρ'=2(1-ρ)/(2-ρ)].

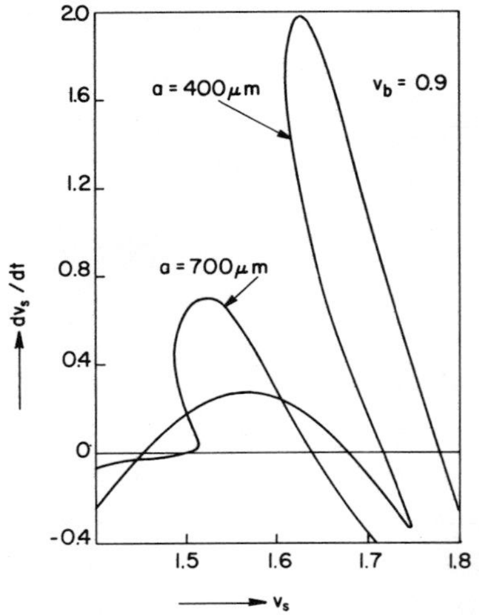

Fig. 16. Effect of the particle radius and temperature on the time derivative of the particle temperature caused by the non-uniqueness of the solution of equation (74).

When multiple solutions of (74) occur, we have three different values of $[\frac{\partial v}{\partial \rho}]_{\rho=\delta}$ to choose from in evaluating χ and Γ. Depending on which is chosen, a different temperature history or "branch" is obtained emanating from a single point of the v_s - z plane. Although each transient solution shares the same particle temperature, because the surface gradient $[\frac{\partial v}{\partial \rho}]_{\rho=\delta}$ differs in each case, so do the surface concentrations, u_{is}.

A typical set of results is shown in Fig. 17 for the 400 μm particle in a $v_b = 0.9$ environment. Choosing $v_{so} = 1.74$ so that the transients at z = 0 are those previously shown in Fig. 15, yields the branches labeled A, B, C. A given transient approaches the steady state that more closely resembles it. This is also illustrated by branches D and E. A third branch (not shown) out of the same initial temperature approaches the uppermost steady state. Branch F which originates from a unique transient solution near the second steady state, displays behaviour similar to that obtained with the one-film model.

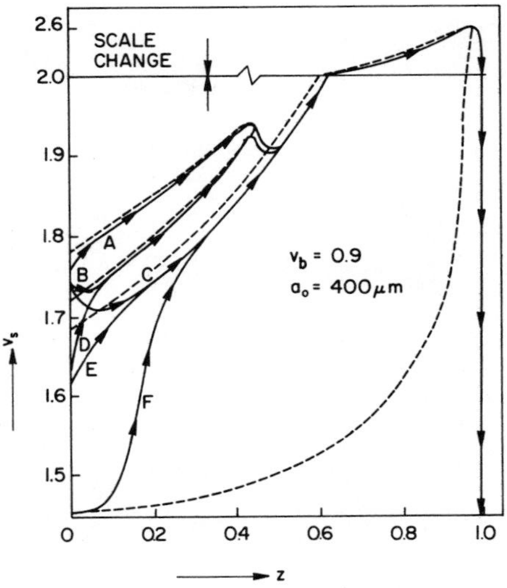

Fig. 17. Select number of temperature histories for a particle with the steady state structure shown by the dotted lines.

In the neighborhood of the bifurcation point of Fig. 9, at z = 0.30 (a = 700 μm), we previously found three transient solutions for $1.49 \leq v_s \leq 1.58$, although only two steady state solutions are possible. The transients associated with these states in terms of resemblance are the first and third. Hence the intermediate transient does not possess an associated steady state. Choosing that transient yields the branch labeled A in Fig. 18 (a blow-up of the region near the bifurcation). As may be observed the temperature history does not approach any of the steady states in the shown size range. However, branch C, out of the same initial v_s - z point but the first instead of the intermediate transient, does not lead to the same sharp initial increase in temperature. Behaviour similar to that of branch A may also be obtained from an intermediate transient near the lower bound of multiplicity (branch B).

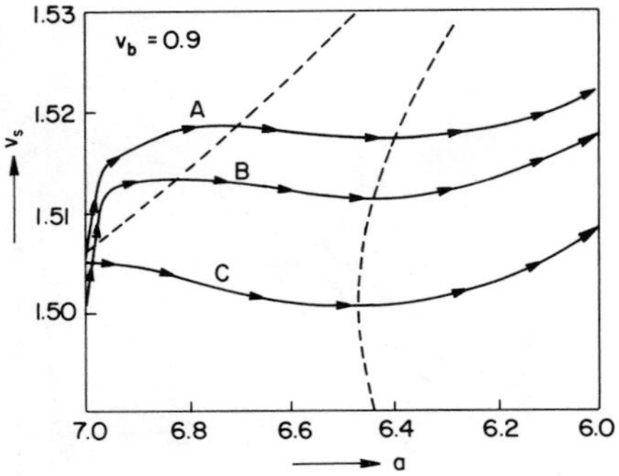

Fig. 18. Blow-up of the region enclosed by the dotted lines in Fig. 9, illustrating the transients that result in the neighborhood of the bifurcation point.

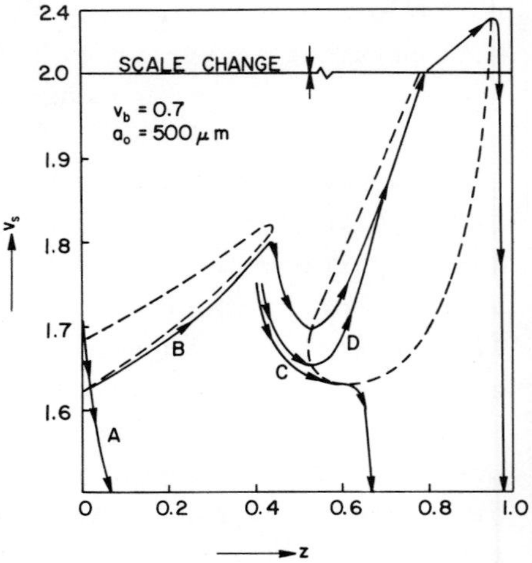

Fig. 19. Select number of transients for a particle with the steady state structure of Fig. 10.

The low temperature case, $v_b = 0.7$ is examined in Fig. 19. Here at z = 0, there is transient multiplicity for $1.575 \leq v_s \leq 1.735$. Branch A, that originates from a transient resembling the third steady state shows how the particle may become unignited from high temperatures. Branch B results from a transient resembling the fourth steady state and burns along that state but drops sharply during the gap between loops, then increases as it approaches the upper state of the second loop. Branches C and D out of $v_{so} = 1.75$, z = 0.41, 0.42, respectively, are presented to illustrate the domain of attraction of the second loop. While branch D is deflected towards the upper state, branch C is able to break away towards the unignited state.

We must finally note that the stability character of the steady states does not translate into the present analysis due to the assumption of a quasi-steady state boundary layer.

NOMENCLATURE

A = $a_o{}^2 c^{3/2} k_3/c$

a = particle radius

b = external edge of the boundary layer

c_i = molar concentration of species i

c = molar concentration of the mixture

C_p = specific heat of the gas

C_s = specific heat of the solid

D_i = diffusivity of the ith species

E_i = activation energy of the ith reaction

f_i = $y \exp(-\varepsilon_i/v_s)$

$(-\Delta H_i)$ = heat of the ith reaction

k_i = frequency factor for the ith reaction

k = thermal conductivity

M_c = molecular weight of carbon

m_j = $(3-j)M_c k_j c a_o/\alpha\rho_c$

n_j = $a_o k_j/\alpha$

P = $\alpha\rho_c C_s/3M_c k$

Q_j = $(-\Delta H_j) k_j a_o c/kT_e$

q_R = radiant heat flux

R_i = rate of the ith reaction

R = universal gas constant

r = radial position variable

s = $\sigma\varepsilon a_o T_e{}^3/k$

T = temperature

T_e = constant parameter, 1000K

T_m = constant parameter

t = time

u_i = dimensionless concentration variable

v = dimensionless temperature variable

y = dimensionless particle radius

z = $1 - y$

Greek Letters

α = thermal diffusivity

β = $(-\Delta H_3)/cC_p T_e$

β_{ij} = stoichiometric coefficient of the jth component in the ith heterogeneous reaction

Γ = a function, defined by equation (17)

γ_i = constant parameter

δ = dimensionless boundary layer thickness

ε = emissivity of carbon

ε_j = E_j/RT_e

θ = dimensionless time

ν_i = stoichiometric coefficient of the ith species in the homogeneous reaction

ρ = dimensionless position variable

ρ_s = density of carbon

ρ_F = location of the flame front

σ = Stefan-Boltzmann constant

τ = particle lifetime

Φ = a function, defined by equation (22)

χ = a function, defined by equation (76)

Ω = a function, defined by equation (74)

Subscripts

s = surface of the particle

b = external edge of the boundary layer

o = initial value

REFERENCES

1. Burke, S.P., Schumann, T.E.W., Ind. Eng. Chem., 23, 406 (1931).
2. Burke, S.P., Schumann, T.E.W., Proc. 3rd Int. Conf. Bituminous Coal, 2, 485 (1931).
3. Caram, H.S., Amundson, N.R., Ind. Eng. Chem. Fundam. 16, 171 (1977).
4. Mon, E., Amundson, N.R., Ind. Eng. Chem. Fundam., 18, 162 (1979).

The early phases of this work were supported by the University of Minnesota and the University of Houston. For the past sixteen months the work has been supported by the Department of Energy, Grant No. ET-78-G-01-2020.

Neal R. Amundson and E. Mon
Department of Chemical Engineering
University of Houston
Houston, Texas 77004

On the Closure and Character of the Balance Equations for Heterogeneous Two-Phase Flow

Paul S. Gough

1. INTRODUCTION

We are concerned here with aspects of the behaviour of heterogeneous multiphase flows that can be described adequately in terms of a macroscopic formulation. We will use the words macroscopic and microscopic to refer to length scales which are respectively large or small by comparison with the scale of heterogeneity. Problems falling into the category of interest here include wave propagation in porous and composite materials, bubbly flows, and the combustion of granular propellants. We present both general and specific results, the latter pertaining particularly to problems of granular propellants.

Equations to describe the macroscopic aspects of heterogeneous multiphase flows have been presented by many authors, using many different approaches. Even confining our attention to approaches based on formal averaging of the microscopic equations, we note the contributions of Anderson and Jackson[1], Slattery[2], Panton[3], Drew[4], Gough[5], and Ishii[6], that of Ishii representing the most comprehensive study to date.

The use of a weighting function to describe the average values of the state variables was introduced by Anderson and Jackson[1] who presented macroscopic balances of mass and momentum for a flow consisting of an incompressible gas and an aggregate of particles. Their approach was extended by Gough [5] to include averaging with respect to time as well as

ISBN 0-12-669550-4

position and to account for compressibility of the gas. Here we extend further our derivation by deducing the energy equation in greater generality than previously and discussing the need for consistent approximation in the neglect of correlation terms.

For those problems in which macroscopic diffusion may be neglected, the balance equations are found to be hyperbolic but not totally hyperbolic. This observation has raised several questions in the literature concerning the well-posedness of the equations. The mathematical character of the balance equations is influenced significantly by superficially minor details of the constitutive assumptions used to close the system. For this reason we compare our approach to closure with that of Ishii[6]. We also review some of the approaches which have been taken to render the equations totally hyperbolic.

2. DEFINITION OF MACROSCOPIC VARIABLES

In order to form macroscopic state variables we suppose that the microscopic quantities will be averaged in some fashion with respect to either space or time or both. We incorporate full generality of the domain over which the average is to be formed by the introduction of a weighting function $g(\vec{r},t)$ whose arguments are understood to be coordinates relative to the point of collocation. We impose on the function g only the weakest of requirements; we assume g is non-negative and that it tends to zero sufficiently rapidly, as its arguments become infinite, that we may ignore sufficiently remote contributions to the average value. In order to maintain simplicity we also assume that g has as many continuous partial derivatives as we may need. Finally, we assume that g is normalized:

$$\int_{-\infty}^{+\infty}\int_{-\infty}^{+\infty}\int_{-\infty}^{+\infty}\int_{-\infty}^{+\infty} g(\vec{r},t)dr_1dr_2dr_3dt = 1 \qquad (2.1)$$

We now recognize explicitly the heterogeneity of the mixture. At each instant of time t we assume that the media occupy disjoint and complementary regions $V_1(t)$ and $V_2(t)$ separated by the interface A(t). It will also be useful to introduce $\mathcal{V}_1$, $\mathcal{V}_2$ and $\mathcal{A}$ as the products of V_1, V_2 and A with time.

Thus V_i is the part of the spacetime continuum occupied by the first medium or phase. The volume fraction associated with each phase is defined formally as:

$$\alpha_i(\vec{x},t) = \int_{V_i} g(\vec{y}-\vec{x}, \tau-t) \, d\vec{y} \, d\tau \tag{2.2}$$

For each microscopic state variable ψ_i defined for phase i we may introduce a macroscopic quantity $<\psi_i>$ according to:

$$\alpha_i(\vec{x},t)<\psi_i(\vec{x},t)> = \int_{V_i} \psi_i(\vec{y},\tau) g(\vec{y}-\vec{x},\tau-t) d\vec{y} \, d\tau \tag{2.3}$$

We note that whereas ψ_i is only defined in V_i, $<\psi_i>$ is defined everywhere. Naturally, we will attempt to formulate the macroscopic theory as consistently as possible, using equation (2.3) to define the state variables. However, it will be the case that certain quantities associated with interphase transfer phenomena are only defined in a meaningful way by reference to an average whose sample set is restricted to the interface. Thus analogously to equation (2.2) we introduce the surface area per unit volume as:

$$S(\vec{x},t) = \int_A g(\vec{y}-\vec{x}, \tau-t) d\vec{y} \, d\tau \tag{2.4}$$

and the surface average of ψ_i is given by $<\psi_i>_s$ as:

$$S(\vec{x},t)<\psi_i(\vec{x},t)>_s = \int_A \psi_i(\vec{y},\tau) g(\vec{y}-\vec{x}, \tau-t) d\vec{y} \, d\tau \tag{2.5}$$

In principle, the foregoing definitions of macroscopic quantities are sufficient for the complete development of the balance equations[5]. However, the formulism can be simplified in regard to the treatment of the correlations of fluctuations about the average values if we introduce a density weighted average [3,4,6]. We define the density weighted average of ψ_i as $<\psi_i>_\rho$ according to:

$$<\rho_i><\psi_i>_\rho = <\rho_i \psi_i> \tag{2.6}$$

where ρ_i is the microscopic density of the ith phase.

3. SMOOTHNESS OF THE AVERAGE AND FLUCTUATIONS

As in the theory of turbulence[7], we will assume that any macroscopic quantity behaves like a constant in any subsequent averaging process. This is assumed to be true not only of the macroscopic quantities themselves but also of their partial derivatives of any order. The significance of this postulate, in the present context, has been discussed in detail by Whitaker[8], Drew[4], Gough[5] and Ishii[6]. In essence we must suppose that g is selected so that we sample over length and time scales which are large compared with the scale of heterogeneity of the mixture and that the significant length and time scales of the macroscopic variables are, in turn, large compared with those over which the average is formed.

With this assumption we may introduce the fluctuation of ψ_i about its average value as:

$$\psi_i' = \psi_i - \langle\psi_i\rangle \tag{3.1}$$

According to the assumption of smoothness of the average we have:

$$\langle\psi_i'\rangle = 0 \tag{3.2}$$

and, for a product, the usual factorization rule follows as:

$$\langle\psi_i\phi_i\rangle = \langle\psi_i\rangle\langle\phi_i\rangle + \langle\psi_i'\phi_i'\rangle \tag{3.3}$$

It should be borne in mind, however, that the three types of average (2.3), (2.5) and (2.6) do not commute. For example, $\langle\langle\psi_i\rangle\rangle_s = \langle\psi_i\rangle$ while $\langle\langle\psi_i\rangle_s\rangle = \langle\psi_i\rangle_s$. Accordingly, we may not assume $\langle\psi_i'\rangle_s = 0$ or $\langle\psi_i'\rangle_\rho = 0$. We also introduce the fluctuation about the density weighted average which we will distinguish by a superscripted asterisk, ψ_i^*; however, we will not consider the influence of fluctuations about the surface averages.

4. COMMUTATION OF DERIVATIVE AND AVERAGE

It is well known that the average of the derivative differs from the derivative of the average. Derivations of the law of commutation have been deduced for various types of space and time averages by Anderson and Jackson[1],

Slattery[2], Drew[4], Gough[5], Ishii[6] and, most recently, by Gray and Lee[9]. In view of the effort exerted by Gray and Lee, it is perhaps worth noting just how tractable the macroscopic theory becomes when a weighting function is used. Take the gradient of both sides of (2.3) with respect to $\vec{x}$. Since the region of integration on the right hand side of (2.3) does not depend on $\vec{x}$, the gradient may be taken inside. Thus:

$$\nabla_x(\alpha_i(\vec{x},t)<\psi_i(\vec{x},t)>) = \int_{V_i} \psi_i(\vec{y},\tau)\nabla_x g(\vec{y}-\vec{x},\ \tau-t)d\vec{y}\ d\tau$$

Now we may use the antisymmetry of g with respect to its arguments $\vec{x}$ and $\vec{y}$ since:

$$\nabla_x g(\vec{y}-\vec{x},\ \tau-t) = -\nabla_y g(\vec{y}-\vec{x},\ \tau-t)$$

Therefore we may write:

$$\nabla_x(\alpha_i(\vec{x},t)<\psi_i(\vec{x},t)>) = \int_{V_i} g(\vec{y}-\vec{x},\ \tau-t)\nabla_y\psi_i(\vec{y},\tau)d\vec{y}\ d\tau$$

$$- \int_{V_i} \nabla_y\{\psi_i(\vec{y},\tau)g(\vec{y}-\vec{x},\ \tau-t)\}d\vec{y}\ d\tau$$

The second term on the right hand side may be transformed to a surface integral by means of the divergence theorem and the contribution at infinity neglected to yield:

$$\nabla(\alpha_i<\psi_i>) = \alpha_i<\nabla\psi_i> - \int_A \psi_i(\vec{y},\tau)g(\vec{y}-\vec{x},\ \tau-t)\vec{n}\ d\vec{y}\ d\tau \quad (4.1)$$

which may be seen to be the familiar result. Similarly:

$$\frac{\partial}{\partial t}(\alpha_i<\psi_i>) = \alpha_i<\frac{\partial\psi_i}{\partial t}> + \int_A \psi_i(\vec{y},\tau)g(\vec{y}-\vec{x},\ \tau-t)\vec{w}\cdot\vec{n}\ d\vec{y}\ d\tau \quad (4.2)$$

where $\vec{w}$ is the velocity of a point on the interface A(t) and $\vec{n}$ is the vector normal to the surface pointing out of V_i. Setting $\psi_i = 1$ in (4.1) and (4.2) provides the derivatives of the volume fraction.

When there is no mass transfer between the phases, we may take A to be a material surface on which the microscopic boundary condition $\vec{u}_i \cdot \vec{n} = \vec{w} \cdot \vec{n}$ is satisfied by the velocity field $\vec{u}_i$ of both phases, even if we consider relative slip. In such a case one may show that:

$$\frac{D\alpha_i}{Dt_i} = \frac{\partial \alpha_i}{\partial t} + \langle \vec{u}_i \rangle_\rho \cdot \nabla \alpha_i = \int_{V_i} g \nabla \cdot \vec{u}_i{}^* d\vec{y}\, d\tau \tag{4.3}$$

Equation (4.3) is useful in the reduction of the mechanical work terms which arise in the macroscopic energy equation. Moreover, it clearly indicates that the derivatives of the fluctuation field must be treated carefully as they may contribute macroscopic terms to the equations.

5. FUNDAMENTAL ASSUMPTIONS

The following assumptions are made concerning the microscopic flow.

Microscopic Balance Equations

Each phase is assumed to behave as a continuum and to be governed by the following balances of mass, momentum and energy throughout V_i:

$$\frac{\partial \rho_i}{\partial t} + \nabla \cdot \rho_i \vec{u}_i = 0 \tag{5.1}$$

$$\frac{\partial}{\partial t} \rho_i \vec{u}_i + \nabla \cdot (\rho_i \vec{u}_i \vec{u}_i - \overleftrightarrow{\sigma}_i) = 0 \tag{5.2}$$

$$\frac{\partial}{\partial t} \left(\rho_i \left(e_i + \frac{u_i^2}{2}\right)\right) + \nabla \cdot \left[\rho_i \vec{u}_i \left(e_i + \frac{u_i^2}{2}\right) - \vec{u}_i \cdot \overleftrightarrow{\sigma}_i + \vec{q}_i\right] = 0 \tag{5.3}$$

Here we have ρ the density, $\vec{u}$ the velocity, $\overleftrightarrow{\sigma}$ the stress tensor, e the internal energy, $\vec{q}$ the heat flux and $u_i^2 = \vec{u}_i \cdot \vec{u}_i$. It will also be useful to represent the stress tensor in terms of the pressure p and the deviator $\overleftrightarrow{\tau}$ as $\overleftrightarrow{\sigma} = \overleftrightarrow{\tau} - p\overleftrightarrow{I}$ where $\overleftrightarrow{I}$ is the unit tensor.

Microscopic Boundary Conditions

We assume that mass transfer may occur between phases and that the following jump conditions apply on A:

$$[\rho(\vec{u} - \vec{w}) \cdot \vec{n}]_2^1 = 0 \tag{5.4}$$

$$[(\rho\vec{u}(\vec{u}-\vec{w}) - \overleftrightarrow{\sigma})\cdot\vec{n}]_2^1 = 0 \qquad (5.5)$$

$$[(\rho(e + \frac{u^2}{2})(\vec{u}-\vec{w}) + \vec{q} - \vec{u}\cdot\overleftrightarrow{\sigma})\cdot\vec{n}]_2^1 = 0 \qquad (5.6)$$

where we have the notation $[\psi]_2^1 = \psi_1 - \psi_2$. It should be noted that these conditions neglect the influence of surface tension. A more general formulation of the jump conditions can be found in the work of Delhaye[10]. We assume continuity of the tangential components of velocity and traction and we introduce the normal pressure jump:

$$\vec{n}\Delta p = [\overleftrightarrow{\sigma}\cdot\vec{n}]_2^1 \qquad (5.7)$$

We will subsequently suppose that when mass transfer occurs, the rate of regression of the interface into phase 2 is available from a macroscopic correlation. Thus we are motivated to write:

$$\vec{w} = \vec{u}_2 + \vec{n}\dot{d} \qquad (5.8)$$

where $\vec{n}$ points into phase 2 and the $\dot{d}$ is the rate of regresion. For the moment, (5.8) may be viewed as a purely formal definition, valid whether or not the desired correlation for $\dot{d}$ is available.

Smoothness of the Average

We take as a postulate that any macroscopic quantity and its derivatives behave as constants in any further averaging operation. Moreover, we assume that correlations of fluctuations about the surface averages can be neglected.

6. MACROSCOPIC BALANCE OF MASS

Taking the average of equation (5.1) and using the laws of commutation expressed by equations (4.1) and (4.2) we have:

$$\frac{\partial}{\partial t}\alpha_i\langle\rho_i\rangle + \nabla\cdot\alpha_i\langle\rho_i\vec{u}_i\rangle = -\int_A g\rho_i(\vec{u}_i - \vec{w})\cdot\vec{n}d\vec{y}d\tau \qquad (6.1)$$

If we were to expand $\langle\rho_i\vec{u}_i\rangle$ as $\langle\rho_i\rangle\langle\vec{u}_i\rangle + \langle\rho_i'\vec{u}_i'\rangle$ we would introduce a correlation term into the balance equation. The formal simplicity of the mass averaged velocity thus becomes apparent. The right hand side of (6.1) vanishes in the absence of a phase change. Otherwise, we make use of (5.4) and (5.8) to deduce the balance of mass in the macroscopic form:

$$\frac{\partial}{\partial t}\alpha_i<\rho_i> + \nabla\cdot\alpha_i<\rho_i><\vec{u}_i>_\rho = \dot{m}_i \qquad (6.2)$$

where the mass transfer term on the right hand side is given by:

$$\dot{m}_1 = -\dot{m}_2 = S(\vec{x},t)<\rho_2>_s<\dot{d}>_s \qquad (6.3)$$

It should be noted that, in accordance with our basic postulate, we have assumed that the correlation of the surface fluctuations of ρ_2 and $\dot{d}$ can be neglected. In the special case when phase 2 is incompressible we may use the obvious relationship $\alpha_2 = 1-\alpha_1$ to recast the mass balance equation for that phase into what is essentially a governing equation for the volume fraction α_1.

7. MACROSCOPIC BALANCE OF MOMENTUM

By averaging equation (5.2) and using equation (6.2) we may deduce the macroscopic equation of motion in the form:

$$\alpha_i<\rho_i>\frac{D}{Dt_i}<\vec{u}_i>_\rho = \nabla\cdot\alpha_i<\overleftrightarrow{\sigma}_i> + \nabla\cdot\alpha_i\overleftrightarrow{\tau}_i^{\,T} - <\vec{u}_i>\dot{m}_i - \int_A g[\rho_i\vec{u}_i(\vec{u}_i-\vec{w}) - \overleftrightarrow{\sigma}_i]\cdot\vec{n}d\vec{y}d\tau \qquad (7.1)$$

where we have introduced the macroscopic convective derivative:

$$\frac{D}{Dt_i} = \frac{\partial}{\partial t} + <\vec{u}_i>_\rho\cdot\nabla \qquad (7.2)$$

and the pseudo-turbulent stress tensor:

$$\overleftrightarrow{\tau}_i^{\,T} = -<\rho_i><\vec{u}_i^{\,*}\vec{u}_i^{\,*}>_\rho \qquad (7.3)$$

In order to complete the macroscopic formulation of the equations of motion we wish to make use of the microscopic boundary condition (5.5) and to extract, from the surface integral on the right hand side of (7.1), the contribution of the average stress. The particular approach to the manipulation of the surface integral which we have previously adopted [5] was predicated on considerations of the closure relations for the macroscopic theory. In the event that phase 1 is connected and phase 2 is dispersed, a common assumption is that the average pressures in the two phases are equal (Kraiko and Sternin[11], Nigmatulin[12], Gough[5], Ishii[6]). The necessity of a macroscopic assumption of this sort follows from the observation that even if we neglect all interphase transfer processes and the pseudo-turbulent stresses, we are still left with one essentially new state variable, namely the volume fraction of one or the other phases. This point has also been discussed by Stuhmiller[13]. On this basis it is convenient to write both equations of motion by reference to the pressure in the first phase, effectively eliminating the pressure in the second phase as an explicit independent variable.

We will follow the same path here while keeping track of the difference in pressure of the two phases so that the assumption of equality can be relaxed. However, the generally valid microscopic boundary conditions will be embedded into the equations so as to reduce the number of unknowns as much as possible.

Thus we may write the two equations of motion in the form:

$$\alpha_1\langle\rho_1\rangle \frac{D}{Dt_1} \langle\vec{u}_1\rangle_\rho = \alpha_1\nabla\cdot\langle\overleftrightarrow{\sigma}_1\rangle + \nabla\cdot\alpha_1\overleftrightarrow{\tau}_1^{\,T} + (\langle\vec{u}_2\rangle_s - \langle\vec{u}_1\rangle_\rho)\dot{m}_1 + \langle\Delta p\rangle_s \nabla\alpha_1 - \vec{f} \tag{7.4}$$

and

$$\alpha_2\langle\rho_2\rangle \frac{D}{Dt_2} \langle\vec{u}_2\rangle_\rho = \alpha_2\nabla\cdot\langle\overleftrightarrow{\sigma}_1\rangle + \nabla\cdot\alpha_2\overleftrightarrow{\tau}_2^{\,T} + \nabla\cdot\alpha_2(\langle\overleftrightarrow{\sigma}_2\rangle - \langle\overleftrightarrow{\sigma}_1\rangle) + (\langle\vec{u}_2\rangle_s - \langle\vec{u}_2\rangle_\rho)\dot{m}_2 + \langle\Delta p\rangle_s\nabla\alpha_2 + \vec{f} \tag{7.5}$$

where we have introduced the surface average of the stress fluctuation:

$$\vec{f} = -\int_A g\overleftrightarrow{\sigma}_1{}' \cdot\vec{n}\; d\vec{y}d\tau \tag{7.6}$$

Subsequently, we will interpret $\vec{f}$ as the interphase drag, a quantity to be determined, in part, by reference to an empirical correlation. We also note that the normal vector $\vec{n}$ in equation (7.6) is directed into phase 2. The representation of the interphase drag according to (7.6) is somewhat different from that of Ishii[6]. Ishii expresses the surface integral in equation (7.1) by reference to the surface average pressure. It is our view that to do so is to impose the requirement that an additional constitutive law be determined so as to relate the surface average pressure to the bulk averaged quantities. Ishii develops equations under the assumption that the surface and bulk average are equal. Stuhmiller[14], on the other hand, attempts to deduce a relationship between the bulk and surface averages on a semitheoretical basis. We shall comment further on this point when we discuss the characteristics of the balance equations.

8. MACROSCOPIC BALANCE OF ENERGY

When equation (5.3) is averaged, the terms expanded, and use is made in the usual way of the continuity equation (6.2) and the equation of motion (7.1) we have the balance expressed by reference to the density weighted internal energy in the form:

$$\alpha_i\langle\rho_i\rangle \frac{D}{Dt_i}\langle e_i\rangle_\rho + \nabla\cdot\alpha_i\langle\vec{q}_i\rangle - \langle\overleftrightarrow{\sigma}_i\rangle : \nabla\alpha_i\langle\vec{u}_i\rangle_\rho$$
$$= \int_A g[\vec{u}_i\cdot\overleftrightarrow{\sigma}_i - \vec{q}_i - \rho_i(\vec{u}_i-\vec{w})(e_i + \frac{u_i^2}{2})]\cdot\vec{n}\, d\vec{y}\, d\tau$$
$$- \langle\vec{u}_i\rangle_\rho\cdot\{\langle\overleftrightarrow{\sigma}_i\rangle\cdot\nabla\alpha_i + \nabla\cdot\alpha_i\overleftrightarrow{\tau}_i^{\,T} - \langle\vec{u}_i\rangle\dot{m}_i$$
$$- \int_A g[\rho_i\vec{u}_i(\vec{u}_i-\vec{w}) - \overleftrightarrow{\sigma}_i]\cdot\vec{n}\, d\vec{y}\, d\tau\}$$
$$- \dot{m}_i[\langle e_i\rangle_\rho + \frac{\langle u_i\rangle_\rho^2}{2}] - \Theta_i \qquad (8.1)$$

and where the correlation term Θ_i is given by:

$$\Theta_i = \frac{\partial}{\partial t}[\alpha_i\langle\rho_i\rangle \frac{\langle u_i^{*2}\rangle}{2}] - \nabla\cdot\alpha_i[\langle\vec{u}_i^{\,*}\rangle\cdot\langle\overleftrightarrow{\sigma}_i\rangle + \langle\vec{u}_i^{\,*}\cdot\overleftrightarrow{\sigma}_i'\rangle]$$
$$+ \nabla\cdot\alpha_i\langle\rho_i\rangle[\langle\vec{u}_i^{\,*}\frac{u_i^{*2}}{2}\rangle_\rho + \langle\vec{u}_i^{\,*}e_i^*\rangle_\rho + \langle\vec{u}_i\rangle_\rho\cdot\langle\vec{u}_i^{\,*}\vec{u}_i^{\,*}\rangle_\rho$$
$$+ \langle\vec{u}_i\rangle_\rho \frac{\langle u_i^{*2}\rangle_\rho}{2}] \qquad (8.2)$$

We now desire to extract the macroscopic state variables, particularly those which are expressed as derivatives, from the integrals on the right hand side of equation (8.1). As the only terms which present any difficulty are those associated with mechanical work, let us ignore, for the moment, heat and mass transfer as well as the correlation terms in (8.1). Combining terms, (8.1) reduces to:

$$\alpha_i\langle\rho_i\rangle \frac{D}{Dt_i}\langle e_i\rangle_\rho - \alpha_i\langle\overleftrightarrow{\sigma}_i\rangle : \nabla\langle\vec{u}_i\rangle_\rho = \int_A g\vec{u}_i^{\,*}\cdot\overleftrightarrow{\sigma}_i\cdot\vec{n}d\vec{y}\, d\tau \tag{8.3}$$

Writing $\overleftrightarrow{\sigma}_i = -\langle p_i\rangle\overleftrightarrow{I} + \langle\overleftrightarrow{\tau}_i\rangle + \overleftrightarrow{\sigma}_i{}'$ in the right hand side, using the divergence theorem on the term involving $\langle p_i\rangle$ and making use of (4.3) yields:

$$\alpha_i\langle\rho_i\rangle \frac{D}{Dt_i}\langle e_i\rangle_\rho - \alpha_i\langle\overleftrightarrow{\sigma}_i\rangle : \nabla\langle\vec{u}_i\rangle_\rho + \langle p_i\rangle\frac{D\alpha_i}{Dt_i} = \int_A g\vec{u}_i^{\,*}\cdot(\overleftrightarrow{\sigma}_i{}' + \langle\overleftrightarrow{\tau}_i\rangle)\cdot\vec{n}d\vec{y}\, d\tau \tag{8.4}$$

Finally, if the macroscopic shear stress is negligible on A as compared with $\overleftrightarrow{\sigma}_1{}'$ and the velocity fluctuations of phase 2 are negligible, as might be the case when phase 2 consists of fairly massive particles, we can put $\vec{u}_1^{\,*} \sim \langle\vec{u}_2\rangle_\rho - \langle\vec{u}_1\rangle_\rho$ on A so that (8.4) becomes, for phase 1:

$$\alpha_1\langle\rho_1\rangle \frac{D}{Dt_1}\langle e_1\rangle_\rho - \alpha_1\langle\overleftrightarrow{\sigma}_1\rangle : \nabla\langle\vec{u}_1\rangle_\rho + \langle p_1\rangle\frac{D\alpha_1}{Dt_1} = (\langle\vec{u}_1\rangle_\rho - \langle\vec{u}_2\rangle_\rho)\cdot\vec{f} \tag{8.5}$$

where $\vec{f}$ is given by equation (7.6). In view of the interpretation of $\vec{f}$ as the interphase drag, the right hand side is seen to be an approximate evaluation of the dissipation in the boundary layer. The dissipation in the bulk of the flow is embedded, principally, in the term $\nabla\cdot\alpha_i\langle\vec{u}_i^{\,*}\cdot\overleftrightarrow{\sigma}_i{}'\rangle$ which contributes to Θ as shown in equation (8.2). It has been shown by Celmins[15] that this term can be comparable in magnitude to the right hand side of (8.5) in the particular case of single phase flow through a cylindrical duct. Thus, while such terms may be neglected in macroscopic models, for reasons of tractability, it should be borne in mind that their influence may be comparable to that of other terms, such as that on the right hand side of (8.5), which are retained for reasons of thermodynamic consistency. Following our discussion of the general forms of the energy equation we shall present an example of the importance of consistent approximation in regard to the neglect of the correlation terms.

As with the momentum equations, further reductions of the general energy equations can be obtained under the assumption that it is useful to eliminate the pressure in the second phase in favor of that in the first. Using the jump conditions (5.5) and (5.6) we have the energy equation for phase 1 in the form:

$$\alpha_1\langle\rho_1\rangle \frac{D}{Dt_1}\langle e_1\rangle_\rho + \nabla\cdot\alpha_1\langle\vec{q}_1\rangle + \alpha_1\langle p_1\rangle\nabla\cdot\langle\vec{u}_1\rangle_\rho + \langle p_1\rangle\frac{D\alpha_1}{Dt_1}$$

$$+ \langle\Delta p\rangle_s\frac{D\alpha_1}{Dt_1} = \alpha_1(\langle\overleftrightarrow{\tau}_1\rangle + \langle\overleftrightarrow{\tau}_1\rangle^T): \nabla\langle\vec{u}_1\rangle_\rho - \int_A g\vec{q}_2\cdot\vec{n}\, d\vec{y}\, d\tau$$

$$+ (\langle\vec{u}_2\rangle_\rho - \langle\vec{u}_1\rangle_\rho)\cdot\int_A g(\langle\overleftrightarrow{\tau}_1\rangle + \overleftrightarrow{\sigma}_1{}')\cdot\vec{n}\, d\vec{y}\, d\tau$$

$$+ \dot{m}_1[\langle e_2\rangle_s - \langle e_1\rangle_\rho + \frac{\langle p_1\rangle + \langle\Delta p\rangle_s}{\langle\rho_2\rangle_s} + \frac{(\langle\vec{u}_2\rangle_s - \langle\vec{u}_1\rangle_\rho)^2}{2}]$$

$$+ \int_A g\vec{u}_2{}^*\cdot(\langle\overleftrightarrow{\tau}_1\rangle + \overleftrightarrow{\sigma}_1{}')\cdot\vec{n}\, d\vec{y}\, d\tau - \theta_1 \qquad (8.6)$$

where $\theta_1 = \Theta_1 + \langle\vec{u}_1\rangle_\rho\cdot\nabla\cdot\alpha_1\langle\rho_1\rangle\langle\vec{u}_1{}^*\vec{u}_1{}^*\rangle_\rho$

We have also used the approximation $\langle u_2{}^2\rangle_s = \langle u_2\rangle_s{}^2$ in evaluating the energy transported with $\dot{m}_i$ in equation (8.6). Similarly, the energy equation for phase 2 takes the form:

$$\alpha_2\langle\rho_2\rangle\frac{D}{Dt_2}\langle e_2\rangle_\rho + \nabla\cdot\alpha_2\langle\vec{q}_2\rangle + \alpha_2\langle p_1\rangle\nabla\cdot\langle\vec{u}_2\rangle_\rho + [\langle p_1\rangle + \langle\Delta p\rangle_s]\frac{D\alpha_2}{Dt_2}$$

$$= \alpha_2(\langle p_1\rangle - \langle p_2\rangle)\nabla\cdot\langle\vec{u}_2\rangle_\rho + \alpha_2(\langle\overleftrightarrow{\tau}_2\rangle + \langle\overleftrightarrow{\tau}_2\rangle^T): \nabla\langle\vec{u}_2\rangle_\rho$$

$$- \dot{m}_1[\langle e_2\rangle_s - \langle e_2\rangle_\rho + \frac{\langle p_1\rangle + \langle\Delta p\rangle_s}{\langle\rho_2\rangle_s} - \frac{(\langle\vec{u}_2\rangle_s - \langle\vec{u}_2\rangle_\rho)^2}{2}]$$

$$- \int_A g\vec{u}_2{}^*\cdot(\langle\overleftrightarrow{\tau}_1\rangle + \overleftrightarrow{\sigma}_1{}')\cdot\vec{n}\, d\vec{y}\, d\tau + \int_A g\vec{q}_2\cdot\vec{n}\, d\vec{y}\, d\tau - \theta_2 \qquad (8.7)$$

and where θ_2 is defined analogously with θ_1. It should be noted that we have not yet assumed $<e_2>_s = <e_2>_\rho$. We also note that $\vec{n}$ is directed into phase 2 in both (8.6) and (8.7).

The generality of equations (8.6) and (8.7) should be noted. They follow in a formal manner from the basic assumptions expressed in equations (5.1) through (5.6) together with the postulate concerning the smoothness of the average and the neglect of certain surface correlations which arise in the context of a phase change between the media.

We now note that by making use of the continuity equation (6.2), expanding the heat flux terms and introducing the additional assumptions $<e_2>_s = <e_2>_\rho$ and $<\vec{u}_2>_s = <\vec{u}_2>_\rho$ appropriate to the case when phase 2 is an aggregate of particles, (8.7) becomes:

$$\alpha_2<\rho_2> \frac{D}{Dt_2} <e_2>_\rho - \alpha_2\{<p_1> - <\Delta p>_s\}\frac{D}{Dt_2} \ln<\rho_2> = - \alpha_2\nabla\cdot<\vec{q}_2>$$
$$+ \int_A g\vec{q}_2{}'\cdot\vec{n}\, d\vec{y}\, d\tau + \{<p_1> - <p_2> - <\Delta p>_s\}\nabla\cdot<\vec{u}_2>_\rho$$
$$+ \alpha_2(<\overleftrightarrow{\tau}_2> + <\overleftrightarrow{\tau}_2>^T) : \nabla<\vec{u}_2>_\rho - \int_A g\vec{u}_2{}^*\cdot(<\overleftrightarrow{\tau}_1> + \overleftrightarrow{\sigma}_1{}')\cdot\vec{n}\, d\vec{y}\, d\tau - \theta_2 \tag{8.8}$$

It is instructive to consider equation (8.8) in the limit as phase 2 becomes incompressible. Let us also neglect heat and mass transfer and the stress deviators $<\vec{\tau}_1>$ and $<\vec{\tau}_2>$ so that (8.8) reduces to:

$$\alpha_2<\rho_2> \frac{D}{Dt_2} <e_2>_\rho = \{<p_1> - <p_2>\}\nabla\cdot<\vec{u}_2>_\rho$$
$$- \int_A g\vec{u}_2{}^*\cdot\overleftrightarrow{\sigma}_1{}'\cdot\vec{n}\, d\vec{y}\, d\tau - \theta_2 + \alpha_2<\overleftrightarrow{\tau}_2>^T : \nabla<\vec{u}_2>_\rho \tag{8.9}$$

We now have the apparent anomaly that if $<p_1> \neq <p_2>$ a macroscopic dilatation of the phase results in work being done. Clearly, if the phase is dispersed we may make $\nabla\cdot<\vec{u}_2>_\rho$ as large as we wish without storing energy in the internal modes. It is tempting to proceed from (8.9) through the second law of thermodynamics to argue, therefore, that $<p_1> = <p_2>$ necessarily. Of course, this is not so and cases in which the average stresses are different are of practical interest as we discuss subsequently. In order to understand equation (8.9) note that θ_2 can be written as:

$$\theta_2 = \phi_2 - \nabla\cdot\alpha_2<\vec{u}_2^{\,*}\cdot\overleftrightarrow{\sigma}_2'> \tag{8.10}$$

where the form of ϕ_2 is evident by inspection of equation (8.2). Also note that if ρ_2 = constant, $<\psi_2> = <\psi_2>_\rho$ for any property of phase 2. Now, using the microscopic boundary condition (5.5) it is easy to show that:

$$\nabla\cdot\alpha_2<\vec{u}_2^{\,*}\cdot\overleftrightarrow{\sigma}'> = \int_V g\nabla\cdot(\vec{u}_2^{\,*}\cdot\overleftrightarrow{\sigma}_2')\,d\vec{y}\,d\tau + \int_A g\vec{u}_2^{\,*}\cdot[\overleftrightarrow{\sigma}_1'\cdot\vec{n} - (<p_1> - <p_2>)\,\vec{n}]\,d\vec{y}\,d\tau \tag{8.11}$$

where the sign of $\vec{n}$ has been chosen to conform with that in equation (8.9). Using (4.3) and (6.2) with $\dot{m}_i = 0$ we see that (8.9) can be written as:

$$\alpha_2<\rho_2>\frac{D}{Dt_2}<e_2>_\rho = \int_{V_2} g\nabla\cdot(\vec{u}_2^{\,*}\cdot\overleftrightarrow{\sigma}_2')d\vec{y}\,d\tau - \phi_2 + \alpha_2<\overleftrightarrow{\tau}_2>^T : \nabla<\vec{u}_2>_\rho \tag{8.12}$$

By recalling the definitions of ϕ_2 and $<\overleftrightarrow{\tau}_2>^T$, using the constancy of ρ_2 and the microscopic boundary condition $\vec{u}_2\cdot\vec{n} = \vec{w}\cdot\vec{n}$ on A we may recast (8.12) as:

$$\alpha_2<\rho_2>\frac{D}{Dt_2}<e_2>_\rho = \int_{V_2} g\overleftrightarrow{\sigma}' : \nabla\vec{u}^{\,*}\,d\vec{y}\,d\tau - \int_{V_2} g<\rho_2>u_2^{\,*}\cdot\left[\frac{\partial\vec{u}_2^{\,*}}{\partial\tau} + \vec{u}_2\cdot\nabla u_2^{\,*} + u_2^{\,*}\cdot\nabla<\vec{u}_2> - \frac{1}{\rho_2}\nabla\cdot\overleftrightarrow{\sigma}'\right]d\vec{y}\,d\tau \tag{8.13}$$

By expanding the microscopic equation of motion into mean and fluctuating components it is easy to see that the integral of the second term on the right hand side of (8.13) is the product of a fluctuation and a macroscopic quantity. The second integral therefore vanishes leaving us with:

$$\alpha_2 \langle\rho_2\rangle \frac{D}{Dt_2} \langle e_2 \rangle_\rho = \int_{V_2} g \overleftrightarrow{\sigma}_2{}' : \nabla \vec{u}_2{}^* \, d\vec{y} \, d\tau \tag{8.14}$$

We can reduce this further by writing $\overleftrightarrow{\sigma}_2{}' = \overleftrightarrow{\tau}_2{}' - p_2{}'\overleftrightarrow{I}$. The contribution of $p_2{}'$ is seen to vanish since the condition of incompressibility implies $\nabla\cdot\vec{u}_2{}^* = -\nabla\cdot\langle\vec{u}_2\rangle_\rho$. Accordingly, we have the physically correct result that the changes in internal energy are only due to shear work. Of course, if phase 2 is rigid we have $\nabla\vec{u}_2{}^* = -\nabla\langle\vec{u}_2\rangle_\rho$ and the right hand side of (8.14) vanishes altogether.

Apart from providing some insights into the relationship between the correlation terms and certain of the macroscopic state variables, the foregoing discussion of the energy equation for an incompressible phase demonstrates the importance of consistency in regard to the neglect of terms. If the correlation terms were dropped from equation (8.8), unphysical behaviour would be predicted in the limit $\langle\rho_2\rangle$ = constant.

9. CLOSURE AND CONSTITUTIVE LAWS

The balance equations have been put into a macroscopic form on the basis of certain broad assumptions. In particular we note, in retrospect, that the neglect of correlations about the surface average has been used only to simplify certain terms associated with mass transfer between the phases. We now introduce some additional assumptions appropriate to the case of convective flamespreading in a granular bed. Phase 1 is assumed to be a compressible gas and phase 2 is assumed to be an aggregate of solid, incompressible particles, packed or dispersed, with constant density ρ_p. We neglect macroscopic shear stress and the pseudo-turbulent Reynolds stress in both phases. We also neglect macroscopic heat conduction through each phase as well as the correlation term Θ_1 for the gas phase. The deformation of the solid phase is assumed to be sufficiently small and the particles sufficiently massive that $\langle\vec{u}_2\rangle_\rho = \langle\vec{u}_2\rangle_s$ and moreover, that $\vec{u}_2{}^*$ is much smaller in absolute value than $\langle\vec{u}_2\rangle_\rho - \langle\vec{u}_1\rangle_\rho$. Finally, we assume that $\langle\Delta p\rangle_s$ is negligible by comparison with $\langle p_1\rangle$. With these assumptions the macroscopic balance equations reduce to the forms we have previously used [5]. In the following text we will no longer distinguish the density weighted average velocities by a subscript ρ. Furthermore, we denote the gas phase properties by unsubscripted variables and those of the solid phase by a subscript p.

Balance of Mass of Gas Phase

$$\frac{\partial}{\partial t}\,\alpha\langle\rho\rangle + \nabla\cdot\alpha\langle\rho\rangle\langle\vec{u}\rangle = \dot{m} \qquad (9.1)$$

Balance of Mass of Solid Phase

$$\frac{\partial\alpha}{\partial t} - \nabla(1-\alpha)\langle\vec{u}_p\rangle = \dot{m}/\rho_p \qquad (9.2)$$

Balance of Momentum of Gas Phase

$$\alpha\langle\rho\rangle\,\frac{D}{Dt}\,\langle\vec{u}\rangle + \alpha\nabla\langle p\rangle = (\langle\vec{u}_p\rangle - \langle\vec{u}\rangle)\dot{m} - \vec{f} \qquad (9.3)$$

Balance of Momentum of Solid Phase

$$(1-\alpha)\rho_p\,\frac{D}{Dt_p}\,\langle\vec{u}_p\rangle + (1-\alpha)\nabla\langle p\rangle - \nabla\cdot(1-\alpha)(\langle p\rangle\overleftrightarrow{I} + \langle\overleftrightarrow{\sigma}_p\rangle) = \vec{f} \qquad (9.4)$$

Balance of Energy of Gas Phase

$$\alpha\langle\rho\rangle\,\frac{D}{Dt}\,\langle e\rangle + \langle p\rangle[\frac{D\alpha}{Dt} + \alpha\nabla\cdot\langle\vec{u}\rangle] = [\langle\vec{u}\rangle - \langle\vec{u}_p\rangle]\cdot\vec{f} - q + \dot{m}[\langle e_p\rangle - \langle e\rangle + \frac{\langle p\rangle}{\rho_p} + \frac{(\langle\vec{u}_p\rangle - \langle\vec{u}\rangle)^2}{2}] \qquad (9.5)$$

The terms $\dot{m}$, $\vec{f}$ and q are macroscopic quantities describing interfacial phenomena according to:

$$\dot{m} = \rho_p S(\vec{x},t)\langle\dot{d}\rangle_s \qquad (9.6)$$

$$\vec{f} = -\int_A g\overleftrightarrow{\sigma}'\cdot\vec{n}\;d\vec{y}\;d\tau \qquad (9.7)$$

$$q = \int_A g\vec{q}_p\cdot\vec{n}\;d\vec{y}\;d\tau \qquad (9.8)$$

These have the interpretations of interphase mass transfer, interphase drag and interphase heat transfer respectively. Closure requires that these be related to the macroscopic state variables. In practical cases this will necessitate the use of empirical correlations. As would be expected, we require an equation of state for the gas phase in the form $\langle e\rangle = \langle e\rangle(\langle p\rangle,\langle\rho\rangle)$. Also, as expected intuitively and supported by the discussion of the previous section, the assumption of incompressibility of the solid phase eliminates the need for a macroscopic energy equation. However, closure is not effected until some constitutive relation is provided for $<\sigma_p>$, so that an equation of state is required for the solid phase, even when it is microscopically incompressible.

The particular choices of constitutive laws for problems of convective flamespreading have been discussed elsewhere [6]. Here we simply remark that the relationships introduced to close the equations have the effect of adding algebraic terms to the balance equations, with the exception of the equation of state for $<\overleftrightarrow{\sigma}_p>$ and the constitutive law for $\vec{f}$. Regarding the latter we note that many investigators have expressed $\vec{f}$ as a steady state component superimposed upon a virtual mass component. The virtual mass component may be expressed in differential form in any one of several ways (Gough[5], Lyczkowski [17], Drew[18]. In the present study we will neglect the virtual mass effect except for occasional comments on the nature of its influence on the results summarized herein. Thus we have only the equation of state for $<\overleftrightarrow{\sigma}_p>$ to consider.

In fact, the constitutive law for $<\overleftrightarrow{\sigma}_p>$, in the case of incompressible particles, has been established by most authors in such a form as to express a principle of stress equilibrium at the microscopic level. Thus Kraiko and Sternin[11], Gough [5] have used, in dispersed flow:

$$<\overleftrightarrow{\sigma}_p> = - <p> \overleftrightarrow{I} \tag{9.9}$$

which states that the average pressure is the same in both phases. The same assumption has been used by Ishii[6] and by Stuhmiller[14] in addressing the problem of bubble flow in which the dispersed phase may be viewed as either compressible or incompressible. When compressible, the isothermal relation between $<\rho_p>$ and $<\overleftrightarrow{\sigma}_p>$ is required.

There can be little doubt concerning the correctness of (9.9) when the particles are dispersed and at rest. When they are in contact we have considered that forces may be transmitted from particle to particle and that when the particles are microscopically incompressible we may write:

$$<\overleftrightarrow{\sigma}_p> = - \{<p> + R(\alpha)\} \overleftrightarrow{I} \tag{9.10}$$

where $R(\alpha)$ represents the intergranular stress and is, according to (9.10), assumed to be isotropic. We also define, for future reference, the quantity a according to:

$$a^2 = -\frac{1}{\rho_p}\frac{d}{d\alpha}(1-\alpha)R(\alpha) \qquad (9.11)$$

and which can be shown to represent the rate of propagation of intergranular stresses [5]. It is not difficult to extend (9.10) to account for compressibility of the particles and to include macroscopic shear stresses. This problem has been addressed by Aguirre-Ramirez and Saxe[19], by Morland[20] and by Garg and Nur[21].

However, the validity of (9.9) for finite values of the Mach number based on relative velocity and the speed of sound in the gas has not been established. The fundamental correctness of (9.9) stems from the idea of local equilibration of the macroscopic stress field. When the relative flow velocity is negligibly small it is reasonable to argue, following Nigmatulin[12], that (9.9) is valid provided that the characteristic time of the macroscopic process is much greater than the characteristic time required to establish mechanical equilibrium over microscopic regions. Central to this argument is the possibility of transmitting information through a macroscopically infinitesimal control volume in a negligibly short time. Yet, as the Mach number based on the relative flow velocity approaches unity, the possibility of transmitting information through the control volume in the upstream direction will disappear altogether. Thus the physical basis for (9.9) becomes increasingly obscure with increasing Mach number.

However, as the Mach number increases we may expect that the velocity dependent drag will be very much larger in magnitude than the pressure gradient term in the solid phase momentum equation. In such cases our inability to resolve the exact contributions of $\langle p \rangle$ and $\langle \overleftrightarrow{\sigma}_p \rangle$ to the motion of the solid phase may become immaterial and a sufficiently accurate equation may be the approximation:

$$(1-\alpha)\,\rho_p \frac{D}{Dt_p}\langle \vec{u}_p \rangle = \vec{f} \qquad (9.12)$$

Such an equation is frequently used in studies of highly dispersed dusty flows at high Mach numbers such as those which occur in rocket nozzles or in the relaxation zone behind a shock.

10. CHARACTERISTICS OF BALANCE EQUATIONS

The concept of a characteristic surface arises most naturally from a consideration of the Cauchy problem for a system of partial differential equations with initial data on an arbitrary surface. Following Petrovsky[22] we may consider the quasilinear system:

$$\sum_{j,k_o,k_1\ldots k_n} A_{ij}^{(k_o,\ldots,k_n)} \frac{\partial^{n_j} u_j}{\partial x_o^{k_o} \partial x_1^{k_1} \ldots \partial x_n^{k_n}} + \ldots + y_i = 0 \qquad (10.1)$$

$$i,j = 1,2,\ldots,N$$

Thus we have N equations, labelled by the subscript i, in N unknowns u_j. In each equation we write explicitly only the highest order term for each u_j that appears in the system as a whole. This maximum order is designated by n_j for each variable u_j. Consider an arbitrary surface whose intrinsic coordinates are $\xi_1\ldots\xi_n$ and let ξ_o be a direction normal to the surface. Cauchy data for this surface consist of the values of u_j, $j = 1,..,N$ together with all the normal derivatives of order less than n_j. Then the surface is said to be free if the system (10.1) enables the determination of $\partial^{n_j}u_j/\partial\xi_o^{n_j}$, $j = 1,\ldots,N$ and characteristic if it does not. Evidently, when the surface is free the possibility exists for the determination of an analytic solution in some neighborhood by an iterative determination of the Taylor coefficients.

It follows from an application of a point transformation $\{x_i\}\to\{\xi_i\}$ that the values of $\partial^{n_j}u_j/\partial\xi_o^{n_j}$ are given by a linear system whose matrix of coefficients is:

$$\{\Delta_{ij}\} = \left\{ \sum_{k_o + k_1 + \ldots + k_n = n_j} A_{ij}^{(k_o\ldots k_n)} \mu_o^{k_o} \mu_1^{k_1} \ldots \mu_n^{k_n} \right\} \qquad (10.2)$$

Thus the i-j element is a sum over all the contributing terms of the ith equation which involve a derivative of order n_j of u_j, no matter how mixed. The quantities μ_i are given by $\mu_i = \partial\xi_o/\partial x_i$, $i = 0,1,\ldots n$ and may be interpreted as the components of a vector normal to the surface ξ_o = constant. A necessary and sufficient condition for the surface to be characteristic is that the rank of Δ be less than N. When the surface is characteristic the data cannot be wholly independent if (10.1) is to be satisfied. They are constrained according to the condition of solvability for the linear system governing the values of $\partial^{n_j} u_j / \partial \xi_o^{n_j}$. The resulting constraints are called the conditions of compatibility corresponding to the characteristic surface. Finally, we note that if the system (10.1) is first order and depends only on $x_o = t$ and $x_1 = x$ in the form

$$A_{ij} \frac{\partial u_j}{\partial t} + B_{ij} \frac{\partial u_j}{\partial x} = C_{ij}, \tag{10.3}$$

then $\Delta_{ij} = \mu_t A_{ij} + \mu_x B_{ij}$ and we have the slope of the characteristic lines in the form $dx/dt = -\mu_t/\mu_x$. A system of N partial differential equations is said to be hyperbolic if it has real characteristic surfaces and totally hyperbolic if there are N distinct such surfaces.

We have previously established the characteristic surfaces of (9.1)-(9.5) in the case of one dimensional unsteady flow [5] taking into account the influence of the effect of virtual mass. More recently we have derived results for two dimensional unsteady flow, neglecting virtual mass [23]. In the latter case we introduced a parameter λ into the solid phase equation so that when $\lambda = 1$ the pressure gradient was treated as a differential term, as expressed by (9.4), and when $\lambda = 0$, it was treated as a non-homogeneous term.[†] Thus we rewrite (9.4) as:

$$(1-\alpha)\rho_p \frac{D}{Dt_p} \langle \vec{u}_p \rangle + \lambda(1-\alpha)\nabla\langle p \rangle - \nabla(1-\alpha)R(\alpha) = \vec{f} - (1-\lambda)(1-\alpha)\nabla\langle p \rangle \tag{10.4}$$

and we have made use of (9.10). In the case of one dimensional unsteady flow the characteristics of the system (9.1), (9.2), (9.3), (10.4) and (9.5) may be expressed as:

$$\frac{dx}{dt} = \langle u \rangle \tag{10.5}$$

on which the condition of compatibility is:

$$\frac{d}{dt}\langle p \rangle - c^2 \frac{d}{dt}\langle \rho \rangle = h_5 \tag{10.6}$$

where c is the isentropic sound speed in the gas phase, and:

$$\frac{dx}{dt} = \langle u_p \rangle \tag{10.7}$$

[†]The pseudo-characteristics based on $\lambda \neq 1$ are useful in the determination of numerical results and also provide insight into the role played by the pressure gradient term.

on which the condition of compatibility is:

$$\frac{d}{dt}(1-\alpha)R(\alpha) = -\rho_p a^2 \frac{d\alpha}{dt} \tag{10.8}$$

and, finally, as the roots of the quartic equation:

$$[(\frac{dx}{dt} - \langle u\rangle)^2 - c^2][(\frac{dx}{dt} - \langle u_p\rangle)^2 - a^2] = \lambda(\frac{1-\alpha}{\alpha})\frac{\langle\rho\rangle}{\rho_p} c^2 [\frac{dx}{dt} - \langle u\rangle]^2 \tag{10.9}$$

on which the condition of compatibility is:

$$k_\alpha \frac{d\alpha}{dt} + k_u \frac{d}{dt}\langle u\rangle + k_p \frac{d}{dt}\langle p\rangle + k_{u_p} \frac{d}{dt}\langle u_p\rangle + k_R \frac{d}{dt}(1-\alpha)R + k_h = 0 \tag{10.10}$$

Here we use h_i, i = 1,....,5 to designate the right hand sides of (9.1), (9.2), (9.3), (10.4) and (9.5) respectively and the coefficients in (10.10) may be tabulated as:

$$k_\alpha = \langle\rho\rangle[(\frac{dx}{dt} - \langle u_p\rangle)^2 - a^2][\frac{\langle u\rangle}{\frac{dx}{dt} - \langle u\rangle} - \frac{\langle u_p\rangle}{\frac{dx}{dt} - \langle u_p\rangle}](\frac{dx}{dt} - \langle u\rangle) \tag{10.11}$$

$$k_u = \alpha\langle\rho\rangle[(\frac{dx}{dt} - \langle u_p\rangle)^2 - a^2][1 + \frac{\langle u\rangle}{\frac{dx}{dt} - \langle u\rangle}] \tag{10.12}$$

$$k_p = \frac{\alpha}{c^2}[(\frac{dx}{dt} - \langle u_p\rangle)^2 - a^2]\frac{dx}{dt} \tag{10.13}$$

$$k_{u_p} = \langle\rho\rangle(1-\alpha)(\frac{dx}{dt} - \langle u\rangle)\frac{dx}{dt} \tag{10.14}$$

$$k_R = \frac{\langle\rho\rangle}{\rho_p}(\frac{dx}{dt} - \langle u\rangle)(1 + \frac{\langle u_p\rangle}{\frac{dx}{dt} - \langle u_p\rangle}) \tag{10.15}$$

$$k_h = -[(\frac{dx}{dt} - \langle u_p\rangle)^2 - a^2][h_1 + \frac{\alpha h_5}{c^2} + \frac{h_3}{\frac{dx}{dt} - \langle u\rangle}]\frac{dx}{dt} + \langle\rho\rangle(\frac{dx}{dt} - \langle u\rangle)[(\frac{dx}{dt} - \langle u_p\rangle)h_2 - \frac{h_4}{\rho_p}]\frac{dx}{dt} \tag{10.16}$$

We note that the coefficients k_α and k_R, which are individually singular as $dx/dt \to \langle u_p\rangle$, combine to yield a finite result due to (10.8). The material characteristics (10.5) and (10.7) and their conditions of compatibility (10.6) and (10.8) are straightforward in meaning and will not concern us henceforth. Our attention is therefore confined to (10.9) and (10.10).

The value $\lambda = 1$ which corresponds to the adoption of (9.9) in dispersed flow or (9.10) in packed flow yields the characteristic condition which we have discussed elsewhere in detail [5,22]. In particular, when $a = 0$ so that the particles are non-interacting, (10.9) reduces to the form first analyzed by Kraiko and Sternin[11]. In this case (10.9) yields only two real roots if the relative Mach number $|\langle u\rangle - \langle u_p\rangle|/c$ is less than unity. At values somewhat larger than unity four real roots may appear. Thus the equations for dispersed flow, (9.1) through (9.5) are always hyperbolic but not necessarily totally hyperbolic. When $a \neq 0$ in (10.9), corresponding to packed flow, the equations are, in general, totally hyperbolic although the minimum value of a to ensure four real roots of (10.9) increases with the value of the relative Mach number and also the value of $(1-\alpha)\langle\rho\rangle/\alpha\rho_p$.

When the particles are packed so that $a \neq 0$ and the velocities of the phases sufficiently low that there are four real roots of (10.9) the balance equations may be said to be well posed. In the sense in which we have used this term, it refers to the possibility of verifying the completeness of initial and boundary data. In the nuclear engineering literature, however, it has been the possibility of unstable solutions which has led to the use of "ill posed" to describe the non-totally hyperbolic system. The connection between the characteristics of a quasi-linear system and the stability of small perturbations has been discussed by Ramshaw and Trapp[24]. The distinction between the meaning of the expression "ill posed" as used here and as used in the nuclear engineering literature should therefore be noted.

Concern over the lack of total hyperbolicity of the balance equations appears to have arisen in the nuclear engineering literature as a consequence of a theorem of Lax relating to the impossibility of obtaining stable numerical solutions for such a scheme (Lyczkowski et al[17]). However, Stuhmiller [14] pointed out that the conclusions of a linear theory might not necessarily be true in the context of a quasi-linear system of equations. The inability to generate stable numerical solutions has been offered in certain cases as evidence of the correctness of the position that only the well posed systems can be solved (Lyczkowski et al, [17], Chao et al[25]). We, on the other hand, have found no detectable numerical difficulty associated with the solution of the ill posed system which arises in dispersed flow. Possibly, however, the difference between our own experience and that of Lyczkowski et al is related to the fact that whereas we were concerned with a dispersed phase consisting of solid particles, Lyczkowski et al were concerned with a bubbly flow.

Let us consider briefly the well-posedness of the equations in the sense used here. Suppose we have flow adjacent to a boundary through which both phases pass at a subsonic velocity. Then if $a \neq 0$, we may expect that two characteristic lines will pass through both the boundary and a line bearing initial data within the region for which the solution is sought. To simplify the discussion we will assume that the gas is isentropic, so that (10.6) is not of interest, and that the granular stress does not depend on the history of loading so that (10.8) need not be considered and, moreover, in (10.10) we may combine k_R with k_α so that k_R disappears and k_α becomes:

$$k_\alpha' = \langle\rho\rangle[\frac{dx}{dt} - \langle u\rangle]\{[(\frac{dx}{dt} - \langle u_p\rangle)^2 - a^2]\frac{\langle u\rangle}{\frac{dx}{dt} - \langle u\rangle} - a^2$$

$$- \langle u_p\rangle(\frac{dx}{dt} - \langle u_p\rangle) \qquad (10.17)$$

Our independent variables are therefore just α, $\langle p\rangle$, $\langle u\rangle$ and $\langle u_p\rangle$. Because of the constraints imposed by the conditions of compatibility (10.10) on each of the characteristic lines, at most two boundary data can be specified independently of arbitrary initial data. Thus if we specify $\langle u\rangle$ and $\langle u_p\rangle$ on the boundary, unique values of α and $\langle p\rangle$ are determined as consistent with these data and with the flow in the interior.

On the other hand, if $a = 0$, corresponding to dispersed flow, and the Mach number is sufficiently low, only one characteristic line will exist to connect the boundary and initial data. There is no explicit denial of the possibility of specifying three independent quantities on the boundary. Physically, this conclusion is not unreasonable. It implies, for example, that we may arbitrarily specify the pressure, the volume fraction and the slip velocity at the entrance to a duct, provided that the particles are dispersed. When the particles are in contact with one another corresponding to two independent data, adjustments in the volume fraction cannot be made without influencing the particle velocity and thus only the pressure and volume fraction, say, can be independently set.

Thus, the lack of total hyperbolicity in the case of dispersed flow may be viewed as physically reasonable by reference to the number of degrees of freedom it admits in the boundary conditions for entering flow. We regard as more disturbing the totally hyperbolic behaviour exhibited by the equations when the relative Mach number sufficiently exceeds unity. However, given the uncertainties which arise in the validity of the pressure gradient term at finite values of the Mach number, the existence of the additional characteristic roots may well be a mathematical anomaly due to extrapolation of the constitutive law (9.9) beyond its regime of validity. Such a view seems also to be expressed by Kraiko and Sternin[11].

We can therefore accept the lack of total hyperbolicity as an attribute consistent with the possible physical behavior of dispersed flow. We nevertheless refer to the equations as ill posed, in our sense, for the following reason. Consider dispersed flow adjacent to an impermeable wall. The only physical boundary conditions to be imposed are the vanishing of the normal components of each of the phase velocities. According to our previous discussion it is possible to demonstrate from the conditions of compatibility only that the values of pressure and volume fraction are related. We are unable to assert that they are uniquely determined. This neither proves nor disproves uniqueness of the solution. However, we are evidently on less firm theoretical ground than in the case when the equations are totally hyperbolic. As we have stated elsewhere [16] we have assumed nevertheless that the initial data in combination with the conditions on the normal velocity component are indeed sufficient to ensure uniqueness so as to proceed with the determination of numerical results.

Consider now the consequences of setting $\lambda = 0$ in equation (10.4), corresponding to the neglect of the pressure gradient in the solid phase momentum equation. In this case (10.9) obviously has four real roots namely $dx/dt = \langle u \rangle \pm c$ and $\langle u_p \rangle \pm a$ under all circumstances. It might therefore be thought that the neglect of the pressure gradient term, when justified by reason of its smallness relative to the interphase drag or the intergranular stress term, might lead to a better posed system of equations. In fact, this is not so. For, if a = 0, corresponding to dispersed flow, it is easy to see that for the double characteristic root $dx/dt = \langle u_p \rangle$, equation (10.10) reduces to $\rho_p(1-\alpha)du_p/dt = h_4$ which is just the solid phase momentum equation. Accordingly, the conditions of compatibility do not constrain the data at either a permeable or impermeable boundary in any essentially new fashion relative to the conditions which arise in the case $\lambda = 1$.

It might also be thought that the ill-posedness is associated with the assumed microscopic incompressibility of the particles. In fact, this is not so. Including the compressibility of the solid phase leads to the characteristic equation:

$$\alpha\frac{\langle\rho_p\rangle}{c^2}\left[\frac{dx}{dt} - \langle u_p\rangle\right]^2\left[\left(\frac{dx}{dt} - \langle u\rangle\right)^2 - c^2\right] + \frac{\langle\rho\rangle(1-\alpha)}{c_p^{\,2}}\left[\frac{dx}{dt} - \langle u\rangle\right]^2\left[\left(\frac{dx}{dt} - \langle u_p\rangle\right)^2 - c_p^{\,2}\right] = 0 \qquad (10.18)$$

where c_p is the isentropic sound speed in the solid phase. As $c_p \to \infty$ we recover (10.9) with $\lambda = 1$. It is not difficult to show (Lyczkowski et al[17]) that (10.18) does not have real roots under all conditions.

Ramshaw and Trapp[24] consider the possibility of introducing additional terms to ensure the totally hyperbolic character of the equations. For separated flow of two compressible phases they reasoned that the condition of local equilibrium (9.9) could be replaced by a statement that the average pressures differed by an amount due to surface tension. Designating the properties of the phases by the subscripts 1 and 2 and evaluating the curvature of the interface under the assumption $\partial\alpha/\partial x \ll 1$ they deduced:

$$\langle p_1\rangle - \langle p_2\rangle = -\sigma H\frac{\partial^2\alpha_1}{\partial x^2} \qquad (10.19)$$

where H is the duct width and σ the surface tension. Considering only the continuity and momentum equations for the two phases and writing them in terms of the variables $\langle p_1\rangle$, $\langle u_1\rangle$, $\langle u_2\rangle$ and α_1 leads to the characteristic conditions $dx/dt = 0$ and:

$$\left[\left(\frac{dx}{dt} - \langle u_1\rangle\right)^2 - c_1^{\,2}\right]\left[\left(\frac{dx}{dt} - \langle u_2\rangle\right)^2 - c_2^{\,2}\right] = 0 \qquad (10.20)$$

so that the roots are always real. It has not been shown, however, that the resulting system is well posed in the sense used here. In addition to boundary values of $<p_1>$, $<u_1>$, $<u_2>$ and α we now require derivatives of α. It is not clear to what extent the derivatives of α can be chosen independently of the other data. However, we do not pursue this question by investigating the conditions of compatibility. The constitutive law (10.19) is simply too unmotivated physically in the applications involving interpenetrating phases which are of interest to us.

It has been claimed by Stuhmiller[14] that the balance equations for dispersed flow can be made totally hyperbolic by considering the details of the interfacial stress even when both media are microscopically incompressible. In fact, we disagree with Stuhmiller's conclusion for reasons which we shall now describe.

In order to do so we shall transform his notation into our own. We assume that phase 1 is an incompressible fluid and that phase 2 is an aggregate of dispersed spherical droplets of radius R in which the bulk average pressure exceeds that in phase 1 by an amount $\sigma/2R$ due to surface tension. With these assumptions the balance equations are (Stuhmiller[14], equations (7) - (12)).

$$\frac{\partial \alpha_1}{\partial t} + \frac{\partial}{\partial x} \alpha_1 <u_1> = 0 \tag{10.21}$$

$$\frac{\partial \alpha_2}{\partial t} + \frac{\partial}{\partial x} \alpha_2 <u_2> = 0 \tag{10.22}$$

$$\alpha_1 \rho_1 [\frac{\partial <u_1>}{\partial t} + <u_1> \frac{\partial <u_1>}{\partial x}] = - \frac{\partial}{\partial x} \alpha_1 <p> + M_x \tag{10.23}$$

$$\alpha_2 \rho_2 [\frac{\partial <u_2>}{\partial t} + <u_2> \frac{\partial <u_2>}{\partial x}] = - \frac{\partial}{\partial x} \alpha_2 <p> - M_x \tag{10.24}$$

$$\alpha_1 + \alpha_2 = 1 \tag{10.25}$$

$$\vec{M} = \int_A \dot{g} \overleftrightarrow{\sigma} \cdot \vec{n} \, d\vec{y} \, d\tau \tag{10.26}$$

These equations agree with our own. We may compare (10.21) and (10.22) with (6.2) in which we set $\dot{m}_i = 0$ and $<\rho_1>$ = constant. Equations (10.23) and (10.24) agree with our (7.1) when reactions and shear stress terms are neglected. Whereas we separated the quantity M, in equations (10.23) and (10.24), into bulk average and fluctuating parts, Stuhmiller takes a different path. First he considers dispersed flow so that we may view the pressure $<p>$ as representative of the pressure at infinity for the case of flow about an isolated sphere. As his interest is with relatively large Reynolds numbers, the contribution of shear stresses in (10.26) is neglected and only the normal pressure component is considered. The pressure on the surface of the sphere is assumed to have the form (Stuhmiller[14], equation (16)):

$$p_i = <p> + \rho_2 U^2 F(\theta) + \tfrac{1}{2} \rho_2 R \frac{dU}{dt} \cos\theta \qquad (10.27)$$

where $U = <u_1> - <u_2>$ is the relative velocity, $F(\theta)$ defines the pressure distribution on the surface and the third term accounts for unsteadiness of the flow. We will neglect this term which simply leads to a virtual mass effect.

Referring back to our own approach we see that, neglecting the contribution of the shear stresses on the surface, the quantity $\rho_2 U^2 F(\theta)$ is the fluctuation p' about the bulk average. Thus the use of (10.27) is consistent with our approach and leads directly to (7.4) and (7.5) when reactions and shear terms are neglected.

Stuhmiller, however, separates p_i into a surface average and fluctuations as:

$$p_i = <p_i>_s + {}^*p_i \qquad (10.28)$$

and we place the superscripted asterisk to the left of the fluctuation so as to distinguish it from a fluctuation about the density weighted bulk average used in the derivation of our balance equations. As the average value of $F(\theta)$ on the surface does not vanish when flow separation is considered we have from (10.27), neglecting dU/dt:

$$<p_i>_s = <p> + \rho_2 U^2 <F(\theta)>_s$$

$${}^*p_i = {}^*F(\theta)\rho_2 U^2$$

Evidently:

$$-\int_A g <p_i>_s \vec{n}_x d\vec{y} d\tau = \{<p> + \rho_2 U^2 <F(\Theta)>_s\} \frac{\partial \alpha_1}{\partial x}$$

and the quantity M_x may be resolved as:

$$M_x = \{<p> + \rho_2 U^2 <F(\theta)>_s\} \frac{\partial \alpha_1}{\partial x} - \int_A g^* F(\theta) \rho_2 U^2 d\vec{y} d\tau \qquad (10.29)$$

Stuhmiller now concludes that the integral on the right hand side of (10.29) is just the total force due to form drag on the particles which is resolved as an algebraic quantity. Thus, in general, M_x is resolved as:

$$M_x = (<p> - \xi \rho_2 U^2) \frac{\partial \alpha_1}{\partial x} \pm \eta \frac{\rho_2 U^2}{L_s} \qquad (10.30)$$

(Stuhmiller[14], equation (18) with dU/dt = 0) where L_s is the total interfacial area per unit volume and ξ and η have been introduced as dimensionless parameters. The use of (10.30) in (10.23) and (10.24) leads to characteristic roots which are always real if ξ is sufficiently large and Stuhmiller advances arguments to demonstrate the plausibility of this assumption.

Now equation (10.29) is still consistent with our formulation. In effect (10.29) states:

$$M_x = <p>_s \frac{\partial \alpha_1}{\partial x} - \int_A g\{(<p> - <p>_s + p')\} \vec{n}_x d\vec{y} d\tau \qquad (10.31)$$

Our results diverge from those of Stuhmiller in passing to equation (10.30). In order to evaluate the surface integral in (10.29) or its equivalent in (10.31), Stuhmiller decomposes it into a sum over individual particles. As a uniform pressure distribution $<p> - <p>_s$ does not contribute to the drag per particle it follows that the contribution of each particle is just the velocity dependent drag. However, while the presence of a uniform pressure distribution does not influence the drag per particle it does influence the total force of interaction between the phases. The drag is seen quite clearly in (10.27) to be identified with p' and not $^*p_i$. In fact, the argument employed by Stuhmiller to pass from equation (10.29) to (10.30) admits the result that by adding and subtracting a uniform value of pressure to the integrand of M_x one can make the coefficient of $\partial\alpha_1/\partial x$ as large or as small as one pleases.

Although we disagree with Stuhmiller's specific conclusion we are left with the following question. In essence we have split the surface integral of $\overleftrightarrow{\sigma}\cdot\vec{n}$ into $\langle\overleftrightarrow{\sigma}\rangle\cdot\vec{n} + \overleftrightarrow{\sigma}'\cdot\vec{n}$ whereas Stuhmiller has split it as $\langle\overleftrightarrow{\sigma}\rangle_s\cdot\vec{n} + {}^*\overleftrightarrow{\sigma}\cdot\vec{n}$. Such splitting is arbitrary from a mathematical point of view. Is it therefore possible that the appropriate macroscopic part of $\overleftrightarrow{\sigma}$ is neither $\langle\overleftrightarrow{\sigma}\rangle$ nor $\langle\overleftrightarrow{\sigma}\rangle_s$ but some other quantity and that it is the surface integral of the fluctuation about this new macroscopic term which actually embeds the velocity dependent drag?

Such a decomposition would lead to the constitutive law:

$$\int_A g\overleftrightarrow{\sigma}\cdot\vec{n}\ d\vec{y}d\tau = -\langle\overleftrightarrow{\sigma}\rangle\cdot\nabla\alpha - \overleftrightarrow{\Phi}\cdot\nabla\alpha + \text{Drag}$$

and would account for the postulated interaction term studied by Lyczkowski et al[17] in their attempts to render the equations totally hyperbolic. Their choice corresponds to:

$$\overleftrightarrow{\Phi} = 2\,\frac{\alpha_1\alpha_2\rho_1\rho_2}{(\alpha_1\rho_2 + \alpha_2\rho_1)}\,[\langle u_1\rangle - \langle u_2\rangle]^2\ \overleftrightarrow{I}$$

However, a fundamental derivation of such a term has not been provided. In all likelihood, the investigation of the origin of such a term should begin with the general momentum equation (7.1) so that the contribution of τ^T is not overlooked. To be meaningful, the investigation should also assess the validity of (9.9). For, it is plausible that $\langle\overleftrightarrow{\sigma}_p\rangle$ is correlated better with $\langle\overleftrightarrow{\sigma}\rangle_s$ than $\langle\overleftrightarrow{\sigma}\rangle$ and Stuhmiller[14] makes it quite clear that the average surface pressure is lower than the bulk average, at least in well dispersed flow.

We also note that Soo(26] has reported results which render the equations for dispersed flow totally hyperbolic. However, these stem from the assumption that the mixture is Newtonian. As this assumption is without apparent foundation for problems of interest to us we do not consider it any further. We also reject, on physical grounds, the totally hyperbolic system which arises if the mechanical interaction term is not split and the volume fraction is retained inside the pressure gradient term. Such systems do not admit an equilibrium solution with a non-uniform distribution of dispersed particles suspended in a fluid.

11. CONCLUDING REMARKS

The use of a formal averaging technique to derive the equations of macroscopic flow of a heterogeneous mixture has the advantage of providing direct analytical access to the underlying microflow and its influence on the macroflow. This aspect of the formulism provides guidance in the formulation and interpretation of constitutive laws and provides a rational path for the derivation of equations which are thermodynamically consistent as well as properly behaved in limiting cases.

The importance of a rational derivation of the balance equations is underscored by the fact that the equations for dispersed flow, as deduced by many authors [1,5,6,11,12,27], fail to be totally hyperbolic.

The failure of the equations to be totally hyperbolic may, in fact, be consistent with considerations of the number of independent data which can be specified at a permeable boundary and may therefore represent physical rather than aphysical behaviour. It has been suggested by Harlow et al[27] that the potential for unstable solutions implied by the lack of total hyperbolicity may be physically meaningful. We might also note that many of the initial attempts to derive the balance equations did so for the precise purpose of studying the observed physical instabilities of fluidized beds [28,29,30].

The attempts to render the equations totally hyperbolic, or well posed, that we have noted here, have been of two types. Either they have involved errors of derivation or they have involved the deliberate introduction of terms which are physically unsupported as regards either phenomenology or magnitude. The latter approach may, however, provide a satisfactory circumvention of the problem of ill-posedness if it can be shown that the predictions of interest are indifferent to the magnitude of the spurious term [31].

Whatever the precise mathematical implications may be regarding the lack of total hyperbolicity, it is certainly not true that one cannot obtain stable numerical solutions of such equations. We have obtained stable numerical results for both one and two dimensional flows of granular propellants using accepted methods of computational fluid dynamics with no explicit diffusive or damping terms [23,31]. From this point of view, therefore, it is the case that the aspect of the non-totally-hyperbolic system requiring fundamental attention is that of uniqueness of the solutions rather than boundedness.

We close with the observation that while the present state of understanding of the theoretical structure of the equations is incomplete, not only have numerical solutions been obtained, but predictions of macroscopic features have been found to accord very well with experimental observations [32,33,34].

REFERENCES

1. Anderson, T. B. and Jackson, R. "A Fluid Mechanical Description of Fluidized Beds"
I and EC Fund, vol.6 No.4 November p.527 1967
2. Slattery, J. C. "Flow of Viscoelastic Fluids Through Porous Media"
A.I. Ch. E. J. p.1066 1967
3. Panton, R. "Flow Properties for the Continuum Viewpoint of a Non-Equilibrium Gas-Particle Mixture"
J.Fluid Mech. vol.31, part 2, pp.273-303 1968
4. Drew, D. A. "Averaged Field Equations for Two-Phase Media"
Stud. Appl. Math. vol.L, No.2, pp.133-166 1971
5. Gough, P. S. "The Flow of a Compressible Gas Through an Aggregate of Mobile Reacting Particles"
Ph.D Thesis, Department of Mechanical Engineering, McGill University, Montreal, Canada 1974
6. Ishii, M. "Thermo - Fluid Dynamic Theory of Two-Phase Flow"
Eyrolles, Paris 1975
7. Hinze, J. O. "Turbulence"
McGraw-Hill 1959
8. Whitaker, S. "The Transport Equations for Multi-Phase Systems"
Chem. Eng. Sci. vol.28, pp.139-147 1973
9. Gray, W. G. and Lee, P. C. Y. "On The Theorems for Local Volume Averaging of Multiphase Systems"
Int. J. Multiphase Flow, vol.3, pp.333-340 1977

10. Delhaye,J. M. "Jump Conditions and Entropy Sources in Two-Phase Systems: Local Instant Formulation" Int. J. Multiphase Flow, v.1, pp.395-409 1974

11. Kraiko, A.N. and Sternin, L. E. "Theory of Flows of a Two-Velocity Continuous Medium Containing Solid or Liquid Particles" PMM, vol.29, No.3, pp.418-429 1965

12. Nigmatulin, R. I. "Methods of Mechanics of a Continuous Medium for the Description of Multiphase Mixtures" PMM, vol.34, No.6, pp.1097-1112 1970

13. Stuhmiller, J. H. "A Review of the Rational Approach to Two-Phase Flow Modeling" Electric Power Research Institute NP-197 1976

14. Stuhmiller, J. H. "The Influence of Interfacial Pressure Forces on the Character of Two-Phase Flow Model Equations" Int. J. Multiphase Flow, v.3, n.6,pp.551-560 1977

15. Celmins, A. K. R. "Critical Review of One-Dimensional Tube Flow Equations" BRL Report 2025 1977

16. Gough, P. S. and Zwarts, F. J. "Modeling Heterogeneous Two-Phase Flow" AIAA J. v.17, n.1, pp.17-25 1979

17. Lyczkowski, R. W.; Solbrig, C. W.; Gidaspow, D. and Hughes, E. D. "Characteristics and Stability Analyses of Transient One-Dimensional Two-Phase Flow Equations and Their Finite Difference Approximations Proc. ASME Winter Annual Meeting, Houston, Texas, Nov. 30-Dec. 4 1975

18. Drew, D. A. "A Framework for Dispersed Two-Phase Flow" Proc. ARO Workshop on Multiphase Flows BRL 1-2 Feb. 1978

19. Aguirre-Ramirez, G. and Saxe, H. C. "Phenomenological Theory for the Granular Mass" Isr. J. Technol, vol.9, n.5, pp.499-506 1971

20. Morland, L. W. " A Simple Constitutive Law for a Fluid Saturated Porous Solid" J. Geophys, Res. vol.77, n.5, p.890 1972

21. Garg, S. K. and Nur, A. "Effective Stress Laws for Fluid Saturated Porous Rocks" J. Geophys. Res. vol.78, n.26 1973

22. Petrovsky, I. G. "Partial Differential Equations" Interscience 1964

23. Gough, P. S. "Two Dimensional Convective Flamespreading in Packed Beds of Granular Propellant" BRL Report ARBRL-CR-00404 1979

24. Ramshaw, J. D. and Trapp, J. A. "Characteristics, Stability and Short-Wavelength Phenomena in Two-Phase Flow Equation Systems" ANCR-1272 1976

25. Chao, B. T.; Sha, W. T.; and Soo, S. L. "On Inertial Coupling in Dynamic Equations of Components in a Mixture" Int. J. Multiphase Flow, vol.4, pp.219-223 1978

26. Soo, S. L. "Multiphase Mechanics and Distinctions from Continuum Mechanics" Two-Phase Flow and Heat Transfer Symposium - Workshop, NSF and University of Miami, Fort Lauderdale, Florida 1976

27 Harlow, F. H. and Amsden, A. A. "Numerical Calculations of Multiphase Fluid Flow" J. Comp. Phys. vol.17, pp.19-52 1975

28. Jackson, R. "The Mechanics of Fluidised Beds: Part I: The Stability of the State of Uniform Fluidisation" Trans. Inst. Chem. Engrs. vol.41, pp.13-21 1963

29. Pigford, R. L. and Baron, T. "Hydrodynamic Stability of a Fluidized Bed" I. and E.C. Fund vol.4, n.1, pp.81-87 1965

30. Murray, J. D. "On The Mathematics of Fluidization Part 1. Fundamental Equations and Wave Propagation" J. Fluid, Mech. vol.21,pt.3,pp.465-493 1965

31. Gough, P. S. "The Influence of an Implicit Representation of Internal Boundaries on the Ballistic Predictions of the NOVA Code" Proc. 14th JANNAF Combustion Meeting 1977

32. Gough, P. S. "The Predictive Capacity of Models of Interior Ballistics" Proc. 12th JANNAF Combustion Meeting 1975

33. Horst, A. W.; Smith, T. C.; and Mitchell, S. E. "Key Design Parameters in Controlling Gun-Environment Pressure Wave Phenomena - Theory versus Experiment" Proc. 13th JANNAF Combustion Meeting 1976

34. Kuo, K. K.; Koo, J. H.; Davis, T. R. and Coates, G.R. "Transient Combustion in Mobile Gas-Permeable Propellants" Acta Astron. vol.3 pp. 573-591 1976

This work was supported by the Naval Ordnance Station, Indian Head, Maryland under contract N00174-78-M-8024.

Paul Gough Associates, Inc.
Portsmouth, NH 03801

Index

U

V

W

Z